A Professional Approach

Microsoft® Office
Word®

Specialist

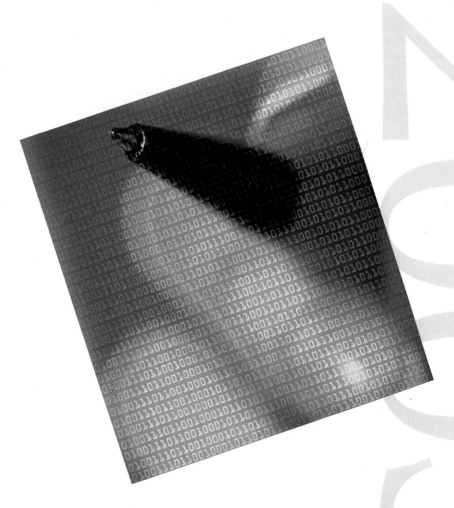

Deborah Hinkle

 **Technology
Education**

Boston   Burr Ridge, IL   Dubuque, IA   Emeryville, CA   Madison, WI   New York   San Francisco   St. Louis
Bangkok   Bogotá   Caracas   Kuala Lumpur   Lisbon   London   Madrid   Mexico City
Milan   Montreal   New Delhi   Santiago   Seoul   Singapore   Sydney   Taipei   Toronto

 **Technology Education**

1333 Burr Ridge Parkway
Burr Ridge, Illinois 60527
U.S.A.

Microsoft® Office Word® 2003: A Professional Approach, Specialist
Student Edition

For information on translations or book distributors outside the U.S.A.,
please see the International Contact Information page immediately following
the index of this book. Some ancillaries, including electronic and print
components, may not be available to customers outside the United States.

1 2 3 4 5 6 7 8 9 0 QPD QPD 0 1 9 8 7 6 5 4

Book p/n 0-07-223213-7 and CD p/n 0-07-223214-5
parts of
ISBN 0-07-223212-9

This book was composed with Corel VENTURA™ Publisher.

www.mhteched.com

Sponsoring Editor
Gareth Hancock

Developmental Editor
Lisa Chin-Johnson

Project Editor
Patty Mon

Technical Editors
Eileen Mullin, Darren Choong, EK Choong

Copy Editor
Marcia Baker

Proofreader
Susie Elkind

Indexer
Valerie Robbins

Composition
Elizabeth Jang, John Patrus,
Kelly Stanton-Scott

Illustrators
Kathleen Edwards, Melinda Lytle

Interior Design
Leggitt Associates, Creative Ink, Inc.,
Peter F. Hancik

Contents

UNIT 2   PARAGRAPH FORMATTING, MARGINS, AND TABS

UNIT 3   ADVANCED EDITING

UNIT 4   PAGE FORMATTING

UNIT 5   TABLES AND COLUMNS

UNIT 6   ADVANCED TOPICS

What does this logo mean?

It means this courseware has been approved by the Microsoft® Office Specialist Program to be among the finest available for learning *Microsoft® Word 2003*. It also means that upon completion of this courseware, you may be prepared to take an exam for Microsoft Office Specialist qualification.

What is a Microsoft Office Specialist?

A Microsoft Office Specialist is an individual who has passed exams for certifying his or her skills in one or more of the Microsoft Office desktop applications such as Microsoft Word, Microsoft Excel, Microsoft PowerPoint, Microsoft Outlook, Microsoft Access, or Microsoft Project. The Microsoft Office Specialist Program typically offers certification exams at the "Specialist" and "Expert" skill levels. *The Microsoft Office Specialist Program is the only program in the world approved by Microsoft for testing proficiency in Microsoft Office desktop applications and Microsoft Project. This testing program can be a valuable asset in any job search or career advancement.

More Information:

To learn more about becoming a Microsoft Office Specialist, visit www.microsoft.com/officespecialist. To learn about other Microsoft Office Specialist-approved courseware from McGraw-Hill/ Technology Education, visit http://www.mhteched.com.

Preface

Microsoft® Office Word® 2003: A Professional Approach, Specialist Student Edition is written to help you master Microsoft Word for Windows. The text takes you step-by-step through the Word features that you're likely to use in both your personal and business life.

Case Study

Learning the features of Word is one component of the text, and applying what you learn is another. That is why a Case Study was created and appears throughout the text. The Case Study offers the opportunity to learn Word in a realistic business context. Take the time to read the Case Study near the front of this book. The Case Study is about Duke City Gateway Travel, a fictional business located in Albuquerque, New Mexico. All the documents for this course involve Duke City Gateway Travel.

Organization of the Text

The text includes six units. Each unit is divided into lessons, and there are 18 lessons, each building on previously learned procedures. This building block approach, together with the Case Study and the following features, enable you to maximize the learning process.

Features of the Text

- ☑ *Objectives* are listed for each lesson.
- ☑ Required skills for the *Microsoft Office Specialist Exam* are listed for each lesson.
- ☑ The *estimated time* required to complete each lesson (up to the "Using Online Help") is stated.
- ☑ Within a lesson, each *heading* corresponds to an objective.
- ☑ Easy-to-follow *exercises* emphasize "learning by doing."
- ☑ *Key terms* are italicized and defined as they are encountered.
- ☑ Extensive *graphics* display screen contents.
- ☑ *Toolbar buttons* and *keyboard keys* are shown in the text when used.
- ☑ *Large toolbar buttons in the margins* provide easy-to-see references.
- ☑ Lessons contain important *Notes*, useful *Tips*, and helpful *Reviews*.
- ☑ *Using Online Help* introduces you to a Help topic related to lesson content.
- ☑ A *Lesson Summary* reviews the important concepts taught in the lesson.
- ☑ A *Command Summary* lists the commands taught in the lesson.

- ☑ *Concepts Review* includes true/false, short answer, and critical thinking questions that focus on lesson content.
- ☑ *Skills Review* provides skill reinforcement for each lesson.
- ☑ *Lesson Applications* ask you to apply your skills in a more challenging way.
- ☑ *On Your Own* exercises let you apply your skills creatively.
- ☑ *Unit Applications* give you the opportunity to use the skills you learn in a unit.
- ☑ Includes an Appendix of Proofreaders' Marks, Standard Forms for Business Documents, and Microsoft Office Specialist Certification standards, a Glossary, and an Index.

Microsoft Office Specialist Certification Program

The Microsoft Office Specialist certification program offers certification at two skill levels—"Specialist" and "Expert." This certification can be a valuable asset in any job search. For more information about this Microsoft program, go to www.microsoft.com/officespecialist. For a complete listing of the skills for the Word 2003 "Specialist" certification exam (and a correlation to the lessons in the text), see Appendix C: "Microsoft Office Specialist Certification."

Professional Approach Web Site

Visit the Professional Approach Web Site at www.mhteched.com/pas to access a wealth of additional materials including tutorials, additional projects, online quizzes, and more!

Conventions Used in the Text

This text uses a number of conventions to help you learn the program and save your work.

- ☑ Text to be keyed appears either in **boldface** or as a separate figure.
- ☑ Filenames appear in **boldface**.
- ☑ Options that you choose from menus and dialog boxes appear in a font that is similar to the on-screen font; for example, "Choose Print from the File menu." (The underline means you can press Alt and key the letter to choose the option.)
- ☑ You are asked to save each document with your initials followed by the exercise name. For example, an exercise might end with this instruction: "Save the document as *[your initials]*5-12." Documents are saved in folders for each lesson.

If You Are Unfamiliar with Windows

If you are unfamiliar with Windows, review the "Windows Tutorial" available on the Professional Approach web site at www.mhteched.com/pas before beginning Lesson 1. This tutorial provides a basic overview of the operating system and shows you how to use the mouse. You might also want to review "File Management" (also on the Professional Approach web site) to get more comfortable with files and folders.

Screen Differences

As you practice each concept, illustrations of the screens help you follow the instructions. Don't worry if your screen is slightly different from the illustration. These differences are due to variations in system and computer configurations.

Acknowledgments

We thank the technical editors and reviewers of this text for their valuable assistance: Eileen Mullin, EK Choong, Darren Choong, Susan Olson, Northwest Technical College, East Grand Forks, MN; John Walker, Doña Ana Community College, Las Cruces, NM; Mary Davey, Computer Learning Network, Camp Hill, PA.

Installation Requirements

You will need Microsoft Word 2003 to work through this textbook. Word needs to be installed on the computer's hard drive or on a network. Use the following checklist to evaluate installation requirements.

Hardware

- Computer with 233 MHz or higher processor and at least 128MB of RAM
- CD-ROM drive and other external media (3.5-inch high-density floppy, ZIP, etc.)
- 400MB or more of hard disk space for a "Typical" Office installation
- Super VGA (800 × 600) or higher-resolution video monitor
- Printer (laser or ink-jet recommended)
- Mouse
- Optional: Modem or other Internet connection (required for Using Online Help exercises)

Software

☑ Word 2003 (from Microsoft Office System 2003)

☑ Windows 2000 with Service Pack 3 or later, or Windows XP or later operating system

☑ Optional: Browser (and Internet access)

Internet Access

Access to the Internet is required for most of the Using Online Help exercises. Many of the help features are only available online. Microsoft Office Online is also a valuable resource for additional clip art, photographs, templates, and other Word resources.

CASE STUDY

Duke City
Gateway Travel Agency

There is more to learning a word processing program like Microsoft Word than simply pressing keys. You need to know how to use Word in a real-world situation. That is why all the lessons in this book relate to everyday business tasks.

As you work through the lessons, imagine yourself working as an intern for Duke City Gateway Travel, a fictional travel agency located in Albuquerque, New Mexico.

Duke City Gateway Travel

It was 1977. Jimmy Carter was president. The first *Star Wars* was showing in movie theaters. People were buying vinyl albums like the Eagle's "Hotel California." And, Duke City Travel began doing business.

Based in Albuquerque, New Mexico, Duke City Travel took its name from Albuquerque's nickname, "Duke City." (The city was named after the Duke of Alburquerque. The first "r" was dropped in the 1880s.) Originally, Duke City Travel was just a small one-person travel agency, but it was an immediate success and began to grow.

Within a few years, the company joined with six other agencies from the United States, Canada, England, and France. They formed the "Gateway Travel Group." (See Figure CS-1 for these Gateway Travel "Flagship Offices.")

The goal of each of the agencies within the Gateway Travel Group is to offer "one-stop shopping" for travelers. However, every company can interpret this goal in its own way. Duke City Gateway Travel has chosen to create four "Specialty Areas." This has the added advantage of increasing the efficiency of the individual travel agents. The four Specialty Areas are as follows:

- International Travel
- Corporate Travel
- Group Travel
- Family Travel

FIGURE CS-1   Gateway Travel Flagship Offices

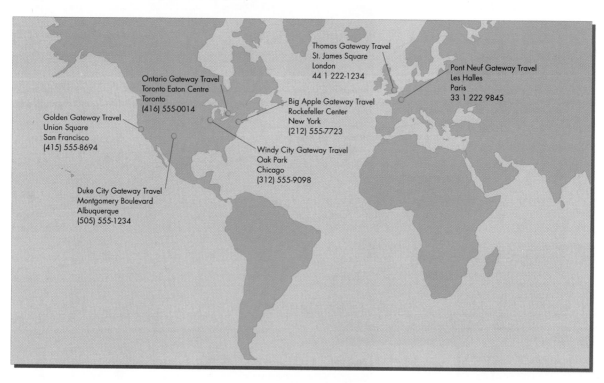

Currently, Duke City Gateway Travel has 12 employees, ranging from Susan Allen, the President/Owner, to Valerie Grier, the Office Manager. You can see a picture of each of the staff members on the next page. It might be helpful to associate the picture of the employee with their name as you work through the course. Also notice that Duke City Gateway Travel refers to travel agents as "Travel Counselors."

To understand the organization of Duke City Gateway Travel, take a look at Figure CS-2. Notice that each of the Specialty Areas is headed by a Senior Travel Counselor. A Travel Counselor or an Associate Travel Counselor works for each of the Senior Travel Counselors.

All of the documents you will use in this text relate to Duke City Gateway Travel. As you work through the documents in the text, take the time to notice the following:

- How the employees interact and how they respond to customers' queries.
- The format and tone of the business correspondence (if you are unfamiliar with the standard formats for business documents, refer to Appendix B).
- References to *The Gregg Reference Manual*, a standard reference manual for business writing and correspondence.
- The content of the correspondence (and its relation to Duke City Gateway's business).

As you use this text and become experienced with Microsoft Word, you will also gain experience in creating, editing, and formatting the type of documents that are generated in a real-life business environment.

FIGURE CS-2   Organization Chart

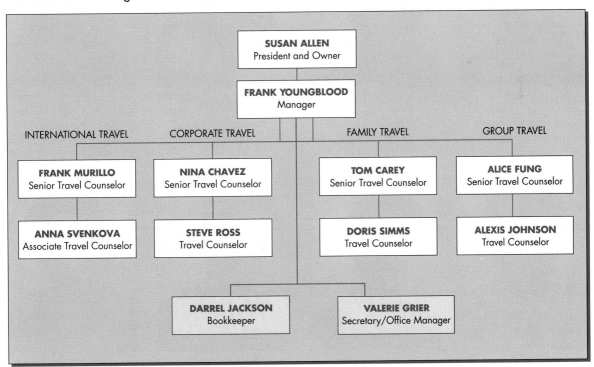

Duke City Gateway Travel

The Staff

SUSAN ALLEN
President and Owner

FRANK YOUNGBLOOD
Manager

FRANK MURILLO
Senior Travel Counselor,
International Travel
Specialist

NINA CHAVEZ
Senior Travel Counselor,
Corporate Travel
Specialist

TOM CAREY
Senior Travel Counselor,
Family Travel Specialist

ALICE FUNG
Senior Travel Counselor,
Group Travel Specialist

DORIS SIMMS
Travel Counselor

ALEXIS JOHNSON
Travel Counselor

STEVE ROSS
Travel Counselor

ANNA SVENKOVA
Associate Travel
Counselor

DARREL JACKSON
Bookkeeper

VALERIE GRIER
Secretary/
Office Manager

Basic Skills

Creating a Document

MICROSOFT OFFICE SPECIALIST ACTIVITIES

In this lesson:
WW03S-1-1
WW03S-5-3

See Appendix C.

OBJECTIVES

After completing this lesson, you will be able to:

1. Start Word.
2. Identify parts of the Word screen.
3. Key text into a document.
4. Perform basic text editing.
5. Name and save a document.
6. Print a document.
7. Close a document and exit Word.

 Estimated Time: 1 hour

Microsoft Word is a versatile, easy-to-use word processing program that helps you create letters, memos, reports, and other types of documents. This lesson begins with an overview of the Word screen. Then you learn how to create, edit, name, save, print, and close a document.

Starting Word

There are several ways to start Word, depending on your system setup and personal preferences. For example, you can use the Start button on the Windows taskbar or double-click a Word shortcut icon that might be on your desktop.

NOTE: Windows provides many ways to start applications. If you have problems, ask your instructor for help.

FIGURE 1-1
Starting Word
from the
Windows taskbar

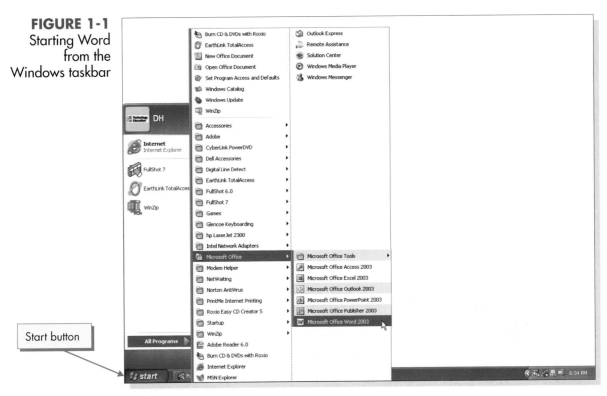

 NOTE: Your screen will differ from the screen shown in Figure 1-1 depending on the programs installed on your computer.

EXERCISE   1-1   Start Word

1. Turn on your computer. Windows loads.

 2. Click the Start button on the Windows taskbar and point to All Programs.

3. On the All Programs menu, click Microsoft Office, and then click Microsoft Office Word 2003. In a few seconds, the program is loaded and the Word screen appears. (See Figure 1-2 on the next page.)

 NOTE: Make sure you are in Normal view. If you're not, click the Normal View button ▤ in the lower left corner of your screen.

FIGURE 1-2
Word screen

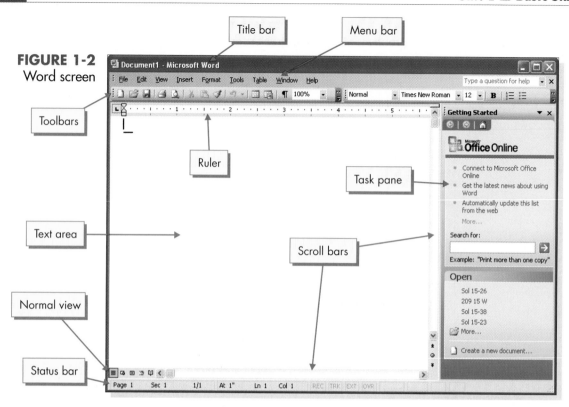

TABLE 1-1   Parts of the Word Screen

PART OF SCREEN	PURPOSE
Title bar	Displays the name of the current document. The opening Word screen is always named "Document1."
Menu bar	Contains the menus you use to perform various tasks. You can open menus by using the mouse or the keyboard.
Toolbars	Contain buttons you click to initiate a wide range of commands. Each button is represented by an icon. Word opens with the Standard and Formatting toolbars displayed in abbreviated form on one row.
Ruler	Shows placement of margins, indents, and tabs.
Text area	Displays the text and graphics in the document.
Scroll bars	Used with the mouse to move right or left and up or down within a document. Buttons to change the way you view a document appear to the left of the horizontal scroll bar.
Status bar	Displays information about the task you are performing, shows the position of the insertion point, and shows the current mode of operation.
Getting Started task pane	Displays options for opening recently used documents and creating new documents. This task pane appears when you start Word.

Identifying Parts of the Word Screen

To become familiar with Word, start by identifying the parts of the screen you'll work with extensively in this course, such as menus and toolbar buttons.

EXERCISE   1-2   Identify Menus and Menu Commands

Each of the items on the menu bar is the name of a menu that contains a list of commands.

1. Move the mouse pointer to Edit on the menu bar. Click the left mouse button to open the menu. Word displays a short version of the Edit menu with the most commonly used Edit menu commands.

2. With the Edit menu open, wait five seconds and Word displays the expanded Edit menu with additional commands. You can also click the arrows at the bottom of the short menu to expand the menu.

    NOTE: Word's short menus are adaptive—they change as you work, listing the commands you use most frequently.

3. Without clicking the mouse button, move the pointer to View on the menu bar. Continue moving the pointer slowly across the menu bar until you display the Help menu.

4. Click Help on the menu bar to close the menu. You can also close a menu by clicking within the text area of the screen or by pressing Esc.

   TIP: The menus show commands with corresponding toolbar buttons and keyboard shortcuts. For example, you can save a document by choosing Save from the File menu, by clicking the Save button on the Standard toolbar, or by pressing Ctrl+S.

5. Double-click View on the menu bar to open the View menu. Double-clicking a menu name opens the expanded menu.

6. Without clicking the mouse button, move the pointer to the Toolbars command. The right-pointing arrow indicates a submenu that shows which toolbars are currently displayed. The submenu also indicates whether the task pane is displayed. (See Figure 1-3 on the next page.)

7. Make sure Task Pane is checked on the View menu (it will also appear on the Toolbars submenu).

8. Click outside the menu in a blank area of the screen to close the menu. Be careful not to click a menu option or toolbar button.

FIGURE 1-3
Displaying
menu options

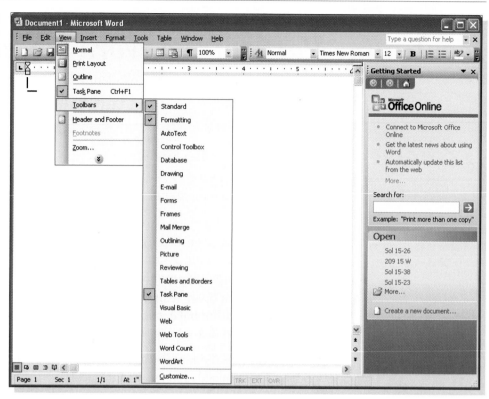

EXERCISE **1-3** **Identify Buttons**

When you start Word, the Standard and Formatting toolbars appear side by side below the menu bar. Only the basic buttons and those that were used most recently are displayed. To see more buttons for either toolbar, click the Toolbar Options button ⚏ at the end of the toolbar. To identify a toolbar button or any other onscreen button by name, point to it with the mouse and Word will display a *ScreenTip*—a box with the button name.

FIGURE 1-4
Identifying a
toolbar button

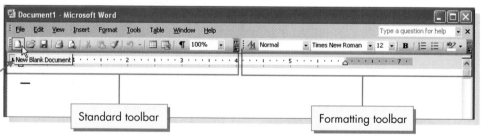

1. Position the pointer over the New Blank Document button on the Standard toolbar. The button appears as a shaded button when you point to it and a ScreenTip appears below the button with the button name. You click this button to start a new blank document.

2. Click the Toolbar Options button at the end of the Standard toolbar to see additional toolbar buttons. Move the mouse pointer over any button to identify it.

FIGURE 1-5
Displaying
additional
toolbar buttons

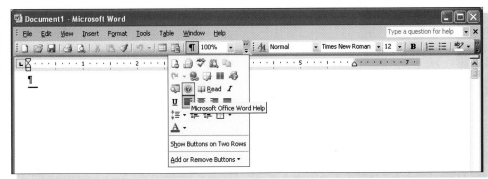

3. Click the Show/Hide ¶ button on the Standard toolbar. Click it again. Notice that when you click (or "press") a button, it appears outlined and shaded in a brighter color. (If you had to click Toolbar Options to locate the Show/Hide ¶ button, notice that it now appears on your Standard toolbar.)

NOTE: Any toolbar button that is currently not available can be identified by pointing to it with the mouse.

4. At the bottom left of the screen, point to the various buttons above the Status bar. You use these buttons to choose various ways to view a document. Check that the Normal View button is selected.

5. Point to the Next Page button at the bottom right of the screen (located on the vertical scroll bar). You use this button, along with the Previous Page button to scroll forward and backward in a document.

NOTE: You might notice the floating Language toolbar on your screen. This toolbar may be used for *speech recognition* if a microphone is connected to your computer.

EXERCISE **1-4** **Identify Task Panes**

The task pane is a screen area that makes many functions easier to access. You can display different task panes, depending on the task at hand, or hide the task pane to make more room for your document.

1. In the Getting Started task pane, point to the down arrow, which is the Other Task Panes button ▾.

FIGURE 1-6
Task pane buttons

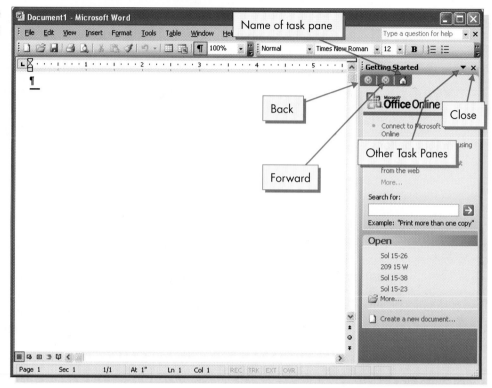

2. Click the button to display a list of the other available task panes; then open one of the other panes.

3. Click the Back button ⊚ in the new task pane to return to the Getting Started task pane.

4. Click the Close button ✕ in the task pane to hide the pane.

NOTE: To redisplay the task pane, open the View menu and choose Task Pane. Task panes will have more meaning as you become more familiar with Word.

Keying Text

When keying text, you'll notice various shapes and symbols in the text area. For example:

- The *insertion point* is the vertical blinking line that marks the position of the next character to be entered.
- The *end mark* is a short horizontal line that moves as you key text and indicates the end of the document.

- The mouse pointer takes the shape of an *I-beam* I when it's in the text area. It changes to an arrow ↳ when you point to a toolbar button or menu outside the text area.
- The *paragraph mark* ¶ marks the end of a paragraph. The paragraph mark displays when Show/Hide ¶ [¶] is selected.

EXERCISE 1-5 Key Text and Move the Insertion Point

1. Before you begin, make sure the Show/Hide ¶ button [¶] on the Standard toolbar is selected. When this feature is "turned on," you can see paragraph marks and spacing between words and sentences more easily.

2. Key the words **Duke City Gateway** (don't worry about keying mistakes now—you can correct them later). Notice how the insertion point and paragraph mark move as you key text. Notice also how a space between words is indicated by a dot.

FIGURE 1-7
The insertion point marks the place where you begin keying.

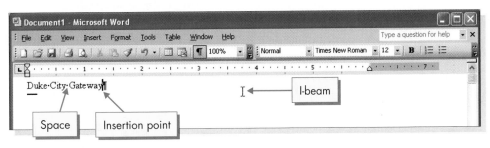

3. Move the insertion point to the left of the word "Duke" by positioning the I-beam and clicking the left mouse button.

4. Move the insertion point back to the right of "Gateway" to continue keying.

 NOTE: The documents you create in this course relate to the Case Study about Duke City Gateway Travel, a fictional travel agency (see pages 1 through 4).

EXERCISE 1-6 Wrap Text and Correct Spelling

As you key more text, you will notice Word performs several tasks automatically. For example, Word does the following by default:

- Wraps text from the end of one line to the beginning of the next line
- Alerts you to spelling and grammatical errors

- Corrects common misspellings, such as "teh" for "the" and "adn" for "and"
- Suggests the completed word when you key the current date, day, or month

 NOTE: Word's capability to correct spelling and grammar, and to complete words for you is discussed thoroughly in Lesson 4: "Writing Tools."

1. Continue the sentence you started in Exercise 1-5, this time keying a misspelled word. Press [Spacebar], and then key **is New Mexico's leeding travel agency** (don't key a period). Word recognizes that "leeding" is misspelled and applies a red, wavy underline to the word.

 TIP: The Spelling and Grammar Status icon at the right side of the Status bar displays an "x" instead of a check mark when it detects an error.

2. To correct the misspelling, use the mouse to position the I-beam anywhere in the underlined word and click the *right* mouse button. A shortcut menu appears with suggested spellings. Click "leading" with the *left* mouse button and Word makes the correction. Notice the change in the Spelling and Grammar Status icon on the Status bar.

FIGURE 1-8
Choose the correct spelling from the shortcut menu.

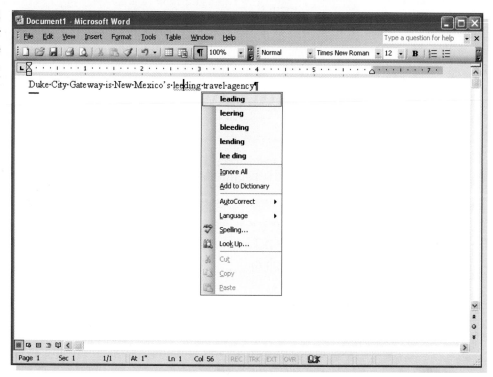

3. Move the insertion point to the right of "agency" and press [Spacebar]. Continue the sentence with another misspelled word by keying **adn** and press [Spacebar]. Notice that "adn" is automatically corrected to "and" when you press [Spacebar].

4. Complete the sentence by keying **is located in Albuquerque on Montgomery Boulevard.** Notice how the text automatically wraps from the end of the line to the beginning of the next line. (If you spell "Albuquerque" incorrectly, right-click it and apply the correct spelling.)

 NOTE: The text "Montgomery Boulevard" may appear with a purple, dotted underline. This is called a *smart tag,* which Word applies to data such as dates and addresses.

5. Verify that the insertion point is to the immediate right of the period following Boulevard, and then press the `Spacebar` once. Key the following text:

It is a franchise division of Gateway Travel, an international travel agency with 49 offices throughout the United States, Canada, and Europe.

 NOTE: Throughout this text, one space is used after a period to separate sentences. This is the standard format for word processing and desktop publishing.

6. Press `Enter` once to start a new line. Press `Enter` again to leave a blank line between paragraphs.

7. Key the second paragraph shown in Figure 1-9. When you key the first four letters of "Monday" in the first sentence, Word suggests the completed word in a small box. Press `Enter` to insert the suggested word, and then press `Spacebar` before you key the next word. Follow the same procedure for "Saturday."

FIGURE 1-9

```
For more information about Duke City Gateway, stop by our office
Monday through Saturday, or visit our Web site anytime. Our travel
counselors will be happy to help you plan your next trip.
```

NOTE: When Word suggests a completed word as you key text, you can ignore the suggested word and continue keying or insert it by pressing `Enter`.

Basic Text Editing

The keyboard offers many options for basic text editing. For example, you can press `Backspace` to delete a single character or `Ctrl`+`Delete` to delete an entire word.

TABLE 1-2   **Basic Text Editing**

KEY	RESULT
Backspace	Deletes characters to the left of the insertion point.
Ctrl + Backspace	Deletes the word to the left of the insertion point.
Delete	Deletes characters to the right of the insertion point.
Ctrl + Delete	Deletes the word to the right of the insertion point.

EXERCISE   1-7   Delete Text

1. Move the insertion point to the right of the word "It" in the second sentence of the first paragraph. (Use the mouse to position the I-beam and click the left mouse button.)
2. Press Backspace twice to delete both characters and key **Duke City Gateway**.
3. In the same sentence, move the insertion point to the left of the number "49."
4. Press Delete twice and key **50**.
5. Move the insertion point to the left of the word "information" in the second paragraph.
6. Hold down Ctrl and press Backspace. The word "more" is deleted.

NOTE: When keyboard combinations (such as Ctrl+Backspace) are shown in this text, hold down the first key as you press the second key. Release the second key, and then release the first key. An example of the entire sequence is this: Hold down Ctrl, press Backspace, release Backspace, release Ctrl. With practice, this sequence becomes easy.

7. Move the insertion point to the right of "Albuquerque" in the first sentence of the first paragraph.
8. Hold down Ctrl and press Delete to delete the word "on." Press Ctrl+Delete two more times to delete the words "Montgomery Boulevard."

EXERCISE   1-8   Insert Text

When editing a document, you can insert text or key over existing text. When you insert text, Word is in regular *Insert mode*. To key over existing text, you switch to *Overtype mode*. Double-click the Status bar indicator OVR to turn Overtype mode on and off, or press Insert on the keyboard.

1. In the first sentence of the first paragraph, move the insertion point to the left of the "A" in "Albuquerque." Key **downtown** and press (Spacebar) once to leave a space between the two words.

2. Move the insertion point to the beginning of the document, to the left of "Duke."

3. Point to OVR on the Status bar and double-click. You know you're in Overtype mode when OVR appears in black. (You're in regular Insert mode when OVR is dimmed.)

4. Press (CapsLock). When you key text in Caps Lock mode, the keyed text appears in all uppercase letters.

 TIP: Most keyboards have a light indicator to show when Caps Lock is turned on.

5. Key **duke city gateway** over the old text. Repeat the process for "Duke City Gateway" in the second sentence.

6. Press (CapsLock) to turn off Caps Lock mode. Double-click OVR (or press (Insert)) to turn off Overtype mode.

 TIP: Always remember to turn off Overtype mode as soon as you're done with it to avoid accidentally keying over text.

FIGURE 1-10
Edited document

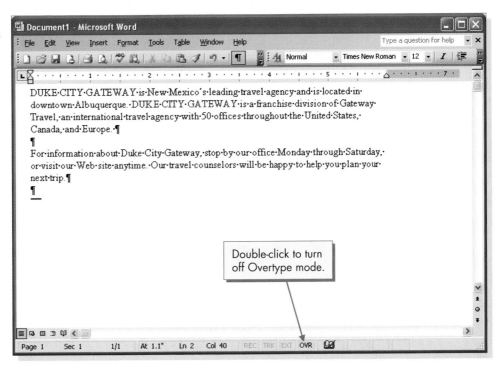

EXERCISE **1-9** **Combine and Split Paragraphs**

1. At the end of the first paragraph, position the insertion point to the left of the paragraph mark (after the period following "Europe").

2. Press (Delete) twice—once to delete the end-of-paragraph mark and once to delete the blank line between paragraphs. The two paragraphs are now combined, or merged, into one.

3. Press (Spacebar) once to insert a space between the sentences.

4. With the insertion point to the left of "For" in the combined paragraph, press (Enter) twice to split the paragraphs again.

Naming and Saving a Document

Your document, called "Document1," is stored in your computer's temporary memory. Until you name it and save it to disk, it can be lost if you have a power failure or computer hardware problem. It's always a good idea to save your work frequently.

The first step in saving a document for future use is to give it a *filename*. Here are some rules about naming documents:

- Filenames can be up to 255 characters long, including the drive letter and the folder name. The following characters cannot be used in a filename: / \ > < * ? " : ;

- Filenames can include uppercase letters, lowercase letters, or a combination of both. They can also include spaces. For example, a file can be named "Business Plan."

- Throughout this course, document filenames will consist of *[your initials]* (which might be your initials or the identifier your instructor asks you to use, such as **rst**), followed by the number of the exercise, such as **4-1**. The filename would, therefore, be **rst4-1**.

You can use either the <u>S</u>ave command or the Save <u>A</u>s command to save a document. Here are some guidelines about saving documents:

- Use Save <u>A</u>s when you name and save a document the first time.

- Use Save <u>A</u>s when you save an existing document under a new name. Save <u>A</u>s creates an entirely new file and leaves the original document unchanged.

- Use <u>S</u>ave to update an existing document.

- Before you save a new document, decide where you want to save it. Word saves documents in the current drive and folder unless you specify otherwise. For example, to save a document to a floppy disk, you need to change the drive to A: or B:, whichever is appropriate for your computer.

 NOTE: Your instructor will advise you on the proper drive and folder to use for this course.

EXERCISE 1-10 Name and Save a Document

1. Click <u>F</u>ile to open the File menu and choose Save <u>A</u>s. The Save As dialog box appears.

2. In the File <u>n</u>ame text box, a suggested filename is highlighted. Replace this filename by keying *[your initials]***1-10**.

3. At the top of the dialog box, click the down arrow in the Save <u>i</u>n box and choose the appropriate drive for your data disk (3 1/2 Floppy (A:), for example). Make sure you have a formatted disk in the drive.

NOTE: The default document type is Word Document. You can specify other file types such as RTF (Rich Text Format, which is a format used to exchange text documents between applications and operating systems) and TXT (Plain Text, which contains no formatting). To change the file type, simply click the down arrow beside the Save as <u>t</u>ype text box.

4. Click <u>S</u>ave. Your document is named and saved for future use.

FIGURE 1-11
Save As dialog box

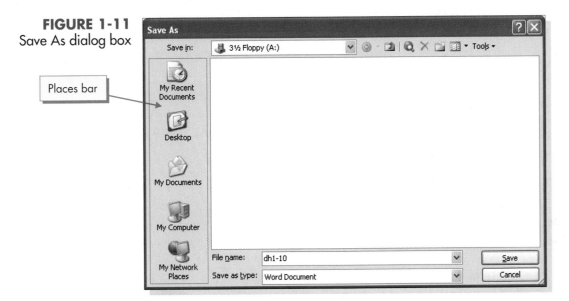

NOTE: If you were working at home and saving files to your computer's hard drive, you might use the *Places bar* on the left side of the Save As dialog box. The Places bar lists locations for commonly used files and folders. For example, "My Documents" is a folder created by Windows to help you organize your files. Click "My Recent Documents" to list the most recently opened files and folders. Use "Desktop" to save a file on your desktop for easy access. "My Computer" displays files or folders available on your computer.

Printing a Document

After you create a document, printing it is easy. You can use any of the following methods:

- Click the Print button 🖨 on the Standard toolbar.
- Choose P̲rint from the File menu.
- Press Ctrl+P.

The menu and keyboard methods open the Print dialog box, where you can select printing options. Clicking the Print button 🖨 sends the document directly to the printer, using Word's default settings.

NOTE: If you share a printer with your classmates, your instructor might ask you to key your name on all documents before printing.

EXERCISE **1-11** **Print a Document**

1. Choose P̲rint from the File menu to open the Print dialog box. The dialog box displays Word's default settings and shows your designated printer.

FIGURE 1-12
Print dialog box

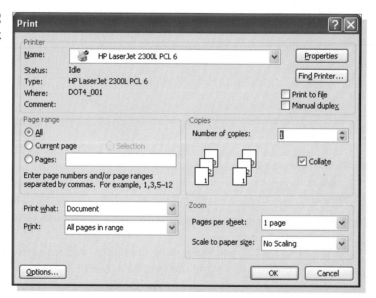

2. Click OK or press Enter to accept the settings. A printer icon appears on the taskbar as the document is sent to the printer.

Closing a Document and Exiting Word

When you finish working on a document and save it, you can close it and open another document or you can exit Word.

The easiest ways to close a document and exit Word include using the following:

- The Close buttons ⊠ and ⊠ in the upper right corner of the window

FIGURE 1-13
Close buttons

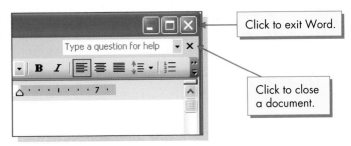

Click to exit Word.

Click to close a document.

- The Close command from the File menu
- Keyboard shortcuts: [Ctrl]+[W] closes a document and [Alt]+[F4] exits Word

EXERCISE **1-12** **Close a Document and Exit Word**

1. Choose Close from the File menu to close the document.

 NOTE: When no document is open, the document window is gray. If you want to start a new document, click the New Blank Document button ▢.

 2. Click the Close button ⊠ in the upper right corner of the screen to exit Word and display the Windows desktop.

USING ONLINE HELP

Online Help is available to you as you work in Word. Just type a question in the Ask a Question box at the top right of the screen and you'll access Microsoft Word's Help system.

Find out more about using Help:

1. Start Word.
2. Locate the Ask a Question box in the upper right corner of the screen. Click the box to activate it; then key **get help**.

3. Press Enter. A list of Help topics appears in the Search Results task pane. Scroll to see additional topics. Notice the mouse pointer is shaped like a hand when you point to a topic.

FIGURE 1-14
Using the
Ask a Question box

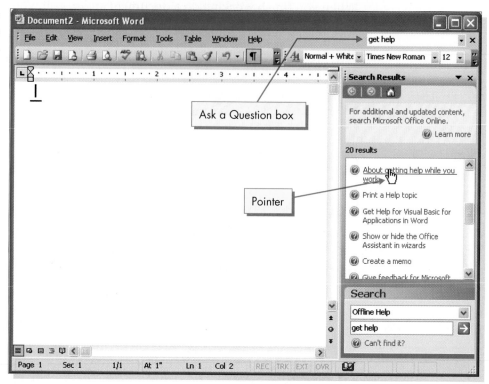

4. Scroll through the list of topics.

5. Click the topic About getting help while you work. The Microsoft Word Help dialog box opens.

6. Click each of the topics listed in blue text (or click Show All in the upper right corner of the Help screen). Review the information.

7. Close Help by clicking the Help window's Close button.

LESSON Summary

➤ To start Microsoft Word, click the Start button on the Windows taskbar, point to All Programs, click Microsoft Office, and click Microsoft Office Word 2003.

➤ The Title bar is at the top of the Word screen and displays the current document name.

➤ The menu bar appears below the Title bar. Each menu name, when clicked, displays a list of commands to perform various tasks.

➤ To open a menu, point to the menu name and click. To display all options on a menu, click the arrows at the bottom of the menu.

➤ The Standard and Formatting toolbars appear below the menu bar. The toolbars contain buttons you click to initiate commands. Each button is represented by an icon.

➤ Identify a button by name by pointing to it with the mouse. Word displays a ScreenTip with the button name. Display hidden buttons by clicking the Toolbar Options button ⸬ at the end of either the Standard or Formatting toolbar.

➤ The ruler appears below the Standard and Formatting toolbars.

➤ Scroll bars appear to the right and bottom of the text area. They are used to view different portions of a document. Buttons to change the way you view a document appear to the left of the horizontal scroll bar.

➤ The Status bar is a gray bar below the horizontal scroll bar. It displays such information as the current position of the insertion point and whether certain modes are active.

➤ The Getting Started task pane appears when you start Word. It displays options for recently used documents and for creating new documents.

➤ To close the task pane, click its Close button. To display the task pane, choose Tas__k__ Pane from the View menu.

➤ The blinking vertical line is called the insertion point. It marks the position of the next character to be entered. The end mark is a short horizontal line that moves as you key text and indicates the end of the document.

➤ The mouse pointer displays on the screen as an I-beam I when it's in the text area and as an arrow ▷ when you point to a toolbar button or menu outside the text area.

➤ When the Show/Hide ¶ button ¶ is turned on, a paragraph mark symbol appears at the end of every paragraph. A dot between words represents a space.

➤ If your computer is appropriately equipped, speech recognition is available for menu commands, toolbar buttons, and text entry.

➤ Word automatically wraps text to the next line as you key text. Press Enter to start a new paragraph or to insert a blank line.

➤ Word flags spelling errors as you key text by inserting a red, wavy line under the misspelled word. To correct the spelling, point to the underlined word, click the right mouse button, and choose the correct spelling.

➤ Word automatically corrects commonly misspelled words for you as you key text. Word can automatically complete a word for you, such as the name of a month or day. Word suggests the completed word and you press Enter to insert it.

➤ Delete a single character by using Backspace or Delete. Ctrl+Backspace deletes the word to the left of the insertion point. Ctrl+Delete deletes the word to the right of the insertion point.

➤ To insert text, click to position the insertion point and key the text.

➤ To enter text over existing text, turn on Overtype mode by double-clicking OVR on the Status bar or pressing Insert on the keyboard.

➤ Insert one space between words and between sentences.

➤ Document names, or filenames, can contain 255 characters, including the drive letter and folder name, and can contain spaces. The following characters cannot be used in a filename: / \ > < * ? " : ;

➤ Save a new document by using the Save As command and giving the document a filename. Use the Save command to update an existing document.

➤ To start a new blank document, click the New Blank Document button 🗋 on the Standard toolbar or choose Blank Document from the New Document task pane.

➤ To use Word Help, key a question or subject in the Ask a Question box, press Enter, and then choose a topic.

LESSON 1	Command Summary		
FEATURE	**BUTTON**	**MENU**	**KEYBOARD**
Save As		File, Save As	F12
Print		File, Print	Ctrl + P
Close a document	x	File, Close	Ctrl + W or Ctrl + F4
Exit Word	X	File, Exit	Alt + F4

 NOTE: Word provides many ways to accomplish a particular task. As you become more familiar with Word, you'll find the methods you prefer.

Concepts Review

TRUE/FALSE QUESTIONS

Each of the following statements is either true or false. Indicate your choice by circling T or F.

T F **1.** You can use the Standard toolbar to start or exit Word.

T F **2.** You can double-click OVR on the Status bar to turn Overtype mode on or off.

T F **3.** You can click ▾ on the Standard toolbar to see more buttons.

T F **4.** The mouse pointer takes the shape of an arrow when it appears in the menu bar.

T F **5.** A red, wavy line appears under words that are misspelled in a document.

T F **6.** Pressing Delete deletes characters to the left of the insertion point.

T F **7.** Ctrl+Delete deletes the word to the right of the insertion point.

T F **8.** You can save a document by choosing Save from the Edit menu.

SHORT ANSWER QUESTIONS

Write the correct answer in the space provided.

1. What menu and menu option open the Print dialog box?

2. Which toolbar contains the Show/Hide ¶ button?

3. If you begin keying a word such as "January" or "Thursday," how can you have Word complete the word for you automatically?

4. What area of the Word screen shows the position of the insertion point and displays indicators that show the current mode of operation?

5. Which toolbar contains the Print button?

6. What shape is the pointer when it appears in the text area of the screen?

_____

7. Which command is used to save a document under a different filename?

_____

8. What happens when you point to the down arrows at the bottom of a menu?

_____

CRITICAL THINKING

Answer these questions on a separate page. There are no right or wrong answers. Support your answers with examples from your own experience, if possible.

1. You can use the Show/Hide ¶ button to hide paragraph marks and space characters. When might it be useful to show these characters? When would you want to hide them?

2. Word allows great flexibility when naming files. Many businesses and individuals establish their own rules for naming files. What kinds of rules would you recommend for naming files in a business? For personal use?

Skills Review

EXERCISE 1-13

Identify parts of the Word screen.

1. Start Word, if necessary, by following these steps:

   a. Click the Start button on the Windows taskbar.

   b. Point to All Programs, point to Microsoft Office, point to Microsoft Office Word 2003, and click.

2. If the New Document task pane is not displayed, open the View menu and choose Task Pane. Select New Document from Other Task Panes, if necessary.

3. Move the pointer to the Open button on the Standard toolbar to identify it. Point to the Print button to identify it.

4. Click the Toolbar Options button at the end of the Formatting toolbar. Point to the Bullets button to identify it.

5. Point to, and then click File on the menu bar. Expand the menu by clicking the arrows at the bottom of the menu.

6. With the File menu still open, move the pointer up the menu list without clicking the mouse button. Notice the submenu for the Send To option.

7. Close the menu by clicking File on the menu bar or clicking in the blank text area.

8. Close the document by clicking the lower Close button ☒ in the upper right corner of the window located on the Menu bar.

EXERCISE 1-14

Key text, correct the spelling of a word, and save a document.

1. Open a new document window by clicking the New Blank Document button ▢ on the Standard toolbar.

2. Make sure you are in Normal view (open the View menu and choose Normal).

3. Click the Toolbar Options button ⸬ on the Standard toolbar and make sure the Show/Hide ¶ button ¶ is selected.

4. Key the text shown in Figure 1-15, including the intentional misspelling of "sponsored."

FIGURE 1-15

> Become a travel agent and see the world! Come to the Travel Fair and learn how. The fair is sponsord by Duke City Gateway Travel.

5. Correct the spelling of "sponsored" by following these steps:
   a. Move the I-beam anywhere within the word and click the right mouse button.
   b. Choose the correct spelling from the shortcut menu by clicking it.

6. Save the document as *[your initials]*1-14 by following these steps:
   a. Choose Save As from the File menu to open the Save As dialog box.
   b. Key the filename *[your initials]*1-14 in the File name text box.
   c. Choose the appropriate drive for your data disk (for example, 3 1/2 Floppy Drive (A:) or another drive specified by your instructor).
   d. Click Save.

7. Close the document by pressing Ctrl + W.

EXERCISE 1-15

Key, edit, and save a document.

1. Click the New Blank Document button ▢ on the Standard toolbar to start a new document.

2. Key the text shown in Figure 1-16. (Use single spacing in all your documents, unless you are told otherwise.)

FIGURE 1-16

> Duke City Gateway has just celebrated another anniversary in the
> travel business. The agency has proudly served over 45,000 area
> residents and visitors to New Mexico from all over the world. Duke
> City Gateway has been at its current location for all of its 25 years.

   3. Correct any spelling mistakes Word locates.

   4. Delete the text "all of its" in the last sentence by following these steps:

   a. Move the insertion point to the right of the word "for" by positioning the
   I-beam and clicking the left mouse button.

   b. Hold down Ctrl and press Delete three times to delete the words "all of its."

   5. Insert text after the word "residents" in the second sentence by following
   these steps:

   a. Check the Status bar to make sure Overtype mode (OVR) is *not* turned on.

   b. Move the insertion point to the immediate right of the word "residents."
   Key a comma, press Spacebar, and key **business travelers** followed by
   another comma.

   6. Split the paragraph by following these steps:

   a. Move the insertion point to the immediate left of the word "Duke" in the
   last sentence.

   b. Press Enter twice.

   7. Save the document as *[your initials]*1-15 on your student data disk.

   8. Close the document.

EXERCISE 1-16

Key, edit, save, and print a document.

   1. Start a new blank document.

   2. Key the text shown in Figure 1-17. Correct spelling mistakes as you key.

FIGURE 1-17

> When you travel, take advantage of inexpensive attractions in the
> area, such as public parks, outdoor concerts, museums, zoos,
> historical landmarks, and art exhibits.

3. At the end of the paragraph, move the insertion point to the left of the word "exhibits." Double-click OVR on the Status bar to change to Overtype mode and key **galleries** over the word "exhibits." Be sure to key a period.

4. Double-click OVR again to turn off Overtype mode and key **(or free!)** after the word "inexpensive."

5. Use ⌈Delete⌉ to delete the word "zoos" and key **botanical gardens** to replace it.

6. Check the spacing before and after the replacement text.

7. Save the document as *[your initials]***1-16** on your data disk.

8. Click the Print button on the Standard toolbar to print the document.

9. Close the document.

Lesson Applications

Key, edit, and print a document.

1. Start a new document. Turn on Caps Lock mode and key **TO THE STAFF:**.

2. Turn off Caps Lock and press Enter twice to start a new paragraph.

3. Key the text shown in Figure 1-18, including the corrections. Refer to Appendix A: "Proofreaders' Marks," if necessary. *Proofreaders' marks* are handwritten corrections to text, often using specialized symbols.

> **TIP:** When Word suggests the completed word for "September," "November," and "Saturday," you can press Enter to insert the word. Remember to press Spacebar after the completed word.

FIGURE 1-18

The lower airfares are good for travel between ~~June~~ 15 and ~~September~~ [September] [November]
15 only. Travelers must stay over one Saturday night to qualify. The
reservation policy‸is first come‸first served. Remember: Groups booked
on these low fare flights will not neccessarily be booked on the same
carrier and will not necessarily arrive at the same time.

> **NOTE:** When starting a new paragraph, remember to press Enter twice to insert a blank line between paragraphs.

4. In the first sentence, key **new** to the left of the word "lower." Add a comma after "new." Then delete the word "lower" and key **reduced**.

5. Insert a hyphen between "low" and "fare" in the last sentence. No spaces should appear before or after the hyphen.

6. In the same sentence, delete the colon after "Remember," key **that**, and make the "G" in "Groups" lowercase.

7. Save the document as *[your initials]***1-17** on your data disk.

8. Print the document, and then close it.

Key, edit, and print a document.

1. Start a new document and key the two paragraphs shown in Figure 1-19 (be sure to leave a blank line between the paragraphs).

FIGURE 1-19

As its slogan says, Duke City Gateway truly is the gateway to the
Southwest, offering the greatest number of specialized tours to New
Mexico, Arizona, and Colorado.

The agency's tour planners and tour guides are considered experts in
the history and culture of this area.

2. In the first paragraph, key **Travel** after "Gateway."
3. Switch to Overtype mode and key **Colorado, and Arizona** over the text "Arizona, and Colorado."
4. Turn off Overtype mode and move the insertion point immediately after the "s" in the word "agency's" in the second paragraph.
5. Use Ctrl + Backspace to delete "The agency's" and key **Duke City Gateway's** in its place.
6. Before "area" (the last word in the last paragraph), key **remarkable**.
7. Save the document as *[your initials]*1-18 on your data disk.
8. Print, and then close the document.

EXERCISE 1-19

Key, edit, and print a document.

1. Start a new document. Key the text shown in Figure 1-20. Use single spacing and key each sentence on a new line.

FIGURE 1-20

Reminder
Will the last one to leave the office:
Turn off all computers.
Turn down the thermostat.
Turn off the coffee machine.
Turn off the lights.
Thank you!

2. In the third line, change "computers" to **computer equipment**.

3. In the next-to-last line, change "the lights" to **all lighting fixtures**.

4. Before the "Thank you!" line, insert another line by keying **And don't forget to turn on the alarm and lock the door.**

5. Use CapsLock and Overtype mode to change "Reminder" and "Thank you!" to all uppercase letters.

6. Insert a blank line after every line of text by pressing Enter once.

7. Save the document as *[your initials]*1-19 on your data disk.

8. Print, and then close the document.

EXERCISE 1-20 ✚ *Challenge Yourself*

Key, edit, and print a document.

1. Start a new document and key the two paragraphs shown in Figure 1-21, including the corrections. Refer to Appendix A: "Proofreaders' Marks," if necessary.

FIGURE 1-21

Group travel is an imprtant part of the Gateway travel business. Tours have been organized thru the mountains, over the seas, and across the world. Groups have traveled on buses, ships, planes, trains and 4-wheel-drive vehicles.

Group ~~travel is~~ tours are arranged for families, students, singles, senior citizens, ~~disabled people~~, and any group with individual needs. Stet

2. Correct the spelling of "thru" in the second sentence to **through**.

3. In the last sentence, key **type of** after "and any." Delete "individual" and key **special** in its place.

4. Save the document as *[your initials]*1-20 on your data disk.

5. Print, and then close the document

6. Exit Word.

On Your Own

In these exercises you work on your own, as you would in a real-life work environment. Use the skills you've learned to accomplish the task—and be creative.

EXERCISE 1-21
Write a simple paragraph about yourself that includes your first and last name. Switch to Overtype mode, turn on Caps Lock, and then key over your name in uppercase letters. Save the document as *[your initials]*1-21 and print it.

EXERCISE 1-22
Browse menu options and toolbar buttons until you find something that looks intriguing; then find out more about it by using the Ask a Question box. In a new blank document, write a brief paragraph about your findings. Save the document as *[your initials]*1-22 and print it.

EXERCISE 1-23
Log onto the Internet and search for sites that relate to a particular interest of yours. Jot down a few Web addresses, and then key the addresses into a blank document, under an appropriate heading. Save the document as *[your initials]*1-23 and print it.

LESSON 2

Selecting and Editing Text

OBJECTIVES

After completing this lesson, you will be able to:

1. Open an existing document.
2. Enter nonprinting characters.
3. Move within a document.
4. Undo and Redo actions.
5. Repeat actions.
6. Select text.
7. Save a revised document.
8. Work with document properties.

 Estimated Time: 1¼ hours

MICROSOFT OFFICE
SPECIALIST
ACTIVITIES
In this lesson:
WW03S-5-2
WW03S-5-4
WW03E-4-6

See Appendix C.

To edit documents efficiently, you need to learn to select text and move quickly within a document. In this lesson you learn those skills, as well as how to open and save an existing document.

Opening an Existing Document

Instead of creating a new document, you start this lesson by opening an existing document. There are several ways to open a document:

- Click the Open button on the Standard toolbar
- Choose <u>O</u>pen from the File menu

● Press Ctrl+O
● Use the document links on the Getting Started task pane

EXERCISE 2-1 Open an Existing File

1. On the menu bar, click <u>F</u>ile to open the File menu. Expand the menu (click the arrows at the bottom of the menu). The filenames listed at the bottom of the menu are the last four files opened from this computer. If the file you want is listed, you can click its name to open it from this list.

2. Click <u>O</u>pen to display the Open dialog box. You're going to open a student file called **Duke1** but first, you must locate the appropriate drive on your computer.

3. Click the down arrow to the right of the Look <u>i</u>n box and choose the appropriate drive according to your instructor's directions.

FIGURE 2-1
Files listed in the
Open dialog box

Document files listed
on the CD-ROM

NOTE: If your instructor tells you to open a specific folder, double-click the folder in the list box under the Look <u>i</u>n box.

4. After you locate the student files, click the arrow next to the Views button in the Open dialog box to display a menu of view options.

5. Choose <u>L</u>ist to list all files by filename. (See Figure 2-2 on the next page.)

6. Click the arrow next to the Views button again and choose Pre<u>v</u>iew. This view provides a quick look at a document before you open it.

FIGURE 2-2
Views menu in the
Open dialog box

7. From the list of filenames, locate **Duke1** and click it once to select it.

8. Click <u>O</u>pen.

 TIP: You can also double-click a filename to open a file.

TABLE 2-1   Open Dialog Box Buttons

BUTTON	NAME	PURPOSE
	My Recent Documents	Lists the most recently opened files and folders.
	Desktop	Lists items on your computer's desktop.
	My Documents	Opens a folder Windows provides to help you organize your document files.
	My Computer	Lists drives, folders, and files you have available on your computer.
	My Network Places	Lists folders and files you can access on shared network drives and on Web servers.
	Back	Goes to the most recent previous location displayed in the Look in box.
	Up One Level	Moves up one level in the hierarchy of folders or drives on your computer or on computers connected to your network.
	Search the Web	Opens the Search page of your Internet browser so you can search the Web for information.
	Delete	Deletes a selected file or folder.
	Create New Folder	Lets you create a new folder to organize your files.
	Views	Opens a menu of view options for displaying drives, folders, files, and their icons.
	Tools	Opens a menu of other file functions, such as finding a file, renaming a file, and adding a file to the Favorites folder.

NOTE: The buttons in the Open dialog box also appear in the Save As dialog box.

EXERCISE   **2-2**   **Create a New Folder**

Document files are typically stored in folders that are part of a hierarchal structure similar to a family tree. At the top of the tree is a disk drive letter (such as C: or A:) that represents your computer, network, floppy drive, or CD-ROM drive. Under the disk drive letter, you can create folders to organize your files. These folders can also contain additional folders.

Here's a scenario: You store your files on the C: drive of your office computer. You create a folder on this drive named "Word Documents." Within this folder you create folders named "Letters," "Memos," and "Reports," each containing different types of documents.

For this course, you'll create a new folder for each lesson. You'll store your completed exercise documents in these folders.

1. Open the File menu and choose Save As. You're going to save **Duke1** under a new filename, in a new folder that will contain all the files you save in this lesson.

2. Choose the appropriate drive and folder location from the Save in drop-down list. (For example, to save your files to a floppy disk, put a disk in the drive and make sure Save in indicates drive A:.)

 NOTE: Ask your instructor where to create the new folder.

3. Click the Create New Folder button . The New Folder dialog box appears.

4. In the Name box, key the folder name *[your initials]***Lesson2** and click OK. Now your new folder's name appears in the Save in box. Word is ready to save the file in the new folder.

5. In the File name box, make sure the file's original name (**Duke1**) is selected. If not, double-click it.

6. Key the filename *[your initials]***2-2** and click Save.

 NOTE: To rename a folder, locate the folder to rename and right-click the folder. Click Rename on the shortcut menu, key the new name, and press Enter.

Nonprinting Characters

Lesson 1 introduced the Show/Hide ¶ button ¶ on the Standard toolbar, which shows or hides paragraph marks and other *nonprinting characters*. These characters appear on the screen, but not in the printed document. Nonprinting characters are included as part of words, sentences, and paragraphs in a document. Here are some examples:

- A word includes the space character that follows it.
- A sentence includes the end-of-sentence punctuation and at least one space.
- A paragraph is any amount of text followed by a paragraph mark.

The document you opened contains two additional nonprinting characters: *tab characters,* which you use to indent text, and *line-break characters,* which you use to start a new line within the same paragraph. Line-break characters are useful when you want to create a paragraph of short lines, such as an address, and keep the lines together as a single paragraph.

Another nonprinting character is a *nonbreaking space,* which you use to prevent two words from being divided between two lines. For example, you can insert a nonbreaking space between "Mr." and "Smith" to keep the name "Mr. Smith" undivided on one line.

TABLE 2-2

Nonprinting Characters	
CHARACTER	**TO INSERT, PRESS**
Tab (→)	Tab
Space (·)	Spacebar
Nonbreaking space (°)	Ctrl + Shift + Spacebar
Paragraph mark (¶)	Enter
Line-break character (↵)	Shift + Enter

EXERCISE   **2-3**   **Enter Nonprinting Characters**

1. Click the Show/Hide ¶ button ¶ if the nonprinting characters in the document are hidden.
2. Move the insertion point to the end of the document (after "without a fee.").
3. Press Enter twice to begin a new paragraph and key **Duke City Gateway Travel has been a member in good standing of the IATA for 25**. (Do not press Spacebar.)

4. Insert a nonbreaking space after "25" by pressing Ctrl+Shift+Spacebar. Then key **years.** (including the period). Word now treats "25 years" as a single unit.

FIGURE 2-3
Nonprinting
characters

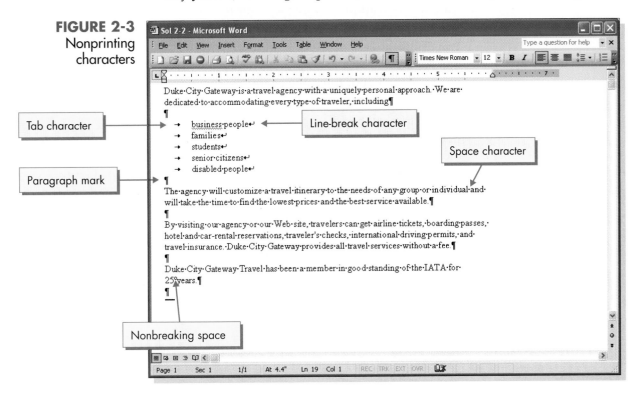

5. Press Enter twice and key the following text as one paragraph at the end of the document, pressing Shift+Enter at the end of the first and second lines instead of Enter.

Duke City Gateway Travel
15 Montgomery Boulevard
Albuquerque, NM 87111

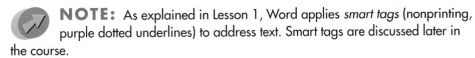

NOTE: As explained in Lesson 1, Word applies *smart tags* (nonprinting, purple dotted underlines) to address text. Smart tags are discussed later in the course.

6. Click the Show/Hide ¶ button ¶ to hide the nonprinting characters and click it again to redisplay them.

Moving Within a Document

You already know how to move around a short document by positioning the I-beam pointer with the mouse and clicking. This is the easiest way to move around a document that fits in the document window. If a document is too long or wide to fit in the window, you need to use different methods.

Word offers two additional methods for moving within a document:

- Using the keyboard
  You can press certain keys on the keyboard to move the insertion point. The Arrow keys, for example, move the insertion point up or down one line or to the left or right one character. Key combinations quickly move the insertion point to specified locations in the document.

- Using the scroll bars
  Use the vertical scroll bar at the right edge of the document window to move through a document. The position of the scroll box indicates your approximate location in the document, which is particularly helpful in long documents. To view and move through a document that's wider than the document window, use the horizontal scroll bar at the bottom of the document window.

NOTE: Scrolling through a document does not move the insertion point. It moves only the portion of the document you are viewing in the document window. When you use the keyboard to move within a document, the insertion point always moves to the new location.

EXERCISE **2-4** **Use the Keyboard to Move the Insertion Point**

1. Press [Ctrl]+[Home] to move to the beginning of the document. Press [End] to move to the end of the first line.

2. Press [Ctrl]+[↓] several times to move the insertion point down one paragraph at a time. Notice how the text with the line-break characters is treated as a single paragraph.

3. When you reach the end of the document, press [Page Up] until you return to the beginning of the document.

TABLE 2-3 **Keys to Move the Insertion Point**

TO MOVE	PRESS
One word to the left	[Ctrl]+[←]
One word to the right	[Ctrl]+[→]
Beginning of the line	[Home]
End of the line	[End]
One paragraph up	[Ctrl]+[↑]
One paragraph down	[Ctrl]+[↓]

continues

TABLE 2-3   **Keys to Move the Insertion Point** *continued*

TO MOVE	PRESS
Up one window	Page Up
Down one window	Page Down
Top of the window	Alt + Ctrl + Page Up
Bottom of the window	Alt + Ctrl + Page Down
Beginning of the document	Ctrl + Home
End of the document	Ctrl + End

> **TIP:** Word remembers the last three locations in the document where you edited or keyed text. You can press Shift + F5 to return the insertion point to these locations. For example, when you open a document you worked on earlier, press Shift + F5 to return to the place where you were last working before you saved and closed the document.

EXERCISE   **2-5**   **Scroll Through a Document**

Using the mouse and the scroll bars, you can scroll up, down, left, and right. You can also set the Previous and Next buttons on the vertical scroll bar to scroll through a document by a specific object, such as tables or headings. For example, these buttons let you jump from one heading to the next, going forward or backward.

1. In the vertical scroll bar, click below the scroll box to move down one window. (See Figure 2-4 on the next page.)

2. Drag the scroll box to the top of the scroll bar.

3. Click the down scroll arrow 🔽 on the scroll bar three times. The document moves three lines.

4. Click the right scroll arrow ▶ on the horizontal scroll bar once, and then click the left scroll arrow ◀ once to return to the correct horizontal position.

5. Click the up scroll arrow 🔼 on the vertical scroll bar three times to bring the document back into full view.
   Notice that as you scroll through the document, the insertion point remains at the top of the document.

6. Click the Select Browse Object button 🔘, located toward the bottom of the vertical scroll bar. A menu of icons appears.

FIGURE 2-4
Using the scroll bars

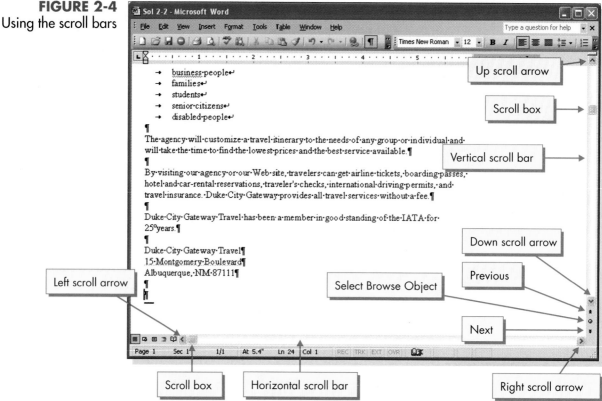

7. Move the pointer over each icon to identify it. These browse options become significant as your documents become more complex. Click the Browse by Page icon.

TABLE 2-4   Scrolling Through a Document

TO MOVE	DO THIS
Up one line	Click the up scroll arrow ▲.
Down one line	Click the down scroll arrow ▼.
Up one window	Click the scroll bar above the scroll box.
Down one window	Click the scroll bar below the scroll box.
To any relative position	Drag the scroll box up or down.
To the right	Click the right scroll arrow ▶.
To the left	Click the left scroll arrow ◀.
Into the left margin	Hold down Shift and click the left scroll arrow ◀.
Up or down one page	Click Select Browse Object ⊙, click Browse by Page ▯, and then click Next ▼ or Previous ▲.

TIP: If you're using a mouse with a wheel, additional navigating options are available. For example, you can roll the wheel forward or backward instead of using the vertical scroll bars, hold down the wheel and drag in any direction to pan the document, or hold down Ctrl as you roll the wheel to change the magnification.

Undo and Redo Commands

Word remembers the changes you make in a document and lets you undo or redo these changes. For example, if you accidentally delete text, you can use the Undo command to reverse the action and restore the text. If you change your mind and decide to keep the deletion, you can use the Redo command to reverse the canceled action.

There are three ways to undo or redo an action:

- Click the Undo button or the Redo button on the Standard toolbar.
- From the Edit menu, choose Undo or Redo.
- Press Ctrl+Z to undo or Ctrl+Y to redo.

EXERCISE 2-6 Undo and Redo Actions

1. Delete the first word in the document, "Duke," by moving the insertion point to the right of the space after the word and pressing Ctrl+Backspace. (Remember that a word includes the space that follows it.)

2. Click the Undo button to restore the word.

3. Move the insertion point to the left of the word "airline" in the paragraph that begins "By visiting our."

4. Key **their** and press Spacebar once. The text now reads "can get their airline tickets."

5. From the Edit menu, choose Undo Typing. (Notice the reminder that the keyboard command Ctrl+Z exists for the same action.) The word "their" is deleted.

6. Click the Redo button to restore the word "their." (You may need to click the Toolbar Options button on the Standard toolbar to locate the Redo button.)

7. Click the down arrow to the right of the Undo button. Word displays a drop-down list of the last few actions, with the most recent action at the top. You can use this feature to choose several actions to undo rather than just the last action. Click the down arrow again to close the list. (See Figure 2-5 on the next page.)

FIGURE 2-5
Undo drop-down list

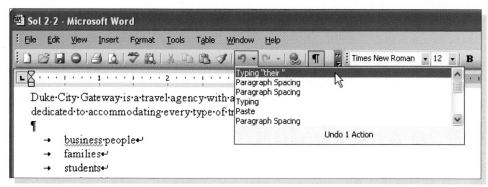

Repeat Command

Suppose you key text you want to add to other areas of a document. Instead of rekeying the same text, you can use the Repeat command to duplicate the text.

There are two ways to use the Repeat command:

● Press F4 or Ctrl+Y.

● Choose Repeat from the Edit menu.

EXERCISE 2-7   Repeat Actions

1. In the paragraph that begins "The agency will," position the insertion point to the left of the word "agency."

2. Key **travel** and press Spacebar once. The sentence now begins "The travel agency will."

3. Move the insertion point to the left of the word "agency" in the next paragraph.

4. Press F4 and the word "travel" is repeated.

 NOTE: If you want to undo, redo, or repeat your last action, do so before you press another key.

Selecting Text

Selecting text is a basic technique that makes revising documents easy. When you select text, that area of the document is called the *selection,* and it appears as a high-lighted block of text. A selection can be a character, group of characters, word, sentence, or paragraph or the whole document. In this lesson, you delete and replace selected text. Future lessons show you how to format, move, copy, delete, and print selected text.

You can select text several ways, depending on the size of the area you want to select.

TABLE 2-5   Mouse Selection

TO SELECT	USE THE MOUSE TO
A series of characters	Click and drag, or click one end of the text block, and then hold down Shift and click the other end.
A word	Double-click the word.
A sentence	Hold down Ctrl and click anywhere in the sentence.
A line of text	Move the pointer to the left of the line until it changes to a right-pointing arrow, and then click. To select multiple lines, drag up or down.
A paragraph	Move the pointer to the left of the paragraph and double-click. To select multiple paragraphs, drag up or down.
The entire document	Move the pointer to the left of any document text until it changes to a right-pointing arrow, and then triple-click (or hold down Ctrl and click).

EXERCISE   2-8   Select Text with the Mouse

1. Select the first word of the document by double-clicking it. Notice that the space following the word is also selected.

FIGURE 2-6
Selecting a word

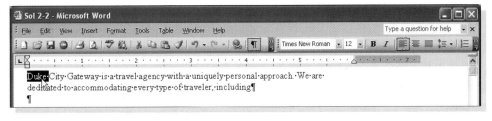

2. Cancel the selection by clicking anywhere in the document. Selected text remains highlighted until you cancel the selection.
3. Select the first sentence by holding down Ctrl and clicking anywhere within the sentence. Notice that the period and space following the sentence are part of the selection. Cancel the selection.
4. Locate the paragraph that begins "By visiting our."
5. To select the text "traveler's checks," click to the left of "traveler's." Hold down the left mouse button and slowly drag through the text, including the comma and space after "checks." Release the mouse button. Cancel the selection.

TIP: When selecting more than one word, you can click anywhere within the first word, and then drag to select additional text. Word will "smart select" the entire first word.

6. To select the entire paragraph by dragging the mouse, click to position the insertion point to the left of "By." Hold down the mouse button, and then drag across and down until all the text and the paragraph mark are selected. Cancel the selection.

7. Select the same paragraph by moving the pointer into the blank area to the left of the text. (This is the margin area.) When the I-beam pointer changes to a right-pointing arrow, double-click. Notice that the first click selects the first line and the second click selects the paragraph, including the paragraph mark. Cancel the selection.

FIGURE 2-7
Selecting text

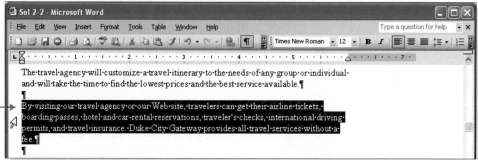

Point and double-click to select paragraph.

 TIP: You can also triple-click within a paragraph to select it.

EXERCISE 2-9 Select Noncontiguous Text

In the previous exercise, you learned how to select *contiguous text*, where the selected characters, words, sentences, or paragraphs follow one another. But sometimes you'd like to select *noncontiguous text,* such as the first and last items in a list or the third and fifth word in a paragraph. In Word, you can select noncontiguous text by using Ctrl and the mouse.

1. Select the first line of the list ("business people").

2. Hold down Ctrl and select the third line of the list ("students"). With these two separate lines selected, you can delete, format, or move them without affecting the rest of the list.

3. Cancel the selection and go to the paragraph that begins "By visiting our."

4. In the first sentence, double-click the word "our" before "Web site." With the word now selected, hold down Ctrl as you double-click the word "their" in the same sentence and "travel" in the next sentence. (See Figure 2-8.) All three words are highlighted.

FIGURE 2-8
Selecting
noncontiguous
words

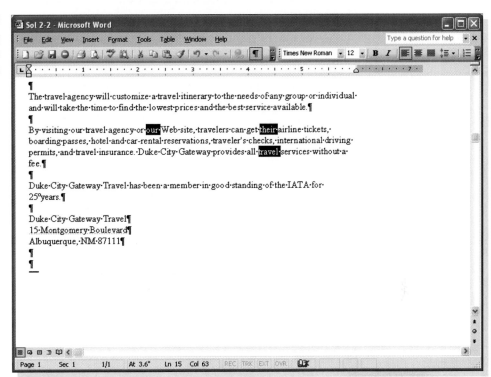

5. Press Delete to delete the selected words.

EXERCISE 2-10 Adjust a Selection Using the Mouse and the Keyboard

1. Select the paragraph beginning "By visiting our travel agency."
2. Hold down Shift and press ← until the last sentence is no longer highlighted. Release Shift.
3. Increase the selection to include the last sentence by holding down Shift and pressing End. Release Shift.
4. Increase the selection to include all the text below it by holding down Shift and clicking at the end of the document (after the ZIP Code).
5. Select the entire document by pressing Ctrl+A. Cancel the selection.

TABLE 2-6   Keyboard Selection

TO SELECT	PRESS
One character to the right	Shift + →
One character to the left	Shift + ←
One word to the right	Ctrl + Shift + →
One word to the left	Ctrl + Shift + ←
To the end of a line	Shift + End
To the beginning of a line	Shift + Home
One line up	Shift + ↑
One line down	Shift + ↓
One window down	Shift + Page Down
One window up	Shift + Page Up
To the end of a document	Ctrl + Shift + End
To the beginning of a document	Ctrl + Shift + Home
An entire document	Ctrl + A

EXERCISE 2-11 Edit Text by Replacing a Selection

You can edit a document by selecting text and deleting or replacing the selection.

1. In the paragraph that begins "By visiting," select the words "travel agency" using the Shift+Click method: click to the left of the word "travel," hold down Shift, and click to the right of the word "agency."

2. Key **office** to replace the selected text.

3. In the paragraph that begins "Duke City Gateway Travel has been," select "IATA" and key **International Air Travel Association**. Notice that, unlike using Overtype mode, when you key over selected text, the new text can be longer or shorter than the selection.

NOTE: Although keying over selected text is an excellent editing feature, it sometimes leads to accidental deletions. Remember, when text is selected in a document (or even in a dialog box) and you begin keying text, Word deletes all the selected text with your first keystroke. If you key text without realizing a portion of the document is selected, use the Undo command to restore the text.

Saving a Revised Document

You already used the Save As command to rename the document you loaded at the beginning of this lesson. Now that you have made additional revisions, you can save a final version of the document by using the Save command. The document is saved with all the changes, replacing the old file with the revised file.

 NOTE: If you wanted to save the current document with a different filename, you would use the Save As command.

EXERCISE **2-12** **Save a Revised Document**

 1. Click the Save button 🖫 on the Standard toolbar. This action does not open the Save As dialog box. Instead, the Save icon flashes quickly at the right end of the Status bar, indicating that Word is saving the document.

 2. Click the Print button 🖨 to print the document.

3. Close the document and exit Word. If you forget to save a document before you close it, Word reminds you to save when you try to close the application.

EXERCISE **2-13** **Use the Web Page File Format**

When you save a file as a Word document, you need Word to open and read the file. However, when you save a file as a Web page, anyone can open and read the document by using a browser. Web pages use the HTML (*Hypertext Markup Language*) file format, which makes a document readable in a browser or on an intranet—without opening Word.

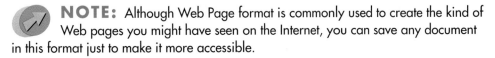 **NOTE:** Although Web Page format is commonly used to create the kind of Web pages you might have seen on the Internet, you can save any document in this format just to make it more accessible.

1. Open the file *[your initials]*2-2.

2. Open the File menu and choose Save as Web Page (you might have to expand the menu). At the bottom of the Save As dialog box, click the down arrow for the Save as type box and choose Web Page.

3. In the File name box, change the name from 2-2 to 2-13.

4. Check that the Save in box shows your Lesson 2 folder, and then click Save. The file is saved in Web Page format and Word changes the view to Web Layout view.

5. To see exactly how your document would look in a browser, open the File menu and choose Web Page Preview. Word opens your browser automatically and displays your document.

6. Close your browser window by clicking its Close button ☒.

EXERCISE **2-14** **Check Word's AutoRecover Settings**

Word's *AutoRecover* feature can automatically save open documents at an interval you specify. However, this isn't the same as saving a file yourself, as you did in the preceding exercise. AutoRecover's purpose is to save open documents "in the background," so a recently saved version is always on disk. Then, if the power fails or your system crashes, the AutoRecover version of the document opens automatically the next time you launch Word. In other words, AutoRecover ensures you always have a recently saved version of your document.

Even with AutoRecover working, you need to manually save a document (by using the Save command) before closing it. AutoRecover documents are not always available; if you save and close your file normally, the AutoRecover version is deleted when you exit Word. Still, a good idea is to make sure AutoRecover is working on your system, and set it to save recovery files frequently.

1. Open the Tools menu and click Options to open the Options dialog box.

2. Click the Save tab.

3. Make sure the Save AutoRecover info every box is checked. If it is not checked, click the box.

4. Click the up or down arrow buttons to set the minutes to 5. Click OK.

FIGURE 2-9
Setting
AutoRecover
options

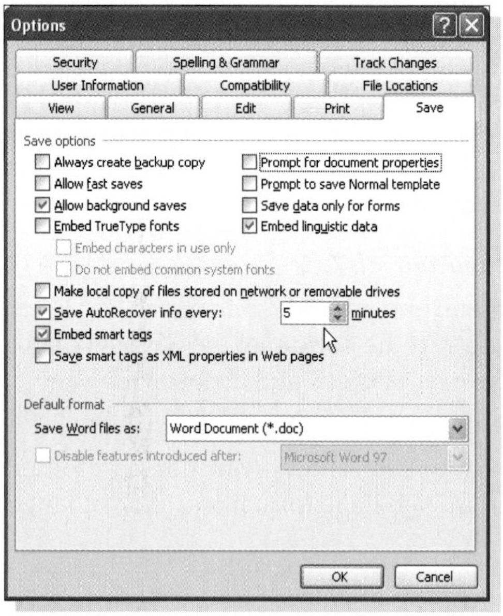

Working with Document Properties

Information that describes your document is called a *property*. Word automatically saves your document with certain properties, such as the filename, the date created, and the file size. You can add other properties to a document, such as the title, subject, author's name, and keywords. This information can help you organize and identify documents.

 NOTE: You can also search for documents based on document properties by using the Search task pane.

EXERCISE **2-15** **Review and Edit Document Properties**

1. With the file *[your initials]*2-13 still open, open the File menu and choose Properties.

2. In the Properties dialog box, click the General tab. This tab displays basic information about the file, such as filename, file type, location, size, and creation date.

3. Click the Statistics tab. This tab shows the exact breakdown of the document in number of paragraphs, lines, words, and so on.

FIGURE 2-10
Entering
Summary
information

Sol 2-13 Properties	⊠

General **Summary** Statistics Contents Custom

Title: Duke City Gateway Introduction

Subject:

Author: Student Name

Manager:

Company:

Category:

Keywords:

Comments:

Hyperlink
base:

Template: Normal

☐ Save preview picture

OK   Cancel

4. Click the Summary tab. Here you can enter specific document property information or change existing information. Notice that Word assigned a temporary title using the document's first sentence.

5. Edit the title to read **Duke City Gateway Introduction** and key your name as the author. Click OK.

6. Save the document, print it, and then close it.

USING ONLINE HELP

You have already learned how to use the Microsoft Word Help feature by using the Ask a Question text box in the upper right corner of the screen. You can also display a Help task pane that provides a Search box or access to a Table of Contents.

Display Microsoft Word Help task pane:

1. Open a new, blank document.
2. Press F1 to display the Microsoft Word Help task pane.

FIGURE 2-11
Help menu

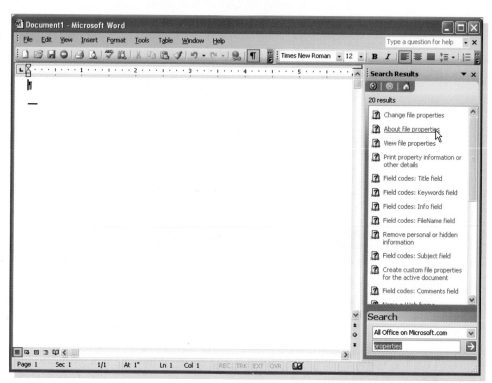

3. Click the Search box and key **properties**. Press Enter or click the Start Searching arrow →.
4. Choose the About file properties link. The Microsoft Word Help window displays.
5. Click the Show All link and read the information about the four types of file properties.
6. Close the Microsoft Word Help window and close the document without saving.

LESSON 2 Summary

- ➤ Use the Open dialog box to open an existing file. Use the Views button ▦ in the dialog box to change the way files are listed.

- ➤ Create folders to organize your files. You can do this in the Save As dialog box, using the Create New Folder button ▣. Rename folders by locating and selecting the folder. Right-click the folder name and choose Rename from the shortcut menu.

- ➤ Nonprinting characters—such as blank spaces or paragraph marks—appear on-screen, but not in the printed document. Insert a line-break character to start a new line within the same paragraph. Insert a nonbreaking space between two words to make sure they appear on the same line.

- ➤ Use the Show/Hide ¶ button ¶ to turn the display of nonprinting characters on and off.

- ➤ When a document is larger than the document window, use the keyboard or the vertical scroll bar to view different parts of the document. Keyboard methods for moving within a document also move the insertion point.

- ➤ Keyboard techniques for moving within a document include single keys (such as (Page Up) and (Home)) and keyboard combinations (such as (Ctrl)+(↑)). See Table 2-3.

- ➤ Scrolling techniques for moving within a document include clicking the up or down scroll arrows on the vertical scroll bar or dragging the scroll box. Scrolling does not move the insertion point. See Table 2-4.

- ➤ If you make a change in a document that you want to reverse, use the Undo command. Use the Redo command to reverse the results of an Undo command.

- ➤ If you perform an action, such as keying text in a document, and you want to repeat that action elsewhere in the document, use the Repeat command.

- ➤ Selecting text is a basic technique for revising documents. A selection is a highlighted block of text you can format, move, copy, delete, or print.

- ➤ There are many different techniques for selecting text, using the mouse, the keyboard, or a combination of both. Mouse techniques involve dragging or clicking. See Table 2-5. Keyboard techniques are listed in Table 2-6.

- ➤ You can select any amount of contiguous text (characters, words, sentences, or paragraphs that follow one another) or noncontiguous text (such as words that appear in different parts of a document). Use (Ctrl) along with the mouse to select noncontiguous blocks of text.

- ➤ When text is selected, Word replaces it with any new text you type, or it deletes the selection if you press (Delete).

- ➤ Use the Save command to save any revisions you make to a document.

➤ Save a document in Web Page file format, known as HTML, as an alternative to Word Document file format. HTML format makes the document readable in a browser, on an intranet, or on the Internet.

➤ Word's AutoRecover feature periodically saves open documents in the background so you can recover a file in the event of a power failure or system crash.

➤ Document properties are details about a file that help identify it. Properties include the filename, file size, and date created, which Word updates automatically. Other properties you can add or change include title, subject, author's name, and keywords. View or add properties for an open document by using the Properties command (File menu).

LESSON 2   Command Summary

FEATURE	BUTTON	MENU	KEYBOARD
Open		File, Open	Ctrl+O or Ctrl+F12
Undo		Edit, Undo	Ctrl+Z or Alt+Backspace
Redo		Edit, Redo	Ctrl+Y or Alt+Shift+Backspace
Repeat		Edit, Repeat	Ctrl+Y or F4
Select entire document		Edit, Select All	Ctrl+A
Save		File, Save	Ctrl+S or Shift+F12
Save as Web Page		File, Save as Web Page	

Concepts Review

TRUE/FALSE QUESTIONS

Each of the following statements is either true or false. Indicate your choice by circling T or F.

T   F   *1.* You can preview a file in the Open dialog box.

T   F   2. A line-break character is used to begin a new paragraph.

T   F   3. A tab mark is a nonprinting character.

T   F   4. The Save As dialog box is used to create a new folder.

T   F   5. Noncontiguous text is text that does not appear consecutively in a Word document.

T   F   6. You can undo only the last change made to a document.

T   F   7. To select a sentence, you can double-click anywhere within the sentence.

T   F   8. You can increase a selection by using Shift+End.

SHORT ANSWER QUESTIONS

Write the correct answer in the space provided.

1. Which menu contains a list of the last four files opened?

_____

2. Which nonprinting character would you insert between two words to keep them together in a sentence?

_____

3. What is the keyboard shortcut to move the insertion point to the beginning of a document?

_____

4. Where does Home move the insertion point?

_____

5. What portion of the vertical scroll bar can you drag up or down?

_____

6. Which keyboard shortcut repeats the text you just keyed?

_____

7. How do you preview a file in the Open dialog box?

_____

8. Which file format is used to save a document as a Web page?

_____

CRITICAL THINKING

Answer these questions on a separate page. There are no right or wrong answers. Support your answers with examples from your own experience, if possible.

1. You can use a nonbreaking space to prevent a line break between two words. Give some examples of word combinations and word-and-number combinations in which you would use a nonbreaking space.

2. Word provides many ways to select text by using the mouse. For example, you can use just the mouse, the mouse in combination with the keyboard, or just the keyboard. When would each of these methods be preferable? Which method do you prefer, and why?

Skills Review

EXERCISE 2-16

Open a document and enter nonprinting characters.

1. Open the document **SanFran** by following these steps:

a. Click the Open button on the Standard toolbar.

b. Click the arrow to the right of the Views button and choose List.

c. Click the down arrow to the right of the Look in box to choose the appropriate drive, according to your instructor's directions. Click the drive. Make sure the appropriate folder is open; if not, double-click it.

d. Scroll to the filename **SanFran** and double-click it.

2. Position the insertion point at the end of the document by pressing `Ctrl`+`End`.

3. Key the text shown in Figure 2-12 as one paragraph, using line-break characters, by following these steps:

a. Press `Enter` twice at the end of the document.

b. Key the first three lines of text, pressing `Shift`+`Enter` at the end of each line.

c. Key the last line (the phone number).

4. Add a new word and a nonbreaking space by following these steps:

a. Move the insertion point to the immediate left of the word "area" in the first paragraph.

b. Key **Bay** and then press `Ctrl`+`Shift`+`Spacebar`.

5. Change "area" to "Area."

FIGURE 2-12

```
Golden Gateway Travel
Union Square
San Francisco, CA 94102
(415) 555-8694
```

6. Save the document as *[your initials]*2-16 in your Lesson 2 folder.

7. Print and close the document.

EXERCISE 2-17

Move within a document.

1. Open the file **Overseas1**.

2. Move the insertion point to the immediate left of "We'll" in the third line and press (Enter) twice to split the paragraph.

3. Move the insertion point to the immediate left of the second sentence in the new paragraph (beginning with "We aim") and press (Enter) twice to split the paragraph.

4. With the insertion point to the left of "We," press ← twice to move the insertion point to the end of the previous paragraph and key **We even have some suggestions about that!**

5. Press (Ctrl)+(End) to move to the end of the document. Press (Spacebar) and key **Call us today or visit our Web site.**

6. Press (Ctrl)+(Home) to move to the beginning of the document. Press (Caps Lock) and key **thinking of overseas travel?** Press (Enter) twice to split the paragraph.

7. Save the document as *[your initials]*2-17 in your Lesson 2 folder.

8. Print and close the document.

EXERCISE 2-18

Undo, redo, and repeat editing actions; save a file in Web page format.

1. Open the file **Business**.

2. Delete the paragraph mark under the heading "Corporate Travel" and click the Undo button to undo the deletion.

3. Move the insertion point to the left of "men" in the first sentence. Key **business** to create the word "businessmen."

4. In the same sentence, move the insertion point to the left of "women." Press (F4) to create the word "businesswomen."

5. In the first part of the next sentence, move the insertion point to the left of "traveler." Key **business** and press (Spacebar).

6. Near the beginning of the next paragraph, use the Repeat command to insert "business" to the left of "travelers."

7. Click the Undo button ⟲ ▾ to undo the text. Then click the Redo button ⟳ ▾ to redo the text.

8. In the first sentence of the second paragraph, replace the text "most efficient" with the word **best** by following these steps:

   a. Select the text "Most efficient."
   b. Key **best**.

9. Save the document, using HTML file format, by following these steps:

   a. Choose Save as Web Page from the File menu.
   b. Key the filename *[your initials]*2-18 in the File name text box.
   c. Choose your Lesson 2 folder from the Save in drop-down list.
   d. Click Save.

   NOTE: Files saved using the .doc extension will not display in the Save as Web Page dialog box unless you change the Save as type option.

10. Print and close the document.

EXERCISE 2-19

Select text, save a revised document, and enter summary information.

1. Start a new document. Key the text shown in Figure 2-13.

FIGURE 2-13

Traveling together with your family can be a rewarding experience, one that will be remembered fondly by all. Duke City Gateway has planned hundreds of trips for families, including both domestic and overseas travel. Travel is educational and stimulates the curiosity of children. Travel brings people closer together.

Family trips are relaxing, adventurous, enriching, and can even include visiting relatives along the way. Whatever your needs, we can create an itinerary to match them. We can recommend the best campsites and the most popular dude ranches. We can help you find a recreational vehicle or a four-wheel-drive vehicle. And, if you're traveling with a feline or canine member of your family, we can tell you where pets are welcome.

2. Save the document as *[your initials]*2-19 in your Lesson 2 folder.

3. Select and replace a word by following these steps:

   a. Place the insertion point in the word "people" in the last sentence of the first paragraph.
   b. Double-click the word to select it and key **families** to replace it.

4. Select and replace the words "are relaxing" in the second paragraph by following these steps:

   a. Move the I-beam pointer just before "are" and click to position the insertion point.

   b. Hold down the mouse button and drag to the right until both words are highlighted. Release the mouse button.

   c. Key **can be for relaxation** in place of the old text.

5. In the same sentence, replace "adventurous" with **for adventure** and "enriching" with **for enrichment**.

6. Select a sentence by following these steps:

   a. Position the I-beam pointer over any sentence.

   b. Hold down Ctrl and click the mouse button. Release Ctrl.

   c. Deselect the sentence.

7. Select the first paragraph by following these steps:

   a. Move the pointer to the left of the paragraph until it changes to a right-pointing arrow.

   b. Double-click the mouse button.

8. Extend the current selection by following these steps:

   a. Hold down Shift and press ↓ twice to extend the selection two lines.

   b. Continue holding down Shift and press End to select the entire line.

   c. Continue holding down Shift and press Page Down to select the rest of the document. Release Shift.

9. Click anywhere to cancel the selection.

10. Select noncontiguous text by following these steps:

    a. Double-click the document's first word ("Traveling").

    b. Move the pointer to the left of the paragraph's third line until the pointer changes to a right-pointing arrow. Hold down Ctrl and click to select the line.

    c. Hold down Ctrl and select the second paragraph by dragging the pointer from the beginning of the paragraph to the end.

11. Click anywhere to cancel the selection.

12. Review the document properties and enter summary information by following these steps:

    a. Open the File menu and choose Properties.

    b. Review the data on the General tab and the Statistics tab and click the Summary tab.

    c. Click in the Title box and press Home to move the insertion point to the beginning of the title. Press Shift+End to select the current title. With the title selected, key **Family Travel** as the new title, replacing the old one.

    d. Click OK.

13. Click the Save button to save the revised document.

14. Print and close the document.

Lesson Applications

Select and edit text and enter a nonbreaking space.

1. Open the file **SanFran**. Make the corrections shown in Figure 2-14 by selecting, and then keying over the selected text.

FIGURE 2-14

> Both tourists and native Californians frequent the San Francisco office of Gateway Travel. The office is in a bustling downtown location and is well known for its wine-tasting tours in Sonoma and Napa Valley. Drop in when you're visiting the area.
>
> *(annotations: Bay Area location; San Francisco; the next time; in)*

2. Press Ctrl+End to move to the end of the document. Press Enter twice to start a new paragraph. Then key:

   For information about monthly travel specials, or for directions to the office, call Rebecca Lee at Golden Gateway Travel. You can also visit us on the Web at www.gwaytravel.com.

3. Insert a nonbreaking space between "Rebecca" and "Lee" in the second paragraph. (Replace the regular space with a nonbreaking space.)

4. Save the document as *[your initials]*2-20 in your Lesson 2 folder.

5. Print and close the document.

Select and repeat text and create a paragraph using line breaks.

1. Open the file **Videos**.

2. In the first sentence, replace "our agency" with **Duke City Gateway**.

3. Use the Repeat command to repeat the keyed text to the left of the word "Customers" in the last sentence. Enter a space after "Gateway" and change the "C" in "Customers" to lowercase.

4. In the second sentence, replace "any video" with **these travelogues** and replace "the agency" with **any of our offices**.

5. Move to the end of the document and press Enter twice.

6. Key the text shown in Figure 2-15 as a single paragraph, using a line break for each new line.

FIGURE 2-15

```
Our most popular travelogues are:
The Grand Canyon
Aloha Hawaii
Paris, City of Light
```

7. Save the document as *[your initials]*2-21 in your Lesson 2 folder.

8. Print and close the document.

EXERCISE 2-22

Select and edit text and insert a nonbreaking space.

1. Open the file **OldTown**.

2. Split the first paragraph at the sentence that begins "For shoppers." Be sure to insert a blank line before the new paragraph.

3. Merge the second paragraph with the third paragraph. Be sure to insert a space between sentences.

4. In the second paragraph, spell out "it's" and "You'll." Replace "the best of" with **superb**.

5. At the end of the second paragraph, key this sentence:

   Choose from the colorful array of T-shirts, turquoise, Apache baskets, Navajo wool rugs, and much more.

6. In the last paragraph, change "278 years" to **284** years and add a nonbreaking space between "284" and "years."

7. Between the first and second sentences of the last paragraph, insert this sentence:

   The church sits at the northern edge of Old Town Plaza and is the scene for such events as the Blessing of the Animals in May and the Old Town Fiesta in June.

8. Save the document as *[your initials]*2-22 in your Lesson 2 folder.

9. Print and close the document.

EXERCISE 2-23 *Challenge Yourself*

Select and edit text, use nonprinting characters, save the document in Web page format, and enter summary information.

1. Open the file **Summer**.
2. Revise the document as shown in Figure 2-16.

FIGURE 2-16

Summer is a great time to visit Albuquerque. ~~In June,~~ the New
Mexico Arts & Crafts Fair in June features more than 150 artists and
craftspeople. Visitors can Enjoy free working exhibitions, live entertainment,
and wonderful ethnic food.

The Fourth of July 4th fireworks in Albuquerque are a special treat. Dozens of fantastic aerial
~~displays~~ and ~~fantastic~~ ground displays are all part of the American
Legion Fireworks Spectacular at University Stadium.

In August, the Cowboy Classic Art Show comes to the New Mexico
State Fairgrounds. Also in August, the Inter-Tribal Ceremonial
presents 50 tribes in a dazzling display of parades and dances at
Red Rock State Park.

3. In the second sentence of the first paragraph, change the ending to **more than 200 New Mexican craftspeople and artists.**
4. In the third sentence, replace "wonderful" with **a wide variety of**.
5. Change the last paragraph so it becomes three lines of text, as follows (use line breaks to start new lines):
   Coming in August:
   The Cowboy Classic Art Show at the State Fairgrounds
   The Inter-Tribal Ceremonial at Red Rock State Park
6. Select the three-line paragraph you just created.
7. Press Delete, and then undo the deletion.
8. At the end of the first paragraph, change "food" to **foods**.
9. Save the document as a Web page named *[your initials]*2-23 in your Lesson 2 folder.

10. Open the Properties dialog box. Key **Summer in Albuquerque** as the title, key your name as the author, and key **summer tourist attractions** in the Comments box.

11. Save, print, and close the document.

On Your Own

In these exercises, you work on your own, as you would in a real-life work environment. Use the skills you've learned to accomplish the task—and be creative.

EXERCISE 2-24
Write a short paragraph about a nearby town or city. Print the document. Edit the document, using the skills you learned in this lesson and changing the information to reflect the town or city where you live. Use nonbreaking spaces, if needed. Save the document in Web page format as *[your initials]*2-24 and print it. View the document in your browser.

EXERCISE 2-25
Open the file **Business**. Use your editing skills to shorten each paragraph, making the document more concise. Save the document as *[your initials]*2-25 and print it.

EXERCISE 2-26
Key a short portion of an historical text or novel (written before the twentieth century) so you have about one-half page of document text. Edit the text to make the language more contemporary. On the Summary tab of the Document Properties dialog box, show the original author under Author. Under Comments, key **Modified by** *[your name]*. Save the document as *[your initials]*2-26 and print it.

LESSON 3

Formatting Characters

OBJECTIVES

After completing this lesson, you will be able to:

1. Apply basic character formatting.
2. Change fonts and font sizes.
3. Choose character formats from the Font dialog box.
4. Repeat and copy character formats.
5. Change case.
6. Highlight text.
7. Create a drop cap effect.
8. Automatically format text and numbers.

MICROSOFT OFFICE SPECIALIST ACTIVITIES

In this lesson:
WW03S-3-1
WW03S-5-7

See Appendix C.

 Estimated Time: 1 hour

Character formatting is used to emphasize text. You can change character formatting by making text bold or italic, for example, or by changing the style of the type. Word also provides special features to copy formats, highlight text, and automatically format text and numbers.

Basic Character Formatting

The basic character formats are bold, italic, and underline. Text can have one or more character formats.

TABLE 3-1

Character Formatting

ATTRIBUTE	EXAMPLE
Normal	This is a sample.
Bold	**This is a sample.**
Italic	*This is a sample.*
Underline	<u>This is a sample.</u>
Bold and italic	***This is a sample.***

The simplest ways to apply basic character formatting are to use:

- Buttons on the Formatting toolbar
- Keyboard shortcuts

You can apply character formatting to existing text, including existing text that is noncontiguous. You can also turn on a character format before you key new text and turn it off after you enter the text. For example, you can click the Bold button **B**, key a few words in bold, and click the button again to turn off the format and continue keying regular text.

EXERCISE **Apply Basic Character Formatting Using the Formatting Toolbar**

1. Open the file **Music**.

2. Click the Show/Hide ¶ button to display paragraph marks and space characters if they are not already showing.

3. In the paragraph that begins "July 1," select "The Marriage of Figaro" (not including the period).

4. Click the Bold button **B** on the Formatting toolbar to make the text bold. (You might need to click the Toolbar Options button 📋 to display the Bold button **B**.)

5. With the text still selected, click the Italic button *I* on the Formatting toolbar to make the text bold and italic.

6. Click the Bold button **B** again to turn off the bold format and make the text italic only. Click the Bold button **B** again to restore the bold italic formatting.

NOTE: If the Bold and Italic buttons did not previously appear on the Formatting toolbar, they do now. Word changes the toolbar display according to your use.

TIP: To display both the Standard and Formatting toolbars in their entirety, open the View menu and choose <u>T</u>oolbars, <u>C</u>ustomize. On the Options tab, click the <u>S</u>how Standard and Formatting toolbars on two rows check box to select it. Click Close.

FIGURE 3-1
Using the Formatting
toolbar to apply
character formatting

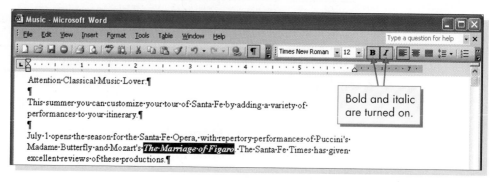

7. Move the insertion point to the end of the same paragraph and press (Spacebar) once.

8. Key **Its production of** and press (Spacebar) once.

9. Click the Bold button **B** and the Italic button **I** and key **La Boheme** in bold italic.

10. Click both buttons to turn the formatting off, press (Spacebar), and complete the sentence by keying **won a rave review from Opera News magazine.** (Key the period.)

11. Select the bold italic text "The Marriage of Figaro" again.

12. Press (Ctrl) and select the bold italic text "La Boheme" as well.

13. Click the Underline button **U** on the Formatting toolbar to underline the noncontiguous selections.

14. Click the Undo button **↺ ▾** to remove the underlines.

EXERCISE **3-2** **Apply and Remove Basic Character Formatting Using Keyboard Shortcuts**

If you like to keep your hands on the keyboard instead of using the mouse, you can use keyboard shortcuts to turn basic character formatting on and off. You can press (Ctrl)+(B) for bold, (Ctrl)+(I) for italic, and (Ctrl)+(U) for underline. To remove character formatting from selected text, press (Ctrl)+(Spacebar).

1. In the same paragraph, select the text "Madame Butterfly."

2. Press (Ctrl)+(B) to make the selected text bold and press (Ctrl)+(I) to add italic.

3. Move the insertion point to the end of the document and press (Enter) twice to start a new paragraph.

4. Press `CapsLock`, press `Ctrl`+`B` to turn on the bold option, and key **jazz fans note:** in bold capital letters.

5. Press `CapsLock` to turn it off, press `Spacebar`, and continue keying in bold:

The first annual Desert Bloom Jazz Festival, featuring some of the world's greatest musicians, will be held in Tucson next May 1-15.

6. Select the bold italic text "Madame Butterfly" again and press `Ctrl`+`Spacebar` to remove the formatting.

7. Click the Undo button `↺ ▾` to restore the bold italic formatting.

Working with Fonts

A *font* is a type design applied to an entire set of characters, including all letters of the alphabet, numerals, punctuation marks, and other keyboard symbols.

FIGURE 3-2
Examples of fonts

Arial is an example of a plain font; Times New Roman (Word's default font) is more ornate; and Monotype Corsiva is an example of a more stylized font. Arial is a *sans serif* font because it has no decorative lines, or serifs, projecting from its characters. Times New Roman is a *serif* font because it has decorative lines. Fonts are also available in a variety of sizes, measured in *points*. There are 72 points to an inch. Like other character formatting, you can use different fonts and font sizes in the same document.

FIGURE 3-3
Examples of
different point sizes

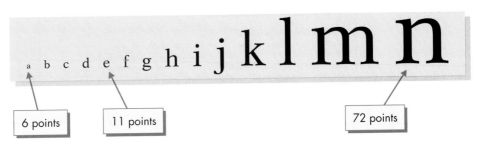

EXERCISE 3-3 **Change Fonts and Font Sizes Using the Formatting Toolbar**

The easiest way to choose fonts and font sizes is with the Formatting toolbar.

1. Move to the beginning of the document and select the first line, which begins "Attention." (Remember, you can press Ctrl+Home to move to the beginning of a document.)

2. Click the down arrow next to the Font box on the Formatting toolbar to open the Font drop-down list. Fonts are listed alphabetically by name and are displayed graphically.

FIGURE 3-4
Choosing a font

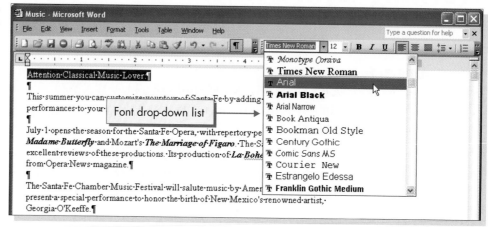

3. Using ↓ or the scroll box on the font list's scroll bar, choose the Arial font.

 NOTE: The fonts you used most recently appear at the top of the Font drop-down list. A divider line separates this list from the alphabetical list.

4. Click the down arrow to open the Font Size drop-down list and choose 16 points. Now the first line stands out as a headline.

FIGURE 3-5
Choosing a font size

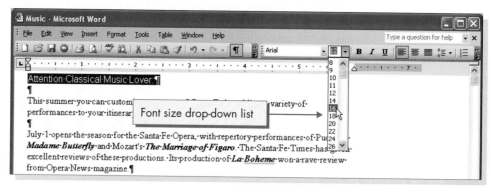

EXERCISE **3-4**   Change Font Size Using Keyboard Shortcuts

If you prefer keyboard shortcuts, you can press Ctrl+Shift+> to increase the font size or Ctrl+Shift+< to decrease the font size.

1. Move the insertion point to the end of the paragraph that begins "For reservations" and press Enter twice to start a new paragraph.

2. Press Ctrl+Shift+> and key **Call Valerie at 555-1234.** The new sentence appears in 14-point type.

 TIP: Sometimes text might appear bold on your screen when it is simply a larger font size.

3. Press Enter twice to begin another paragraph. Press Ctrl+Shift+< to reduce the font size to 12 points and key **Credit card payments are accepted.**

Using the Font Dialog Box

The Font dialog box offers a wider variety of options than those available on the Formatting toolbar. You can conveniently choose several options at one time.

FIGURE 3-6
Shortcut menu

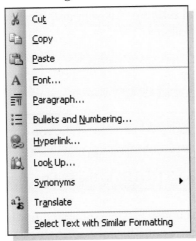

There are several ways to open the Font dialog box:

- Open the Format menu and choose Font.

- Right-click (use the right mouse button) selected text to display a *shortcut menu*, and then choose Font. A shortcut menu shows a list of commands relevant to a particular item you click.

- Display the Reveal Formatting task pane and click the Font link, Font. This is called a *hyperlink*, blue underlined text you click to open a software feature, such as a dialog box.

EXERCISE **3-5**   Choose Fonts and Font Styles Using the Font Dialog Box

1. Select the first line of text, which is currently 16-point Arial.

2. From the Format menu, choose Font to open the Font dialog box. Click the Font tab, if it is not showing.

3. Choose Monotype Corsiva from the Font list, Bold Italic from the Font Style list, and 18 from the Size list. Preview your choices in the Preview box and click OK.

FIGURE 3-7
Using the Font
dialog box

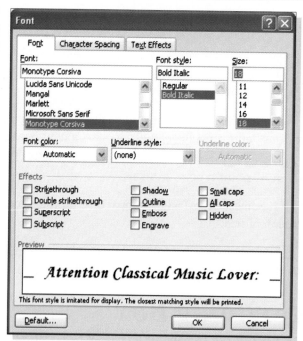

NOTE: Font availability varies, depending on the type of printer you're using and the installed software. Ask your instructor to recommend a substitute font if the specified one is unavailable.

EXERCISE 3-6 Apply Underline Options and Character Effects

In addition to choosing font, font size, and font style, you can choose font color, a variety of underlining options, and special character effects from the Font dialog box.

1. Select the text "JAZZ FANS NOTE" in the last paragraph (do not select the colon).

2. From the Format menu, choose Reveal Formatting to display the task pane. The Reveal Formatting task pane lists the formatting of the selected text.

TIP: The Reveal Formatting task pane allows you to see the formatting that is applied to selected text without having to navigate to individual formatting dialog boxes. You can also open the task pane by choosing Task Pane from the View menu and clicking the Other Task Panes button ▾ to choose from the list of available task panes.

3. Move the mouse pointer over Font under the Font section in the task pane. When the mouse pointer becomes a hand pointer ☝, click to open the Font dialog box.

FIGURE 3-8
Font link in the
Reveal Formatting
task pane

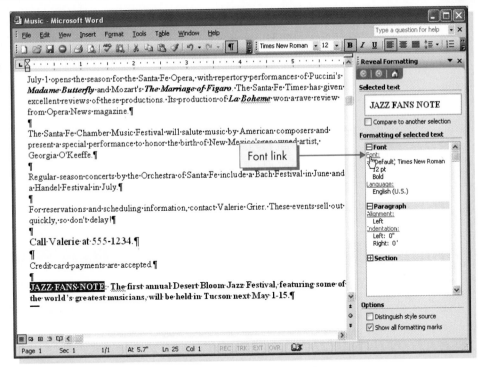

4. Click the down arrow to open the Underline style drop-down list. Drag the scroll box down to see all the available underline styles. Choose one of the dotted line styles.

5. Click the down arrow next to the Font color box and choose "Blue." (Each color is identified by name when you point to it.) Both the text and the underline are now blue in the Preview box.

FIGURE 3-9
Font color options in
the Font dialog box

6. Click the down arrow next to the Underline color box and choose Red.

7. Click OK. The text is blue with a red dotted underline. Notice the description of this formatting in the task pane.

 TIP: As a rule, punctuation such as colons and periods should not be underlined.

8. Select the sentence after the blue, dotted-underlined text "JAZZ FANS NOTE:". Notice that the task pane reflects the newly selected text's formatting.

9. Move the mouse pointer to the Selected text box, which contains "The first annual" at the top of the task pane. The text box becomes a drop-down list box.

10. Click the down arrow. Notice the options available. Choose Clear Formatting from the drop-down list. The bold formatting is removed from the text.

FIGURE 3-10
Selected text drop-down list in Reveal Formatting task pane

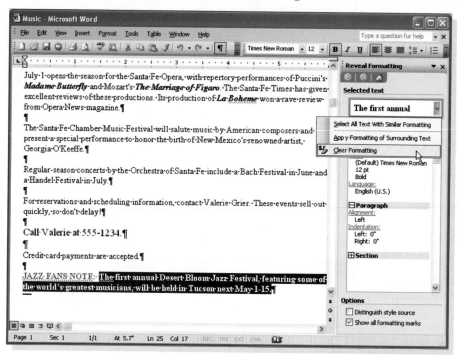

11. In the same sentence, select the text "Desert Bloom Jazz Festival."

12. Click the selected text with the right mouse button and, from the shortcut menu, choose Font to open the Font dialog box. Under Effects, click the Small caps check box and click OK. The text that was formerly lowercase now appears in small capital letters.

13. Select the sentence that begins "Call Valerie."

14. Open the Font dialog box, click the Strikethrough check box, and click OK. The text appears with a horizontal line running through it.

15. Click the Undo button 🔄 ▾ to undo the strikethrough effect.

16. Select the first line of text, which begins "Attention."

TABLE 3-2   Font Effects in the Font Dialog Box

EFFECT	DESCRIPTION AND EXAMPLE
Strikethrough	Applies a ~~horizontal line~~.
Double strikethrough	Applies a ~~double horizontal line~~.
Superscript	Raises text ^{above} other characters on the same line.
Subscript	Places text $_{below}$ other characters on the same line.
Shadow	Applies a **shadow**.
Outline	Displays the inner and outer border of text.
Emboss	Makes text appear raised off the page.
Engrave	Makes text appear imprinted on the page.
Small Caps	Makes lowercase text SMALL CAPS.
All Caps	Makes all text UPPERCASE.
Hidden	Hidden text does not print and appears on-screen only if Word's View options are set to display hidden text. See Tools, Options, View.

17. Open the Font dialog box and click the Text Effects tab. Text effects are intended for screen display, including on-line use.

18. Choose the various options in the Animations list, noticing the effect in the Preview box. Click Cancel to close the dialog box.

FIGURE 3-11
An Animation
text effect

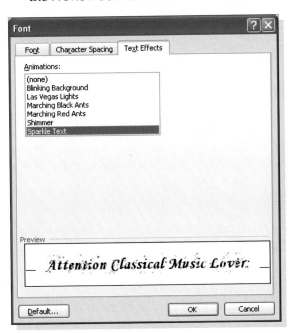

EXERCISE **3-7** Use Keyboard Shortcuts for Underline Options and Font Effects

Word provides keyboard shortcuts for some underlining options and font effects as an alternative to opening the Font dialog box.

 REVIEW: Remember that Ctrl+U turns on and off standard underlining.

1. Start a new sentence at the end of the last paragraph by keying **To beat the heat, bring plenty of H2O.**

2. Select the "2" in H2O.

3. Press Ctrl+= to make it subscript.

4. Select the blue, dotted-underlined text "JAZZ FANS NOTE:". Press Ctrl+Shift+W to change the dotted underlining to words-only underlining.

TABLE 3-3   **Keyboard Shortcuts for Underlining and Character Effects**

KEYBOARD SHORTCUT	ACTION
Ctrl+Shift+W	Turn on or off words-only underlining.
Ctrl+Shift+D	Turn on or off double underlining.
Ctrl+Shift+=	Turn on or off superscript.
Ctrl+=	Turn on or off subscript.
Ctrl+Shift+K	Turn on or off small capitals.
Ctrl+Shift+A	Turn on or off all capitals.
Ctrl+Shift+H	Turn on or off hidden text.

EXERCISE **3-8** Change Character Spacing

The Character Spacing tab in the Font dialog box offers options for changing the space between characters or the position of text in relation to the baseline. Character spacing can be expanded or condensed horizontally, as well as raised or lowered vertically.

1. Select the first line of text, which begins "Attention."

2. Open the Font dialog box and click the Character Spacing tab.

3. Click the down arrow to open the Scale drop-down list. Click 150% and notice the change in the Preview box. Change the scale back to 100%.

4. Click the down arrow to display the Spacing options. Click Expanded, and then click OK. The text appears with more space between each character. Notice that Character Spacing appears in the Reveal Formatting task pane.

TIP: You can increase the space between characters even more by increasing the number in the <u>B</u>y box (click the arrows or key a specific number). Experiment with the <u>S</u>pacing and <u>S</u>cale options on your own to see how they change the appearance of text.

5. Click the Close button ⊠ on the task pane to close it. (This is the Close button ⊠ beside the words "Reveal Formatting" at the top of the task pane.)

6. In the Save As dialog box, create a new folder for your Lesson 3 files and save the document as *[your initials]*3-8.

NOTE: Remember to use this folder for all the exercise documents you create in this lesson.

Repeating and Copying Formatting

Just as you repeated text by using F4 or Ctrl+Y in the previous lesson, you can use F4 or Ctrl+Y to repeat character formatting. You can also copy character formatting with a special tool on the Standard toolbar—the Format Painter button 🖋.

EXERCISE **Repeat Character Formatting**

Before trying to repeat character formatting, keep in mind that you must use the Repeat command immediately after applying the format. In addition, the Repeat command repeats only the last character format applied. (If you apply multiple character formats from the Font dialog box, the Repeat command applies all formatting.)

1. In the paragraph that begins "July 1," select "repertory" and click the Italic button *I* to make it italic.

2. Move the insertion point into the next paragraph, before the word "special."

3. Press F4 to repeat your last action (turning on the Italic button *I*. Key **very** and press Spacebar once. The word "very" appears in italic.

4. Select the sentence that begins "Call Valerie."

5. Open the Font dialog box. Click the Fo<u>n</u>t tab, if it is not already displayed, and choose another font, such as Impact. Make the font size 12 points, and change the font color to red. Click OK. The text appears with the new formatting.

6. Select the text "JAZZ FANS NOTE:" (including the colon) and press F4. Word repeats all formatting you chose in the Font dialog box. If you apply each character format separately, using the Formatting toolbar, the Repeat command applies only the last format you chose.

FIGURE 3-12
Repeating character
formatting

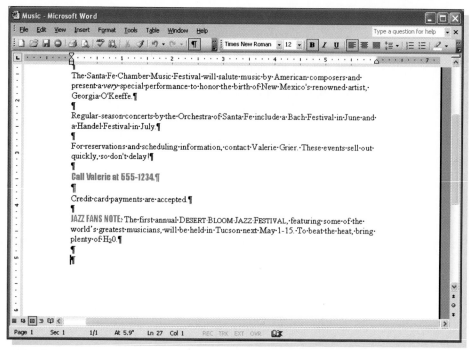

EXERCISE 3-10 Copy Character Formatting

The Format Painter button makes it easy to copy a character format. This is particularly helpful when you copy text with multiple formats, such as bold italic small caps.

To use Format Painter to copy character formatting, first select the text with the formatting you want to copy, and then click the Format Painter button. The mouse pointer changes to a paintbrush with an I-beam pointer. Use this pointer to select the text to which you want to apply the copied formatting.

1. In the paragraph that begins "July 1," select "Santa Fe Times" and click the Underline button to underline the publication name.

2. With the text still selected, click the Format Painter button on the Standard toolbar. When you move the pointer back into the text area, notice the new shape of the pointer.

3. Use the paintbrush pointer to select "Opera News" in the next sentence. This copies the underlining to the selected text and the pointer returns to its normal shape.

4. Select the small caps word "Desert" in the last paragraph.

5. Double-click the Format Painter button. Double-clicking lets you copy formatting repeatedly.

6. Scroll to the top of the document. Notice that the paintbrush pointer becomes an arrow when you move out of the text area to use the scroll bars.

7. Select the sentence that begins "This summer." The small caps formatting is applied to the sentence, and the pointer remains the paintbrush pointer.

8. Scroll down to the line that begins "Call Valerie at" and select the sentence. The paintbrush pointer copies the new formatting over the old formatting.

9. Press Esc or click the Format Painter button to stop copying and restore the regular pointer.

Changing Case

You've used CapsLock to change case and you've seen the Small Caps and All Caps options in the Font dialog box. You can also change the case of characters by using keyboard shortcuts and the Change Case dialog box.

EXERCISE 3-11 Change Case

1. Select the sentence that begins "Credit card payments." Press Shift+F3. This keyboard shortcut changes cases. Now the text appears in all uppercase letters.

2. With the sentence still selected, press Shift+F3 again. Now the sentence appears in all lowercase letters.

3. Press Shift+F3 again and the original case is restored.

4. With the sentence still selected, open the Format menu and choose Change Case. This dialog box offers a few extra case options.

FIGURE 3-13
Change Case
dialog box

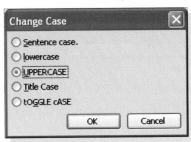

5. Click Title Case and click OK. This option changes the first letter of each word to uppercase, the common format for titles.

6. Click the Undo button to undo the change of case.

7. Click anywhere in the document to deselect the text.

Highlighting Text

To emphasize parts of a document, you can mark text with a color highlighter by using the Highlight button on the Formatting toolbar. As with the Format Painter button, when you click the Highlight button, the pointer changes shape. You then use the highlight pointer to select the text you want to highlight. In addition, you can choose from several highlighting colors.

EXERCISE 3-12 Highlight Text

1. Make sure no text is selected. On the Formatting toolbar, click the down arrow next to the Highlight button to display the color choices. Click "Yellow" to choose it as the highlight color. This turns on the Highlight button, and the color indicator box on the button is now yellow.

FIGURE 3-14
Choosing a
highlight color

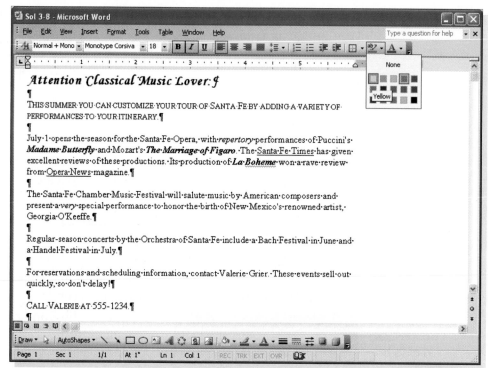

2. Move the highlight pointer ⱥ into the text area.
3. Drag the pointer over the phone number in the paragraph that begins "Call Valerie."
4. Press Esc to turn off the highlighter and restore the normal pointer.
5. Select the first line of text, which begins "Attention."
6. Click the Highlight button to highlight the selection. This is another way to use the highlighter—by selecting the text, and then clicking the Highlight button.
7. Select the first line of text again. Remove the highlight by clicking the down arrow to display the highlight color choices and choosing None.
8. Select the remaining highlighted text and click the Highlight button. Because "None" was last chosen (as shown in the color indicator box on the button), this action removes the highlight from the selected text.

NOTE: You can use highlighting to mark text as you work on a document or to point out text for others opening the same document later. It might not work as well for printed documents because the colors might print too dark. You can use shading, which is discussed in Lesson 5, to emphasize text in a printed document.

Creating a Drop Cap Effect

One way to call attention to a paragraph is to use a dropped capital letter, or a *drop cap.* A drop cap is a large letter that appears below the text baseline. It is usually applied to the first letter in the first word of a paragraph. (For an example of a drop cap, see the first paragraph on the first page of this lesson. The "C" is a drop cap.)

EXERCISE 3-13 **Create a Drop Cap**

1. Delete the paragraph that begins "This summer" (not including the blank line below it).

FIGURE 3-15
Drop Cap
dialog box

2. Place the insertion point within the paragraph that begins "July."
3. From the Format menu, choose Drop Cap to open the Drop Cap dialog box.
4. Under Position, click Dropped. This option makes the paragraph wrap around the letter.
5. Click OK. Click within the document to deselect the "J" of "July," which is the height of three lines. Word is now in Print Layout view. You use this view when you work with graphic objects.

6. Click the Normal View button above the left end of the Status bar to return to Normal view. The "J" appears separate from the paragraph, although it will print as shown in Print Layout view.

NOTE: Print Layout view is explained in detail in Lesson 6: "Margins and Printing Options."

Word's AutoFormat Features

Word has several features that automatically change formatting as you key text or numbers. One of these *AutoFormat* features converts ordinal numbers and fractions into a more readable format, as shown in Table 3-4.

TABLE 3-4

Automatic Formatting of Ordinal Numbers and Fractions	
YOU KEY	**WORD CHANGES THE FORMAT TO**
1st	1st
2nd	2nd
1/2	½
1/4	¼

Another AutoFormat feature changes an Internet address into a hyperlink as you key the text. Clicking on this form of hyperlink takes you to another location, such as an HTML page on the Internet (assuming you were logged on to the Internet).

EXERCISE   **3-14**   **Format Ordinal Numbers and Fractions Automatically**

1. In the paragraph that begins "July 1," move the pointer to the immediate right of the "1" and key **st**.

2. Press [Spacebar] to initiate the automatic formatting. Notice that Word changes the "st" to superscript letters. Now delete the extra space character. (You might want to switch to Print Layout view for a better view of the paragraph. To do this, click the Print Layout View button 回.)

FIGURE 3-16
Formatting
ordinal numbers

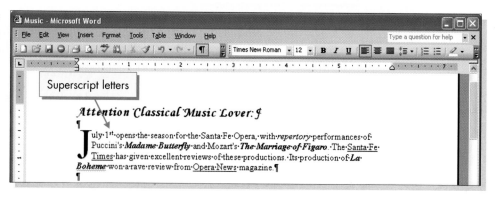

3. Move to the end of the paragraph that begins "Credit card payments." Add a new sentence by keying **Matinee performances are 1/2 price.** Notice that Word automatically converts the numbers and the slash to a fraction symbol.

4. Select the entire line that begins "Call Valerie," remove the small caps formatting, and delete the period. Add the following text after the phone number:

   or send an e-mail message to valeriegrier@gwaytravel.com.

5. Press (Spacebar) at the end of the sentence, after the period, to see the hyperlink formatting. The e-mail address is now blue and underlined. If this were a real e-mail address, any reader of this document could click the text to send an e-mail message to Valerie (providing the reader had an e-mail program installed).

6. Delete one of the blank lines below the first paragraph in the document and save the document as *[your initials]*3-14 in your Lesson 3 folder.

7. Print and close the document.

USING ONLINE HELP

You learned to access the Microsoft Word Help task pane by pressing F1 or choosing Task Pane from the View menu. When the Microsoft Word Help task pane displays, you can explore the Table of Contents link.

Explore Microsoft Word Help Table of Contents:

1. Press F1 and click the Table of Contents link. (See Figure 3-17 on the next page.) A list of topics appears in the task pane, and each is represented by a book icon.

2. Click the book icon beside the Working with Text topic. A list of subtopics appears.

3. Click the topic Formatting Characters.

4. Click the topic to Change the spacing between characters.

5. Click Show All and read the information about kerning. Note the difference between kerning and expanded/condensed spacing.

6. Close the Microsoft Word Help window.

FIGURE 3-17
Exploring Help
Contents

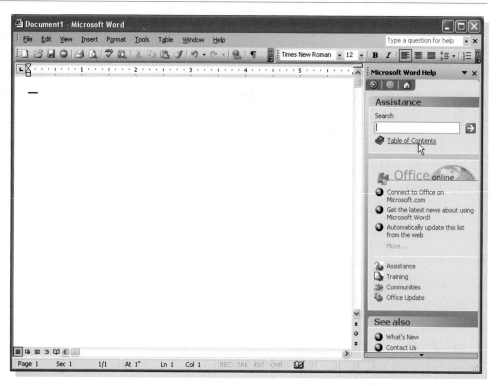

LESSON 3 Summary

➤ Use the Formatting toolbar to apply basic character formatting (for example, bold, italic, and underline) to selected contiguous (text that is together) and noncontiguous text (text that is not together).

➤ Use keyboard shortcuts to apply and remove basic character formatting.

➤ A font is a type design applied to an entire set of characters, including all letters of the alphabet, numerals, punctuation marks, and other keyboard symbols.

➤ A font can be serif (with decorative lines), or sans serif (with no decorative lines).

➤ Fonts are available in a variety of sizes, which are measured in points. There are 72 points to an inch.

➤ You can use the Formatting toolbar to change fonts and font sizes.

➤ Keyboard shortcuts can also be used to change font sizes: Ctrl+Shift+> increases text by one font size and Ctrl+Shift+< decreases text by one font size.

➤ The Font dialog box can be used to change fonts, font sizes, and font styles. The Font dialog box also has settings for underline styles, font and underline colors, effects such as small caps and shadow (see Table 3-2), animated character effects, and character spacing.

➤ Choose Reveal Formatting from the Format menu to display the Reveal Formatting task pane. Use this task pane to see how selected text is formatted, to select similarly formatted text, to clear formatting, and to access the Font dialog box (as well as other formatting dialog boxes).

➤ A hyperlink often appears as blue underlined text you click to open a software feature (such as a dialog box or a Help topic) or to go to an e-mail or Web address.

➤ Keyboard shortcuts are available for some underline styles and font effects (see Table 3-3).

➤ A shortcut menu shows a list of commands relevant to a particular item. To display a shortcut menu, point to the item and right-click the mouse.

➤ Use F4 or Ctrl+Y to repeat character formatting.

➤ Use the Format Painter button 🖌 to copy character formatting. Double-click the button to apply formatting to more than one selection.

➤ To change the case of selected characters, use the keyboard shortcut Shift+F3 or the Change Case dialog box (Format menu, Change Case).

➤ Use the Highlight button 🖉▾ to apply a color highlight to selected text you want to emphasize on-screen.

➤ Use Drop Cap from the Format menu to create a dropped cap. A drop cap is a large letter that appears below the text baseline. It is usually applied to the first letter in the first word of a paragraph.

➤ Take advantage of Word's automatic formatting of ordinal numbers and fractions as you key text. An ordinal number is a number that indicates an order (for example, 1st, 3rd, or 107th).

LESSON 3 Command Summary

FEATURE	BUTTON	MENU	KEYBOARD
Bold	**B**	Format, Font	Ctrl+B
Italic	*I*	Format, Font	Ctrl+I
Continuous, single underlining	U	Format, Font	Ctrl+U
Remove character formatting		Format, Font	Ctrl+Spacebar
Increase font size	12 ▾	Format, Font	Ctrl+Shift+>
Decrease font size	12 ▾	Format, Font	Ctrl+Shift+<
Change case		Format, Change Case	Shift+F3

Concepts Review

TRUE/FALSE QUESTIONS

Each of the following statements is either true or false. Indicate your choice by circling T or F.

T   F   *1.* You can apply single underlining from the Formatting toolbar.

T   F   *2.* To remove character formatting, press [Ctrl]+[Delete].

T   F   *3.* Small Caps is an example of a font.

T   F   *4.* Times New Roman is an example of a sans serif font.

T   F   *5.* You can use [F4] to repeat text or to repeat character formatting.

T   F   *6.* After clicking the Format Painter button [✐], you can press [Esc] to restore the regular pointer.

T   F   *7.* You can use the Font dialog box to change character spacing.

T   F   *8.* You must use the Font tab in the Font dialog box to apply the shadow effect.

SHORT ANSWER QUESTIONS

Write the correct answer in the space provided.

1. Which toolbar contains a button to highlight text?

_____

2. Which dialog box do you use to choose bold italic style?

_____

3. What unit of measurement is used to measure fonts?

_____

4. What keyboard shortcut increases the font size of selected text to the next available size?

_____

5. What character effect places a horizontal line through text?

_____

6. Which button do you use to copy character formatting?

_____

7. What keyboard shortcut do you use to change the case of selected text?

_____

8. What character spacing setting inserts more space between characters?

_____

CRITICAL THINKING

Answer these questions on a separate page. There are no right or wrong answers. Support your answers with examples from your own experience, if possible.

1. Select three examples of effective character formatting in magazine advertisements, articles, or other publications. Describe why you think the character formatting was particularly effective.

2. Using a large font size, key **Fonts & styles** ten times on ten separate lines (you can use the Repeat Typing command). Use a different font for each line. Describe the differences you see among the fonts.

Skills Review

EXERCISE 3-15

Apply basic character formatting. Change font and font size.

1. Open the file **Overseas1**.
2. At the top of the document, press [Enter] twice. Move to the first paragraph mark and key **Become a World Traveler**.
3. Change the font for the entire document by following these steps:
   a. Select the entire document by pressing [Ctrl]+[A].
   b. Open the Font drop-down list on the Formatting toolbar by clicking the down arrow.
   c. Locate Arial and click it.
4. Change the first line you keyed to 16-point bold by following these steps:
   a. Select the text by moving the pointer to the left of the text. When the arrow points to the text, click the left mouse button.
   b. Choose 16 from the Font Size drop-down list on the Formatting toolbar.
   c. Click the Bold button **B** on the Formatting toolbar.
5. Apply italic formatting to noncontiguous text by following these steps:
   a. Move to the end of the document and press [Enter] twice.
   b. Key **Come fly with us!**
   c. Select the sentence you just keyed.

I

   d. Press and hold Ctrl and select the word "Gateway" in the previous paragraph.

   e. Click the Italic button I .

6. Key new bold text by following these steps:

   a. Move to the end of "Come fly with us!" and press Enter.

   b. Click the Italic button I to turn off italic.

B

   c. Click the Bold button B to turn on bold and key **Duke City Gateway Travel**.

   d. Turn off bold.

7. Save the document as *[your initials]*3-15 in your Lesson 3 folder.

8. Print and close the document.

EXERCISE 3-16

Apply formatting options using the Font dialog box and using repeat character formatting.

1. Open the file **Itin1**.

2. Insert a blank line after the first line.

3. Apply character formatting to the first line. Use the Font dialog box and follow these steps:

   a. Select the first line. Click the selected text with the right mouse button and choose Font from the shortcut menu.

   b. Click the Font tab, if it is not already shown. For font, font style, and size, choose Arial, Bold, and 14 points.

   c. Apply the effect Small caps by clicking the check box.

   d. View your options in the Preview box and click OK.

4. Apply and repeat character formatting by following these steps:

   a. Select the text "Day 1:" and choose Font from the Format menu.

   b. Choose the font style Bold Italic.

   c. Open the Font color drop-down list and choose "Blue." Click OK.

   d. Select the text "Day 2:" and press F4. Repeat the formatting through "Day 6:"

5. Apply the Strikethrough and Hidden text effects from the Font dialog box by following these steps:

   a. Select the last line of text (beginning "Day 7:").

   b. From the Format menu, choose Reveal Formatting.

   c. Click Font.

   d. In the Font dialog box, click the Strikethrough check box.

   e. Click OK. Notice the strikethrough effect.

   f. Open the Font dialog box by pressing Ctrl+D, and click the Hidden check box.

g. Click OK. The text appears with a dotted underline.

h. Click the Show/Hide button ¶ to hide the text and nonprinting characters.

6. Save the document as *[your initials]*3-16 in your Lesson 3 folder.

7. Print and close the document.

EXERCISE 3-17

Copy character formatting and change case.

1. Open the file **Offices1**.

2. Change the first line to read **Gateway Travel Agency**.

3. At the end of the first line, press ⟮Enter⟯ and key **seven flagship offices around the United States**.

4. Select the first two lines of text and make them 14-point bold.

5. Use a keyboard shortcut to change the case of the first line to all uppercase by following these steps:

   a. Select the first line of text.

   b. Press ⟮Shift⟯+⟮F3⟯.

6. Change the case of the second line from the Change Case dialog box by following these steps:

   a. Select the second line of text.

   b. From the Format menu, choose Change Case.

   c. Click Title Case, and then click OK.

7. Use the Font dialog box to format the first agency name, "Golden Gateway Travel," as bold italic small caps.

8. Copy the character formatting to the other agency names by following these steps:

   a. With the formatted text selected, double-click the Format Painter button ✍.

   b. Drag the pointer over the next agency name, "Windy City Gateway Travel."

   c. Continue copying the formatting to the other agency names. Use the scroll bar as needed. When you finish copying, click ✍ to restore the normal pointer.

9. In the second line, change "Around The" to lowercase.

10. Save the document as *[your initials]*3-17 in your Lesson 3 folder.

11. Print and close the document.

EXERCISE 3-18

Highlight text, automatically format numbers, and create a dropped capital letter.

1. Start a new document by keying the text shown in Figure 3-18. Use 12-point Arial type. Leave a blank line after the first line.

FIGURE 3-18

Winners of the Albuquerque Bike-a-Thon

Congratulations to the winners of the Bike-a-Thon. More than 300 people of all ages participated and a wonderful time was had by all. A celebration is scheduled for Friday night at the Pueblo Center. Prizes will be awarded to the following winners: Jimmy Ferrini (1st place), Fred Begey (2nd place), and Juanita Hildalgo (3rd place).

2. Highlight part of the document by following these steps:

a. Click the down arrow next to the Highlight button 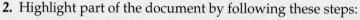 and click the yellow highlight.
b. Use the highlight pointer to select the text "Pueblo Center."
c. Press Esc to restore the normal pointer.

3. Create a dropped capital letter by following these steps:

a. Position the insertion point in the paragraph that begins "Congratulations."
b. From the Format menu, choose Drop Cap.
c. Click the Dropped option and click OK.

4. Click the Normal View button ≡ to return to Normal view.
5. Remove the highlight by following these steps:

a. Select the highlighted text "Pueblo Center."
b. Click the down arrow next to the Highlight button and choose None.

6. Format the first line as 14-point Arial bold.
7. Save the document as *[your initials]*3-18 in your Lesson 3 folder.
8. Print and close the document.

Lesson Applications

EXERCISE 3-19

Apply and copy character formatting. Change font size.

1. Open the file **Visit**.
2. Make the first line ("Visit the Caribbean") 14 points and change the font to Impact. (If Impact is not available, choose another bold-looking font from the Font drop-down list.)

3. Use the Format Painter button ✒ to apply the formatting of the first line to the last line.
4. Apply the shadow effect to the first line.
5. Use the Formatting toolbar to make the text "101 Reasons to Visit the Caribbean" bold. With the text still selected, make it italic as well.
6. Position the insertion point in the next paragraph, before the word "interesting." Press F4 to repeat the last selected character formatting (italic), key **very**, and press Spacebar.
7. Position the insertion point at the beginning of the last line. Key **This book is** and press Spacebar. Make the word "Must" lowercase.
8. Select the entire last line and change the text to 12-point Times New Roman. Add red double underlining to the entire sentence, except the exclamation point.
9. Save the document as *[your initials]*3-19 in your Lesson 3 folder.
10. Print and close the document.

EXERCISE 3-20

Apply and copy basic character formatting. Change font size, case, and character spacing.

1. Start a new document by keying the text shown in Figure 3-19, including the corrections. Use 12-point Arial type.
2. In the first line, change the case of the text "Tourist/Budget" to all capitals.
3. Copy or repeat the all-capitals formatting to the following words in the paragraph that describe hotel classifications: "Moderate," "First Class," "Deluxe," and "Luxury."

FIGURE 3-19

In hotel classificat⌢oin, Tourist/Budget is geared to those seeking
the lowest possible prices. This type of property has limited
amenities
~~services~~, yet offers funct⌃onal accommodations. Moderate means the
hotel is suitable for cost=conscious clients. The rooms are simple
and comfortable. First Class hotels offer a selective variety of
facilities ~~and services~~ (Stet) that will please a majority of travel⌢lers.
Deluxe hotels are outstanding properties, with fine rooms, public
areas, and services. They are the better hotels in the area and
will satisfy most discriminating travel⌢lers. Luxury hotels are ~~the~~
exceptional hotels, offering the highest standard of service and
facilities.ᴧThey are classified as the world's finest.

4. Create a new paragraph for each hotel classification, beginning with
    "MODERATE." (You should have four new paragraphs.) Be sure to include
    a blank line between paragraphs.

5. At the beginning of the document, insert two blank lines (press Enter twice).
    Key the title **Guide to Hotel Classification** at the first paragraph mark.

6. Change the text you just keyed to bold, orange, small caps in 16-point type.

7. Make the text "the lowest possible prices" (in the paragraph under the title)
    and the text "the world's finest" (in the last paragraph) bold at the same time.

8. Select the entire document and make it italic.

9. Save the document as a Web page named *[your initials]*3-20 in your
    Lesson 3 folder.

   NOTE: Because some Web browsers do not support small caps formatting,
   you might discover that Word changes your small caps formatting to all caps.
   If this happens to you, click Tell Me More in the dialog box to learn more about your
   Web browser's capabilities.

10. Print and close the document.

EXERCISE 3-21

Apply and copy character formatting, highlight text, and create a dropped capital letter.

1. Open the file **Special1**.
2. Key the text shown in Figure 3-20 at the beginning of, and as part of, the first paragraph.

FIGURE 3-20

> Traveling alone? Want to experience the wonders of the world? Join one of Duke City Gateway's tours and meet others who share your special interests.

3. Format the "T" of "Traveling" as a dropped capital letter. Switch to Normal view.
4. Highlight the second paragraph (which begins "Our local") in yellow.
5. Format the list of items from "The Great Sphinx" through "The Great Wall of China" as 11-point Arial italic.
6. Repeat the formatting for the second list (from "Hawaii" through "Tahiti").
7. Split the last paragraph so "For more information" starts a new paragraph. Be sure to leave a blank line between paragraphs.
8. Copy the formatting from one of the lists to the new last paragraph.

  REVIEW: You need select only one, or a portion of, the formatted words, click the Format Painter button , and then select the new paragraph.

9. Format the phone number in the last paragraph as bold italic with a dotted underline.
10. Remove the highlight from the second paragraph.
11. Save the document as *[your initials]*3-21 in your Lesson 3 folder.
12. Print and close the document.

EXERCISE 3-22 ✚ *Challenge Yourself*

Apply character formatting, change case, and format numbers automatically.

1. Open the file **Baggage**.
2. At the end of the paragraph that begins "B. Baggage Delay," apply the strikethrough effect to the text "other than your residence."

3. Repeat the effect for the word "Terms" (excluding the colon) in the next paragraph.

4. Place the insertion point after the strikethrough text and before the colon. Key the word **Conditions**. Format the new word in blue text without the strikethrough effect.

5. Using the keyboard shortcut, apply double underlining to the title of the document.

6. Change the case of the title to uppercase.

7. Using the keyboard shortcut, make the text "A. Baggage Loss or Damage" small caps.

8. Repeat this formatting for the text "B. Baggage Delay" in the next paragraph.

9. Key the text shown in Figure 3-21 as a new paragraph at the end of the document. Use line breaks to make the text a single paragraph. Use the formatting shown in the figure.

FIGURE 3-21

```
bf    If you experience baggage problems:
      1st: notify airline personnel at the airport.        Use Arial 11pt.
      2nd: call us at Duke City Gateway Travel.
```

10. Save the document as *[your initials]*3-22 in your Lesson 3 folder.

11. Print and close the document.

On Your Own

In these exercises you work on your own, as you would in a real-life work environment. Use the skills you've learned to accomplish the task—and be creative.

EXERCISE 3-23
Create a list of ten companies in which you are interested. (They could be potential employers, local companies, companies that make products in which you are interested—any companies you want.) Include the companies' addresses. Apply interesting font effects to one of the company names. Copy and repeat the formatting to the other companies in the list. Save the document as *[your initials]*3-23 and print it.

EXERCISE 3-24

Create an itinerary for a long trip you would like to take. (Be imaginative! This could be a real trip or a fantasy trip!) To make the itinerary interesting, use as many as possible of the character formatting features you learned in this lesson. Remember, though, the itinerary must be readable. Save the document as *[your initials]*3-24 and print it.

EXERCISE 3-25

Log onto the Internet and find an interesting Web site. Summarize the information from the site in a Word document at least a half page long. Give the document an interesting title. Using the character formatting you learned in this lesson, try to make your document look as much like the Web site as possible. Use a drop cap in the first paragraph. Save the document as *[your initials]*3-25 and print it.

LESSON 4

Writing Tools

OBJECTIVES

After completing this lesson, you will be able to:

1. **Use AutoComplete, AutoCorrect, and smart tags.**
2. **Work with AutoText.**
3. **Check spelling and grammar.**
4. **Use the Thesaurus and Research task pane.**

 Estimated Time: 1 hour

Word provides several automated features that save you time when keying frequently used text and correcting common keying errors. Word also provides important writing and research tools: a spelling and grammar checker, a Thesaurus, and access to research services. These tools help you create professional-looking documents.

Using AutoComplete and AutoCorrect

By now, you might be familiar with three of Word's automatic features, though you might not know their formal names:

- *AutoComplete* suggests the completed word when you key the first four or more letters of a day, month, or date. If you key "Janu," for example, Word displays a box suggesting the word "January," which you can insert by pressing Enter.

● *AutoCorrect* corrects commonly misspelled words as you key text. If you key "teh" instead of "the," for example, Word automatically changes the spelling to "the." You can create AutoCorrect entries for text you frequently use, and you can control AutoCorrect options.

● *Smart tags* help you save time by performing actions in Word for which you'd normally open other programs (such as Outlook). Word recognizes names, dates, addresses, and telephone numbers, as well as user-defined data types through the use of smart tags, which appear as purple dotted lines.

EXERCISE   4-1   Correct Errors Automatically

1. Open a new document. From the Tools menu, choose AutoCorrect Options to open the AutoCorrect dialog box. Notice the available AutoCorrect options.

2. Scroll down the list of entries and notice the words that Word corrects automatically (assuming the Replace text as you type option is checked).

FIGURE 4-1
AutoCorrect
dialog box

AutoCorrect options

Words that are
corrected
automatically

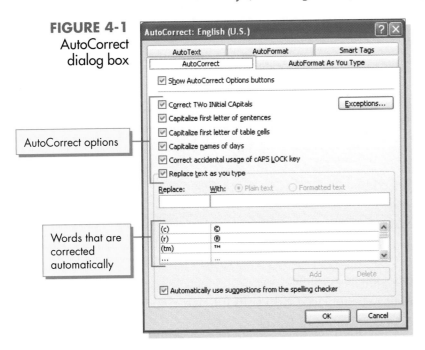

3. Click Cancel to close the dialog box.

4. Key **i am testing teh AutoCorrect feature.** Press [Spacebar]. Word corrects the "i" and "teh" automatically.

5. Try keying another incorrect sentence. Using the exact spelling and case as shown, key **TOdya is**. AutoCorrect corrects the spelling and capitalization of "Today."

6. Key today's date, beginning with the month, and then press (Spacebar). When you see the AutoComplete tip that suggests the current date, press (Enter).

7. Key a period at the end of the sentence.

TABLE 4-1   AutoCorrect Options

OPTIONS	DESCRIPTION
Correct TWo INitial CApitals	Corrects words keyed accidentally with two initial capital letters, such as "WOrd" or "THis."
Capitalize first letter of sentences	Corrects any word at the beginning of a sentence that is not keyed with a capital letter.
Capitalize first letter of table cells	Corrects any word at the beginning of a table cell that is not keyed with a capital letter.
Capitalize names of days	Corrects a day spelled without an initial capital letter.
Correct accidental usage of cAPS LOCK key	If you turn on (Caps Lock) accidentally, and then key "tODAY", AutoCorrect changes the word to "Today" and turns off (Caps Lock).
Replace text as you type	Makes all corrections automatically.

NOTE: AutoCorrect corrects your text only after you complete a word by either pressing (Spacebar) or keying punctuation, such as a period or comma.

EXERCISE   4-2   Create an AutoCorrect Entry

You can create AutoCorrect entries for words you often misspell. You can also use AutoCorrect to create shortcuts for text you use repeatedly, such as names or phrases. Here are some examples of these types of AutoCorrect entries:

- asap for as soon as possible
- Your initials to be replaced with your full name, such as jh for Jill Holmes
- dc for Duke City Gateway Travel

1. Open the AutoCorrect dialog box. In the Replace box, key **fyi**.

2. In the With box, key **For your information**.

3. Click the Add button to move the entry into the alphabetized list. Click OK.

4. Start a new paragraph in the current document and key **fyi, this really works.** Word spells out the entry, just as you specified in the AutoCorrect dialog box.

EXERCISE 4-3 Control AutoCorrect Options

Sometimes you might not want text to be corrected. You can undo a correction or turn AutoCorrect options on or off by clicking the AutoCorrect Options button 🗲 ▾ and making a selection.

1. Move the I-beam over the word "For" until a small blue box appears beneath it.

FIGURE 4-2
Controlling
AutoCorrect options

Small blue box

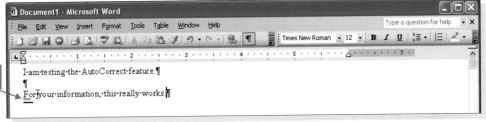

2. Slide the I-beam down over the small blue box until your mouse becomes a pointer and the box turns into the AutoCorrect Options button 🗲 ▾.

3. Click the button and choose Undo Automatic Corrections from the menu list.

FIGURE 4-3
Undoing automatic
corrections

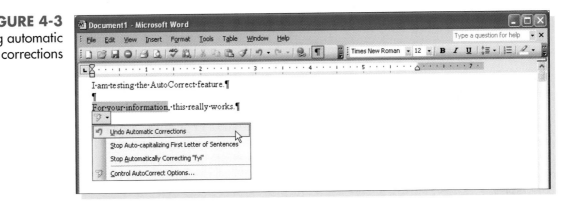

4. Click the AutoCorrect Options button 🗲 ▾ again and choose Redo Automatic Corrections from the menu list. The words "For your information" are restored.

5. Click the button again and choose Control AutoCorrect Options. The AutoCorrect dialog box opens.

6. Remove the "fyi" entry: Choose "fyi," click Delete, and then click OK.

EXERCISE **4-4** **Create an AutoCorrect Exception**

Another way to keep Word from correcting text you do not want corrected is to create an AutoCorrect exception. For example, you might have a company name that uses nonstandard capitalization such as "tuesday's child." In such a case, you can use the AutoCorrect Exceptions dialog box to prevent Word from making automatic changes.

1. In a new paragraph, key the following on two separate lines:

   The ABC of travel:

   ABsolute is a must.

   Notice that AutoCorrect automatically makes the "B" in "ABsolute" lowercase.

2. Open the AutoCorrect dialog box and click Exceptions.

3. In the AutoCorrect Exceptions dialog box, click the INitial CAps tab.

4. Key the exception **ABsolute** in the Don't Correct text box. Click Add. The entry is now in the list of exceptions.

FIGURE 4-4
AutoCorrect
Exceptions
dialog box

5. Click OK, and then click OK again to close both dialog boxes.

6. Select "Absolute" and then key **ABsolute Comfort**.

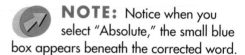 **NOTE:** Notice when you select "Absolute," the small blue box appears beneath the corrected word.

7. Delete the exception: Open the AutoCorrect dialog box, click Exceptions, select "ABsolute" from the list, and click Delete. Click OK, and then click OK again.

TIP: Another good example of an AutoCorrect exception is the use of lowercase initials, which are sometimes entered at the bottom of a business letter as reference initials (see Appendix B: "Standard Forms for Business Documents"). In this case, you would not want Word to capitalize the first letter.

EXERCISE **4-5** **Use Smart Tags**

Just as Word recognizes an e-mail or Web address and automatically creates a hyperlink, it also recognizes names, dates, addresses, and telephone numbers, as well as user-defined data types through the use of smart tags. You can use this feature to perform actions in Word for which you'd normally open other programs, such as Microsoft Outlook. Purple dotted lines beneath text in your document indicate smart tags.

1. From the Tools menu, choose <u>A</u>utoCorrect Options.

2. Click the Smart Tags tab and click the <u>L</u>abel text with smart tags check box if it is not selected.

3. Under <u>R</u>ecognizers, select everything except Person name, Telephone number, and Financial Symbol.

4. Click OK.

5. After the period behind "this really works", press (Enter) twice and key

   Duke City Gateway Travel Agency
   15 Montgomery Boulevard
   Albuquerque, NM 87111.

6. Notice that Word recognizes the text as an address and applies a smart tag indicator (the purple dotted underline).

7. Move the I-beam over the street address, and then move your pointer over the Smart Tag Actions button ⊙ ▾.

8. Click the button to see the list of actions.

FIGURE 4-5
Smart tag list
of actions

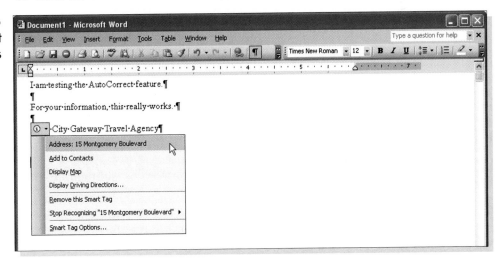

9. Choose <u>A</u>dd to Contacts. Microsoft Outlook launches and an Untitled-Contact dialog box opens. You can add the name and address as well as telephone numbers to the listing.

NOTE: Microsoft Outlook is a program included in the Microsoft Office suite. If it is not set up on your machine, just close the dialog box when it asks you to configure it, or ask your instructor for help. If Outlook does launch, the <u>A</u>dd to Contacts option lets you record information about individuals and businesses. All of this is done from within Word through the use of the smart tag.

10. Look over the dialog box contents and close the dialog box. Click No when you are asked if you want to save the changes.

Working with AutoText

AutoText is another feature you can use to insert text automatically. This feature is extremely versatile. You can use it to:

- Insert over 40 commonly used entries, including letter salutations and closings (such as "To Whom It May Concern," or "Sincerely yours") and mailing instructions (such as "VIA AIRMAIL" or "CONFIDENTIAL").
- Create AutoText entries for text you use repeatedly (the AutoText entry can even include the text formatting). The text for which you create an AutoText entry can be a phrase, a sentence, paragraphs, logos, and so on.

After you create an entry, you can insert it with just a few keystrokes.

 NOTE: You can also create AutoText entries for nontext items such as graphics and tables you use often.

E X E R C I S E   **4-6**   **Insert an Existing AutoText Entry**

You can access Word's existing AutoText entries from the Insert menu, the AutoCorrect dialog box, or a special AutoText toolbar.

1. Delete the first two paragraphs of text you keyed in the practice document, leaving two sentences, the first of which begins "The ABC."

2. Create a blank line above this line, if there is not already one, and position the insertion point in the blank line.

3. From the Insert menu, choose AutoText. Without clicking the mouse, move the pointer slowly down the AutoText submenu to see the existing AutoText entries.

4. Position the pointer over Salutation and click "To Whom It May Concern:". The text is automatically inserted in the document.

5. Click the Undo button 🔄 ▾ to undo the insertion.

6. For better access to Word's AutoText entries, display the AutoText toolbar, if it is not already displayed. From the View menu, choose Toolbars and choose AutoText from the submenu.

7. On the AutoText toolbar, click All Entries. The same list of Word AutoText entries is displayed.

8. Point to Attention Line and click "Attention:". The text is entered in the document.

FIGURE 4-6

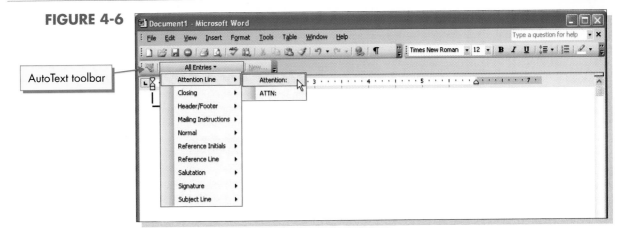

AutoText toolbar

9. Close the document without saving.

EXERCISE 4-7 Create AutoText Entries

1. Open the file **Letter1**. Display the AutoText toolbar if necessary.

2. Select the word "Date" at the top of the letter and enter the date. (Key the month, press Spacebar, and press Enter to accept the AutoComplete date.)

3. At the second paragraph mark below "Dear Mr. Farr," key the opening paragraph: **Enclosed is the information you requested about cruise travel.** Press Enter.

4. Proofread the paragraph you keyed. Then select it along with the blank line below it.

5. Click New on the AutoText toolbar. (See Figure 4-7 on the following page.) The Create AutoText dialog box appears, with a suggested name for the entry.

TIP: When the AutoText toolbar is not displayed, you can create an AutoText entry by choosing Insert, AutoText, New from the menu or by pressing Alt + F3.

6. Replace the suggested name ("Enclosed is") by keying *[your initials]* **cruise**. Click OK.

7. Replace the word "cruise" in the opening paragraph with the word **group**.

8. Select the paragraph along with the blank line below it and press Alt + F3 to open the Create AutoText dialog box.

FIGURE 4-7
Create AutoText
dialog box

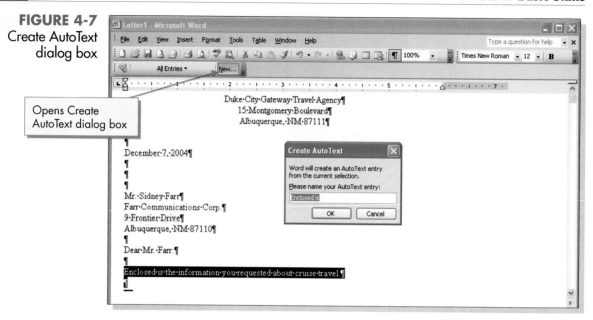

9. Name the AutoText entry *[your initials]***group**. Click OK.

10. Delete the text in this paragraph and replace it with the paragraph shown in Figure 4-8.

FIGURE 4-8

```
Please call with any questions you might have. You can also visit
our Web site at www.gwaytravel.com. We look forward to planning
your trip.
```

11. Proofread the paragraph and save it, with the blank line below it, as an AutoText entry named *[your initials]***closing**.

12. Delete the selected paragraph and the blank line that follows it, so there is one blank line below "Dear Mr. Farr:"

13. Select the first three lines of text, which contain the company letterhead. Format the text as 14-point Arial bold, small caps, and shadowed.

14. Add the two blank paragraph marks below the letterhead to the selection and save the selection as an AutoText entry named *[your initials]***letterhead**.

15. Delete the selected letterhead and the two blank paragraph marks.

EXERCISE   **4-8**   Insert AutoText Entries

To insert an AutoText entry without keying it, key the AutoText entry name or choose the entry from the AutoText toolbar. When you key enough of the entry name, an AutoComplete tip appears with a few lines of the content of the AutoText entry. If you have an entry name that is unique or short, you can key the first couple of letters of the entry name and press F3 to insert the entry without seeing a tip.

1. Position the insertion point at the top of the document, to the left of the date.

2. Begin keying the name of the letterhead AutoText entry, *[your initials]* **letterhead**. After you key the first few letters, an AutoComplete tip appears showing the first two lines of the letterhead. Press Enter to insert the AutoText entry. The letterhead is inserted at the top of the document.

3. Click the Undo button 🔄▾ twice.

4. Press Enter. On the new blank line above the date, key *[your initials]* followed by a lowercase "L".

5. Press F3 to insert the letterhead entry.

6. Press Delete to delete one of the blank lines. There should be only two blank lines above the date line in a block-style letter.

7. Position the insertion point at the paragraph mark under "Dear Mr. Farr" and press Enter to insert a blank line.

8. Click All Entries on the AutoText toolbar. Point to Normal in the list. The submenu shows the AutoText entries you created.

9. Choose the entry that contains the opening paragraph about group travel, *[your initials]***group**. The text is inserted in the document.

10. With the insertion point at the blank paragraph mark at the end of the document, click the AutoText button 🔲 on the AutoText toolbar. The AutoCorrect dialog box opens with the AutoText tab displayed. In this dialog box, you can preview your AutoText entries before you insert them.

11. Scroll the list of AutoText entries and click *[your initials]***closing**.

12. Review the text in the Preview box and click Insert. (See Figure 4-9 on following page.)

> ☀ **TIP:** The dialog box method for inserting AutoText entries is helpful if you forget the name of an entry or just need to preview an entry before you insert it.

13. To complete the letter, click All Entries [All Entries ▾] on the AutoText toolbar and choose the closing text "Sincerely,".

14. Press Enter four times and key your name. On the next line, key the title **Associate Travel Counselor**.

FIGURE 4-9
Previewing an
AutoText entry

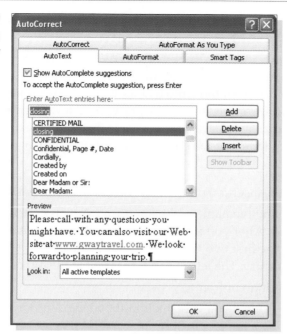

15. Press Enter twice and key **Enclosures (2)**.

16. Split the closing paragraph into two paragraphs by pressing Enter twice before the sentence "We look forward to planning your trip."

17. Insert two more blank lines above the inside address, increasing the number of blank lines to the maximum of five for a block-style letter. (Making this short letter a little longer will help balance it on the printed page.)

 TIP: Check Appendix B: "Standard Forms for Business Documents," to double-check that your letter has the correct number of blank lines between items.

EXERCISE 4-9 Edit and Delete AutoText Entries

After you create an AutoText entry, you might need to change it. If you no longer use an entry, you can delete it.

1. Position the insertion point to the right of the ZIP Code in the letterhead, press Enter, and key the telephone number **(505) 555-1234**.

2. Select the letterhead text and the blank lines below it.

3. Click the AutoText button 🖳 on the AutoText toolbar.

4. Scroll to the existing AutoText name, *[your initials]*letterhead. Select it and click Add.

5. Click Yes to redefine the AutoText entry. The entry now includes the telephone number.

6. To test the change, delete the letterhead text and the two blank paragraph marks below it in the letter and choose the letterhead entry from the AutoText toolbar.

7. Click the AutoText button 🖼 to reopen the dialog box that lists the AutoText entries.

8. Click the first entry that contains your initials and click Delete.

9. Repeat the process for the remaining three entries and click OK to close the dialog box.

10. From the View menu, choose Toolbars, AutoText to close the AutoText toolbar.

11. Save the document as *[your initials]*4-9 in a new Lesson 4 folder.

12. Print and close the document.

Checking Spelling and Grammar

Correct spelling and grammar are essential to good writing. As you've seen, Word checks your spelling and grammar as you key text and flags errors with these on-screen indicators:

- A red, wavy line appears under misspelled words.
- A green, wavy line appears under possible grammatical errors.
- The Spelling and Grammar Status icon at the right end of the Status bar contains an "X".

TABLE 4-2   The Spelling and Grammar Status Icon

ICON	INDICATES
📖	Word is checking for errors as you key text.
📖✗	The document has errors.
📖✓	The document has no errors.

EXERCISE **4-10** **Spell- and Grammar-Check Errors Individually**

You can right-click text marked as either a spelling or a grammar error and choose a suggested correction from a shortcut menu.

1. Open the file **Educate**. This document has several errors, indicated by the red and green wavy lines.

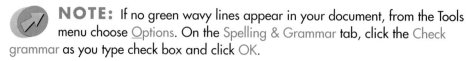 **NOTE:** If no green wavy lines appear in your document, from the Tools menu choose Options. On the Spelling & Grammar tab, click the Check grammar as you type check box and click OK.

2. At the top of the document, press Enter twice and move the insertion point to the first paragraph mark. Notice that the icon on the Status bar now contains an "X".

3. Using 14-point bold type, key a misspelled word by keying the title **Vacatins Can Be Educational**. When you finish, "Vacatins" is marked as misspelled.

4. Right-click the word and choose "Vacations" from the spelling shortcut menu.

5. Right-click the grammatical error "are the range" in the first paragraph under the title. Choose "is the range" from the shortcut menu.

TIP: Word's spelling and grammar tools are not foolproof. For example, it cannot correct a word that is correctly spelled but incorrectly keyed, such as "sue" instead of "use." It might also apply a green, wavy line to a type of grammatical usage, such as the passive voice, which might not be preferred, but is not incorrect.

EXERCISE **4-11** **Spell- and Grammar-Check an Entire Document**

Instead of checking words or sentences individually, you can check an entire document. This is the best way to correct spelling and grammar errors in a long document. Use one of these methods:

- Click the Spelling and Grammar button on the Standard toolbar.
- Press F7.
- From the Tools menu, choose Spelling and Grammar.

 1. Position the insertion point at the beginning of the document and click the Spelling and Grammar button on the Standard toolbar. Word locates the first misspelling, "formaal." (See Figure 4-10 on the next page.)

2. Click Change to correct the spelling to the first suggested spelling, "formal." Next, Word finds a repeated word, "as."

FIGURE 4-10
Checking spelling

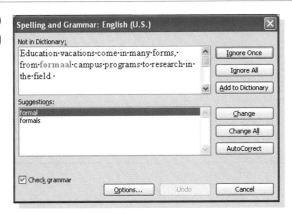

 TIP: To check spelling without also checking grammar, click the Check grammar check box to clear it.

3. Click <u>D</u>elete to delete the repeated word. Next, Word finds a grammatical error—"Their" is used incorrectly.

4. In the dialog box, select "Their" in the Order of Words section and key **There**. Click <u>C</u>hange to change the spelling in the document. Next, Word locates "cavate," which should be "excavate," as you can see from the text.

5. In the dialog box, delete the space between "ex" and "cavate" and click <u>C</u>hange. Word stops on the name "Atacama."

6. Click <u>I</u>gnore Once, because this name is correctly spelled but is not in Word's dictionary. Next, Word flags "desert" and suggests "Desert."

7. Click <u>C</u>hange to accept the suggestion. Next, Word locates "profesionals," which is incorrectly spelled.

8. Click <u>C</u>hange to correct the spelling to "professionals."

9. Click OK when the check is complete. Notice there are no more wavy lines in the document and the Spelling and Grammar Status icon shows a check mark.

TABLE 4-3   Dialog Box Options When Checking Spelling and Grammar

OPTION	DESCRIPTION
<u>I</u>gnore Once	Skips the word.
<u>I</u>gnore All	Skips all occurrences of the word in the document.
<u>A</u>dd to Dictionary	Adds the word to the default dictionary file in Word. You can also create your own dictionary and add words to it.
<u>C</u>hange	Changes the word to the entry in the Change To box or to the word you chose from the Suggestions list.

continues

TABLE 4-3   Dialog Box Options When Checking Spelling and Grammar   *continued*

OPTION	DESCRIPTION
Change All	Same as Change, but changes the word throughout the document.
AutoCorrect	Adds the word to the list of corrections Word makes automatically.
Options	Lets you change the Spelling and Grammar options in Word.
Undo	Changes back the most recent correction made.
Cancel	Discontinues the checking operation.

NOTE: You can create or add a custom dictionary for technical and specialized vocabulary. Choose Options from the Tools menu and click the Spelling & Grammar tab. Click Custom Dictionaries, click New, and key a name for the custom dictionary. To add a custom dictionary that you purchased, follow the steps listed above, except choose Add instead of New. Locate the folder and double-click the dictionary file.

Using the Thesaurus and Research Task Pane

The *Thesaurus* is a tool that can improve your writing. Use the Thesaurus to look up a *synonym* (a word with a similar meaning) for a selected word to add variety or interest to a document. You can look up synonyms for any of these words to get additional word choices. The Thesaurus sometimes displays *antonyms* (words with the opposite meaning) and related words.

After selecting a word to change, you can start the Thesaurus in one of three ways:

- Choose Language from the Tools menu, and then Thesaurus from the submenu.
- Press [Shift]+[F7].
- Right-click the word and choose Synonyms from the shortcut menu.

EXERCISE   4-12   **Use the Thesaurus**

1. Select the word "range" in the first paragraph or place the insertion point in the word.
2. Press [Shift]+[F7]. The Research task pane appears with a list of possible meanings for "range" as a noun and as a verb. It also gives you a list of synonyms, with "variety" as the recommended word.

FIGURE 4-11
Using the Thesaurus

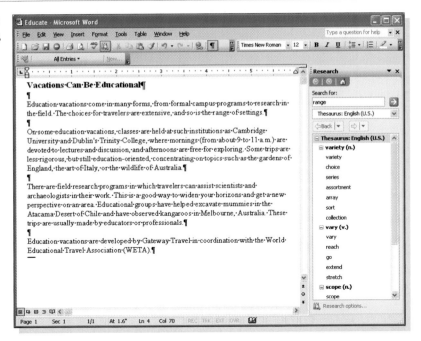

3. Click the word "scope" in the <u>M</u>eanings list. A list of synonyms appears in the task pane.

4. Go back to the word "variety" by clicking the Previous Search button <u>Back</u> ▼ . Point to variety, click the down arrow, and click <u>L</u>ook Up. A list of additional meanings and synonyms appears for "variety."

5. Point to "diversity" (or one of its synonyms), click the down arrow, and choose <u>I</u>nsert. Word replaces "range" with "diversity" (or your chosen word) and returns to the document.

6. Save the document as *[your initials]***4-12** in your Lesson 4 folder.

7. Print the document, but do not close it.

EXERCISE **4-13** **Use References**

If you are connected to the Internet, you can access several research sources, such as a dictionary, an encyclopedia, and research sites such as MSN. You can click the Research button 📖, right-click a word, or press Alt and click a word to open the Research task pane.

1. Press Alt and click the word "extensive" in the first paragraph. (See Figure 4-12 on the following page.)

FIGURE 4-12
Using References

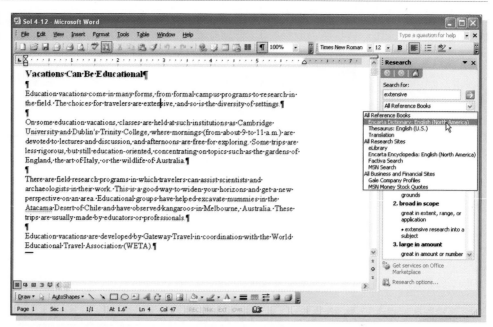

2. Click the drop-down arrow beside the All Reference Books box and choose *Encarta Dictionary*. The task pane indicates the part of speech, syllabication, and several definitions for extensive.

3. Click the drop-down arrow beside the All Reference Books box and choose Translation.

4. Choose English in the From box and French (France) in the To box. The bilingual dictionary displays the French word for extensive.

5. Close the document.

USING ONLINE HELP

Another way to get help on a Word function is to access Microsoft Office Online. This Web site provides current Help topics, templates, training, and product updates for Microsoft Word. Remember, you must have an Internet connection to access Microsoft Office Online.

Use the Office on Microsoft.com to explore topics about grammar-checking:

1. Display the Microsoft Word Help task pane and connect to the Internet.

2. Click the link Connect to Microsoft Office Online.

3. Click the link for Word under Programs.

4. Click the link Word 2003 Help, and click the link Working with Text.

FIGURE 4-13
Using the Help
Task Pane

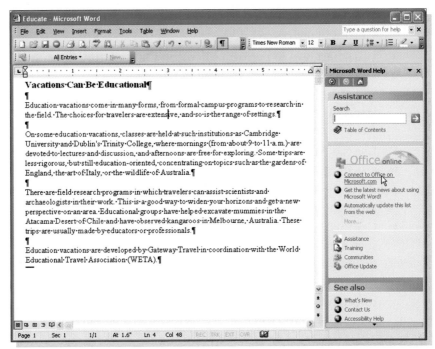

5. Click the link Spelling, Grammar, and Thesaurus. A list of articles appears.

6. Click the About Spelling, Grammar, and Thesaurus link, and then click the link Proofing text in a different language. Read the information in the article.

7. Click the Back button until you return to the list of links for Working with Text.

8. Explore the link Automatic Text Options to learn more about AutoCorrect, AutoComplete, and AutoText.

9. Click the Print button if you want to print a topic.

10. Return to Word and close the Help task pane when you have finished. Disconnect from the Internet.

LESSON 4 Summary

➤ The AutoComplete feature suggests the completed word when you key the first four or more letters of a day, month, or date.

➤ The AutoCorrect feature corrects some misspelled words and capitalization errors for you automatically as you key text.

➤ Use the AutoCorrect dialog box to create entries for words you often misspell and the AutoCorrect Options button to control AutoCorrect options.

➤ Use the AutoCorrect Exceptions dialog box to create an AutoCorrect exception so Word will not correct it.

➤ Use smart tags to perform Microsoft Outlook functions, such as creating entries in Outlook's contact list, without opening Outlook.

➤ AutoText is another versatile feature you can use to insert text automatically. You insert over 40 commonly used entries, as well as creating AutoText entries for text you use repeatedly, including text formatting.

➤ Use the AutoCorrect dialog box to edit and delete AutoText entries.

➤ Use the spelling and grammar checker to correct misspelled words in your document as well as poor grammar usage. Check errors individually or throughout your entire document.

➤ Use the Thesaurus to look up synonyms (words with similar meaning) or sometimes antonyms (words with the opposite meaning) for a selected word to add variety and interest to your document.

➤ Use the Research task pane to look up words or phrases in a dictionary, to research topics in an encyclopedia, or to access bilingual dictionaries for translations. You can also access research sites such as MSN.

LESSON 4 Command Summary

FEATURE	BUTTON	MENU	KEYBOARD
Create AutoText entry	or New (on AutoText toolbar)	Insert, AutoText, New	Alt + F3
Check spelling and grammar		Tools, Spelling and Grammar	F7
Thesaurus		Tools, Language, Thesaurus	SHIFT + F7
Research		Tools, Research	ALT + Click

Concepts Review

TRUE/FALSE QUESTIONS

Each of the following statements is either true or false. Indicate your choice by circling T or F.

T F **1.** AutoCorrect automatically changes "THis" to "This" and "monday" to "Monday."

T F **2.** You can edit an AutoText entry by redefining an existing AutoText entry.

T F **3.** You can start a grammar check from the Formatting toolbar.

T F **4.** To start the Thesaurus, press F7.

T F **5.** The Thesaurus finds synonyms for words.

T F **6.** AutoComplete suggests a complete word or phrase for a date or an AutoText entry.

T F **7.** You can choose to check only the spelling of a document, without checking the grammar.

T F **8.** You can use the View menu to display the AutoText toolbar.

SHORT ANSWER QUESTIONS

Write the correct answer in the space provided.

1. Which function key starts a spell check?

_____

2. Which option in the Spelling and Grammar dialog box lets you skip over an incorrectly spelled word?

_____

3. Which tab, in which dialog box, lets you display and delete AutoText entries?

_____

4. When you click New [New...] on the AutoText toolbar, which dialog box is opened?

_____

5. Which item in the All Entries list on the AutoText toolbar contains such entries as "Sincerely" and "Best Wishes"?

_____

6. Which Word feature corrects accidental usage of Caps Lock ?

_____

7. Which task pane is used to access references such as dictionaries and encyclopedias?

_____

8. What must you do to a word before using the Thesaurus?

_____

CRITICAL THINKING

Answer these questions on a separate page. There are no right or wrong answers. Support your answers with examples from your own experience, if possible.

1. Some educators believe the spell-checking feature in word-processing programs will lead to decreased spelling skills in future generations. Imagine a future in which children learned word processing from the earliest grades. Do you think the students' spelling skills would deteriorate? Explain your answer.

2. Key a sample of your writing that is at least one full page (it could be a paper for another class, a letter you wrote, or anything else you wrote entirely). Grammar-check the writing. Did you find the analysis helpful? What are the advantages and disadvantages of using this tool?

Skills Review

EXERCISE 4-14

Use AutoCorrect and AutoComplete.

1. Start a new document using 12-point Arial.

2. Key the following sentence (including the errors in the first word):

   PEopel visiting Scottsdale, Arizona, enjoy its low humidity, blue skies, and striking sunsets.

3. Press Caps Lock and key these sentences as shown. Be sure to press Shift to capitalize the first word in each sentence.

The average high temperature is 85.9 degrees. The average low temperature is 59.3 degrees.

4. Continue the paragraph by keying the following sentence in all lowercase letters, letting AutoCorrect capitalize the first letter of each sentence. When you see the AutoComplete tip for the months, press Enter and continue keying

   from january through december, you can expect sunshine 86 percent of the time.

5. Start a new paragraph, including a blank line before it. Key the following sentence (including the errors in the first two words):

   thisyear, consider Scottsdale as a vacation destination.

6. Save the document as [your initials]4-14 in your Lesson 4 folder.

7. Print and close the document.

EXERCISE 4-15

Use AutoText and work with smart tags.

1. Open the file **Letter1**. Key today's date in place of "Date."

2. Select only the words "Duke City Gateway Travel" in the letterhead of the document. Format the text as italic, small caps, with a dark blue color.

3. Using the text you just formatted, create an AutoText entry by following these steps:

   a. If it is not shown, display the AutoText toolbar by choosing View, Toolbars, AutoText.

   b. Select the formatted "Duke City Gateway Travel." (Do not select the space after the word "Travel.")

   c. Click New on the AutoText toolbar.

   d. Key the entry name [your initials]**dc**. Click OK.

4. Change the formatting for the entire letterhead to 14 points, small caps, dark blue, regular text (no italic).

5. At the second blank paragraph mark below "Dear Mr. Farr," key the following sentence, substituting your AutoText entry [your initials]dc in place of "dc." When you see the AutoComplete tip, press Enter.

   Thank you for your interest in dc, the leading full-service travel agency in New Mexico.

6. Press Enter twice to start a new paragraph. Click All Entries on the AutoText toolbar, point to Normal, and choose your AutoText entry. Complete the sentence by keying the following text:

   is your gateway to the Southwest, offering the greatest number of specialized tours of this remarkable area.

7. Start a new paragraph and key the following text. Use either of the previous two methods to insert your AutoText entry in place of "dc."

The enclosed material will provide more information about dc. We look forward to helping you plan your next trip.

8. Press Enter twice. Click All Entries from the AutoText toolbar, and then choose Closing and choose an appropriate closing for the letter.

9. Press Enter four times and key the following information:

Nina Chavez
Senior Travel Counselor

10. Press Enter twice, key *[your initials]* in lowercase, and press Enter.

11. Control the AutoCorrect function with the AutoCorrect Options button by following these steps:

   a. Move the I-beam over your first initial until you see the small blue box.
   b. Move the pointer to the small blue box until the AutoCorrect Options button appears.
   c. Click the button icon and choose Undo Automatic Capitalization.

12. On the line below your initials, key **Enclosures (2)**

13. Delete the AutoText entry by following these steps:

   a. Click the AutoText button on the AutoText toolbar.
   b. Select your AutoText entry, *[your initials]*dc, from the list.
   c. Click Delete and click OK.

14. Work with smart tags by following these steps:

   a. Move the I-beam over the date in the date line, and then move your pointer up over the Smart Tag Actions button.
   b. Click the button to see the list of actions.
   c. Choose Schedule a Meeting.
   d. Look over the dialog box contents and close the dialog box.
   e. Click No when you are asked if you want to save the changes.

15. Save the document as *[your initials]*4-15 in your Lesson 4 folder.

16. Print and close the document.

EXERCISE 4-16

Spell-check and grammar-check a document.

1. Open the file **Overseas**.
2. Spell-check and grammar-check the document by following these steps:

   a. With the insertion point at the beginning of the document, click the Spelling and Grammar button on the Standard toolbar.
   b. When Word locates the first misspelled word in the title, choose "Overseas" from the Suggestions list and click Change.

    c. Continue checking the document, changing spelling, deleting words, or correcting grammar as appropriate.

3. Format the title as all uppercase with an extra paragraph mark below it.

4. Save the document as *[your initials]***4-16** in your Lesson 4 folder.

5. Print and close the document.

EXERCISE 4-17

Use the Thesaurus.

1. Open the file **Summer**.

2. Use the Thesaurus to find another word for "wonderful" by following these steps:

    a. Select "wonderful" in the last sentence of the first paragraph.

    b. Press (Shift)+(F7).

    c. Point to a synonym and click the drop-down arrow to the right of the word.

    d. Choose Insert from the drop-down list.

3. Start a new paragraph at the end of the document and key **Come and enjoy the fun!**

4. Select the word "fun." Using the menu, choose Tools, Language, Thesaurus to look for another noun for "fun."

5. Replace "fun" with a noun listed in the Research task pane. Remember to click the down arrow and choose Insert.

6. Save the document as *[your initials]***4-17** in your Lesson 4 folder.

7. Use the Research task pane to define a word by following these steps:

    a. Press (Alt) and click "ethnic" in the first paragraph.

    b. Choose the *Encarta Dictionary* from the All Reference Books drop-down list.

    c. Read the definition.

    d. Close the task pane.

8. Print and close the document.

Lesson Applications

EXERCISE 4-18

Create an AutoText entry and check grammar and spelling.

1. In a new blank document, key the text *[your name]*'s **Good News Café** and press Enter twice. (AutoCorrect applies the accent mark over the "e" of "Café.")

2. Select the text and the paragraph marks and format them as 12-point Arial.

3. Select the text only (excluding the paragraph marks) and create an AutoText entry named *[your initials]*cafe.

4. At the last paragraph mark, key the text shown in Figure 4-14. Include the corrections. Wherever "gn" appears, use the AutoText entry you just created, pressing Enter when you see the AutoComplete tip.

FIGURE 4-14

The best vegetarian restaurant in Tucson is (gn). This popular choice for healthy dining is located at the ^southeast corner of Wilmot Road and East Broadway.

(gn) serves three meals a day, following its own strict rules: no bleached flour, no chemicals, no refined sugar, and no frying. *Many dishes are also dairy-free.*^ The menu at (gn) features pages of unusual dishes, such as Malaysian cashew chicken and veggie nut burgers. Natural fruit shakes, fat ＝ free brownies ^and fresh baked muffins make desserts special.

(gn) is open (7) days a week. Reservations are taken for five or more people, and credit cards are not accepted.

5. Spell-check and grammar-check the document.

6. Format the title as 16-point bold small caps with a text shadow. Add another paragraph mark below the title.

7. Format the restaurant name in small caps throughout the document.

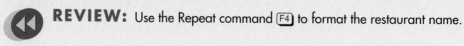 **REVIEW:** Use the Repeat command F4 to format the restaurant name.

8. Delete the AutoText entry you created.

9. Save the document as *[your initials]*4-18 in your Lesson 4 folder.

10. Print and close the document.

EXERCISE 4-19

Spell-check and grammar-check a document and use the Research Task Pane.

1. Open the file **Rockies1**.
2. Spell-check and grammar-check the document, making the appropriate corrections.
3. Use the Thesaurus to look up the word "lovely" in the second sentence of the paragraph that begins "On Day 3."
4. Change the first sentence of the document to title case. Make the sentence into a title with two blank lines below it (delete the period).
5. Format the title as 14-point bold, with expanded character spacing.
6. Save the document as *[your initials]*4-19 in your Lesson 4 folder.
7. Print and close the document.

EXERCISE 4-20

Format a document as a letter, check spelling and grammar, and use the Thesaurus.

1. Open the document **Prices1**.
2. Create a standard business letter in block style: Key the date at the top of the document, followed by the address, salutation, and opening paragraph shown in Figure 4-15. Include five blank lines before and three blank lines after the date, and one blank line each after the address, the salutation, and the opening paragraph. (You can refer to Appendix B: "Standard Forms for Business Documents," for correct letter format.)

 NOTE: If the Office Assistant asks if you want help creating the letter, click Cancel.

FIGURE 4-15

```
Mr. George Coleman

7500 Powder Point Drive

Hickory, NC 28601

Dear Mr. Coleman:

Thank you for your payment for the tour of the Southwest pueblos.
Enclosed you will find an itinerary of events.
```

3. At the end of the document, press (Enter) twice. Choose the AutoText closing "Sincerely," press (Enter) four times, and key **Steve Ross**. On the next line,

key **Travel Counselor**. Press [Enter] twice and key your initials in lowercase. Press [Enter] and key **Enclosures (2)**.

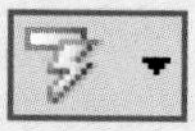

4. Use the AutoCorrect Options button [⟱ ▼] to undo the capitalization of your first initial.

5. Check the spelling and grammar of the document, making the appropriate corrections.

6. Use the Thesaurus to find a synonym for "ensure" in the last sentence of the document.

7. Save the document as *[your initials]*4-20 in your Lesson 4 folder.

8. Print and close the document.

EXERCISE 4-21   ✚   *Challenge Yourself*

Check the grammar and spelling of a document.

1. Open the file **RedRiver**.

2. In the first sentence after "Red River Hotel," key **and Conference Center**. Select the entire name ("Red River Hotel and Conference Center") and create an AutoText entry named *[your initials]*red.

3. After the paragraph that begins "The hotel has two," start a new paragraph by keying the text shown in Figure 4-16. Include the corrections. Replace "red" with your new AutoText entry.

FIGURE 4-16

> The ⟨red⟩ has an̸ indoor pool as well as a heated outdoor pool. It also
> has full health=club facilities∧including ⟨10⟩ racquetball courts, ⟨5⟩
> tennis courts, Nautilus equipment, computerized treadmills,∧steam
> room, and∧sauna. Spa services∧ such as massage∧and facials∧can be
> scheduled/∧ personal trainer.

(handwritten corrections: "heated" above "an"; "a" above treadmills; "a" above "and" before sauna; "s" above "massage"; "as well as the services of a" below "scheduled")

4. In the paragraph that begins "For corporate clients," replace each occurrence of the word "hotel" with your AutoText entry.

5. Replace "Red River Hotel" in the last paragraph with the same AutoText entry.

6. Use the same AutoText entry as a title for the document, leaving two blank lines below it.

7. Add the word "The" to the beginning of the title and format the title as 14-point bold, all capitals, shadowed, with expanded character spacing.

8. Start the spell- and grammar-check from the beginning of the document.

9. Delete the AutoText entry *[your initials]*red.

10. Hide the AutoText toolbar.

11. Save the document as a Web page named *[your initials]*4-21 in your Lesson 4 folder.

> **NOTE:** Some Web browsers might not support formatting that you use in Word, such as the shadow text effect. If a dialog box appears as you save the Web page, click Continue to continue saving.

12. Print and close the document.

On Your Own

In these exercises you work on your own, as you would in a real-life work environment. Use the skills you've learned to accomplish the task—and be creative.

EXERCISE 4-22

Write a summary about a book you have recently read but, before you start the summary, create an AutoCorrect entry for a word you know you often misspell. Use this word in the summary as often as you can. Delete the entry when you have finished. Spell- and grammar-check your document. Make the document attractive, save it as *[your initials]*4-22, and print it.

EXERCISE 4-23

Write a letter to a friend. Insert as many existing AutoText entries into the letter as you can. Create your own AutoText entry and insert it into the letter as well. Delete the AutoText entry when you have finished. Spell- and grammar-check your document. Make the letter attractive (but make sure you follow the standard form for letters), save it as *[your initials]*4-23, and print it.

EXERCISE 4-24

Log onto the Internet and find a Web site about one of your hobbies or interests. Summarize the information from the site in a Word document. Give the document a title and some basic character formatting. Use the Thesaurus to insert synonyms. Spell- and grammar-check the document. Make the document attractive, save it as *[your initials]*4-24, and print it.

Unit 1 Applications

UNIT APPLICATION 1-1

Edit, spell-check, use the Thesaurus, and apply formatting to a document.

1. Open the file **Pueblos1**.
2. Format the entire document as 11-point Arial.
3. Merge the first and second paragraphs.
4. Format the first paragraph with a dropped capital letter that drops three lines and is 0.1 inch from the text.
5. Switch back to Normal view.
6. At the end of the second paragraph, add the sentence **Its curved walls, built out of mud and straw, are symbolic of the architecture of the region.**
7. Spell-check and grammar-check the document. Ignore proper names.
8. In the first paragraph, use the Thesaurus to choose a synonym for the word "worthwhile."
9. Use noncontiguous text selection to format the names "Taos Pueblo" in the second paragraph and "San Juan Pueblo" in the third paragraph as follows:
   - Small caps
   - Bold
   - Italic
   - Expanded character spacing
10. Copy the formatting to the remaining six pueblo names.
11. Save the document as *[your initials]***u1-1** in a new Unit 1 Applications folder.
12. Print and close the document.

UNIT APPLICATION 1-2

Create AutoText entries, use AutoComplete format, spell-check, and grammar-check a document.

1. Open the file **Letter1**.
2. Format the first three lines (the letterhead) as Arial, shadowed.
3. Format the first line of the letterhead as 14-point small caps. Add the phone number **(505) 555-1234** to the letterhead as a fourth line and make sure its formatting matches that of the address lines above it.
4. Display the AutoText toolbar.

5. Select the portion of the document from the letterhead through the blank paragraph mark above "Mr. Sidney Farr." Save the selection as the AutoText entry *[your initials]***letterheading**.

6. Close the document without saving.

7. Start a new document. Insert your AutoText entry **letterheading** by beginning to key its name (and then pressing Enter when AutoComplete suggests the content).

8. Replace "Date" with the current date by using AutoComplete.

9. Address the letter as shown in Figure U1-1.

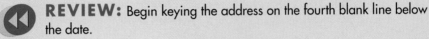 **REVIEW:** Begin keying the address on the fourth blank line below the date.

FIGURE U1-1

```
Ms. Florence Ewing

54 Marble Avenue

Albuquerque, NM 87110

Dear Ms. Ewing:
```

10. For the body of the letter, key the text shown in Figure U1-2. After keying the first sentence of the second paragraph, create an AutoText entry named *[your initials]***ojo** for the name "Ojo Caliente Hot Springs". Use the AutoText entry everywhere "Ojo" appears circled in the figure.

FIGURE U1-2

```
It was nice seeing you last week at the craft show. I received some
information about a new spa, and I thought I'd pass it on to you. I'm
also enclosing a brochure. I know you enjoyed your stay at the Canyon
Ranch Resort last spring, and you said you were interested in finding
a place closer to home.

The new spa is called Ojo Caliente Hot Springs. It is 100 miles from
Albuquerque International Airport and only one hour from Santa Fe. Ojo
is in a very picturesque community, surrounded by rugged mountains and
desert. With three different springs on the grounds, they offer
several of the spa services you enjoy, such as massage and
aromatherapy. They have 25 lovely rooms in the main facility and 5
```

continues

FIGURE U1-2 *continued*

> separate cottages. The restaurant at (Ojo) offers delicious, healthy
> meals, and there is also a gift shop that showcases the work of local
> artists.
>
> If you have any interest in (Ojo), let me know. If you're in our area on
> the weekend, drop by the office and say hello!

11. Use the AutoText toolbar to choose an appropriate closing and identify the writer of the letter as **Anna Svenkova, Associate Travel Counselor**.

12. Refer to Appendix B: "Standard Forms for Business Documents" to check your line spacing.

13. Add your reference initials to the end of the document as well as **Enclosure** below your initials. Control the capitalization of the first initial of your reference initials by using the AutoCorrect Options button .

14. Insert nonbreaking spaces wherever a number appears at the end of a line.

15. Spell-check and grammar-check the document.

16. Delete the AutoText entries you created.

17. Save the document as *[your initials]***u1-2** in your Unit 1 Applications folder.

18. Print and close the document.

UNIT APPLICATION 1-3

Compose a document, apply formatting, and check grammar and spelling.

1. Start a new document. Key just the first paragraph in Figure U1-3, making it into a two-line title.

FIGURE U1-3

```
Morning Flights on Southwest Airlines from Tucson, Arizona, to
Albuquerque, New Mexico

Mon-Sat      Flight 664     Depart 7:20 a.m., Arrive 8:20 a.m.

Sat          Flight 808     Depart 9 a.m., Arrive 10:05 a.m.

Sun          Flight 714     Depart 10:25 a.m., Arrive 11:30 a.m.

Mon-Sun      Flight 968     Depart 10:50 a.m., Arrive 11:55 a.m.
```

2. Format the two-line title as 16-point Arial bold italic.

3. Change the case of the title's first line to all capitals.

4. Insert two blank lines below the title.

5. Translate the flight information that appears in the next four lines of Figure U1-3 into sentences. For each line, create a short paragraph of one to two sentences. Use 12-point Arial and insert a blank line between paragraphs. At this point, the paragraphs should not be bold or italic.

 TIP: Be creative. For example, the first paragraph can begin "Every day except Sunday, Flight 664 departs Tucson at. . . ."

6. In each of the four paragraphs, format the word "Flight" and the flight number in italic with words-only underlining.

7. Write a closing paragraph regarding seat availability for Saturday, August 26. Use the information shown in Figure U1-4.

FIGURE U1-4

Available Seats	Flight Number
12	968
4	808
20	664

8. Insert two blank lines before the closing paragraph. Format the paragraph as 12-point Arial italic with red text and a black dotted underline.

9. Insert nonbreaking spaces in the document, if they are needed.

10. Grammar-check and spell-check the document.

11. Save the document as *[your initials]*u1-3 in your Unit 1 Applications folder.

12. Print and close the document.

UNIT APPLICATION 1-4 *Using the Internet*

Apply character formatting, use AutoFormat features, and check grammar and spelling.

Using the Internet, create a list of five organizations. Be creative. The organizations could be:

- Companies where you would like to be employed
- Schools you'd be interested in attending
- Associations related to your hobbies or interests

Include the organization's name and its Web site address. Include any e-mail addresses, the physical address, and the telephone and fax numbers. Allow AutoFormat to format the Web addresses and e-mail addresses as hyperlinks.

Create a title for the document, followed by a descriptive paragraph that describes the content of the list. Apply appropriate formatting. Check spelling and grammar, watching carefully as Word's spelling and grammar checker moves through the addresses.

Save the document as a Web page named *[your initials]*u1-4 in your Unit 1 Applications folder.

Paragraph Formatting, Margins, and Tabs

Formatting Paragraphs

OBJECTIVES

After completing this lesson, you will be able to:

1. **Align paragraphs.**
2. **Change line spacing.**
3. **Change paragraph spacing.**
4. **Set paragraph indents.**
5. **Apply borders and shading.**
6. **Repeat and copy paragraph formats.**
7. **Create bulleted and numbered lists.**
8. **Insert symbols and special characters.**

 Estimated Time: 1½ hours

In Microsoft Word, a *paragraph* is a unique block of information. Paragraph formatting lets you control the appearance of individual paragraphs within a document. For example, you can change the space between paragraphs. For emphasis, you can indent paragraphs, number them, or add borders and shading.

A paragraph is always followed by a *paragraph mark*. All the formatting for a paragraph is stored in the paragraph mark. Each time you press (Enter), you copy the formatting instructions in the current paragraph to a new paragraph. You can copy paragraph formats to existing paragraphs and view formats in the Reveal Formatting and Styles and Formatting task panes.

Paragraph Alignment

Paragraph alignment determines how the edges of a paragraph appear horizontally. There are four ways to align text in a paragraph, as shown in Figure 5-1.

FIGURE 5-1
Paragraph
alignment options

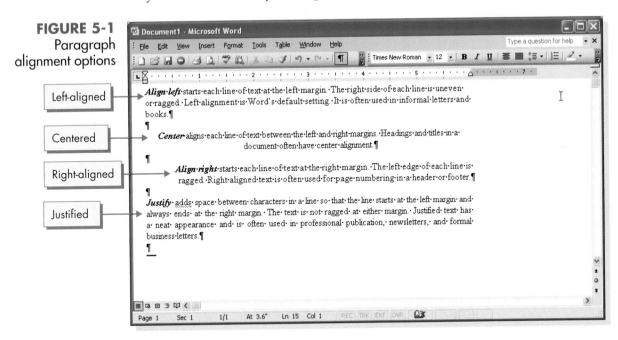

EXERCISE 5-1 Change Paragraph Alignment

The easiest way to change paragraph alignment is to use the alignment buttons on the Formatting toolbar. You can also use keyboard shortcuts: Ctrl+L left-align; Ctrl+E center; Ctrl+R right-align; Ctrl+J justify.

FIGURE 5-2
Alignment buttons
on the Formatting
toolbar

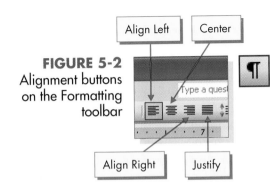

1. Open the file **Turquoise**. Click the Show/Hide button ¶ to display paragraph marks if they are turned off.

NOTE: The documents you create in this course relate to the Case Study about Duke City Gateway Travel, a fictional travel agency (see pages 1 through 4).

2. Position the insertion point anywhere in the first paragraph.

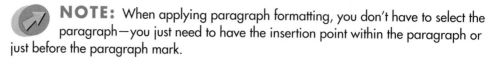 **NOTE:** When applying paragraph formatting, you don't have to select the paragraph—you just need to have the insertion point within the paragraph or just before the paragraph mark.

 3. Click the Center button ≣ on the Formatting toolbar to center the paragraph.

 REVIEW: If you do not see one of the alignment buttons on the Formatting toolbar (or any button used in an exercise), click the Toolbar Options button ⁝ to display more buttons. When you select a button, it is added to the toolbar.

 4. Continue to change the paragraph's formatting by clicking the Align Right button ≣, the Justify button ≣, and the Align Left button ≣. Notice how the lines of text are repositioned with each change.

 5. Position the insertion point in the second paragraph and press Ctrl+E to center the paragraph.

 6. Use the keyboard shortcut Ctrl+R to right-align the third paragraph.

7. Combine the fourth and fifth paragraphs, and use the keyboard shortcut Ctrl+J to justify the merged paragraph.

 TIP: To change the alignment of multiple paragraphs, select them, and then apply the alignment.

Line Spacing

Line space is the amount of vertical space between lines of text in a paragraph. Line spacing is typically based on the height of the characters, but you can change it to a specific value. For example, some paragraphs might be single-spaced and some double-spaced.

EXERCISE   **5-2**   **Change Line Spacing**

You can apply the most common types of line spacing by using keyboard shortcuts: single space, Ctrl+1; 1.5 line space, Ctrl+5; double space, Ctrl+2. Additional spacing options, as well as other paragraph formatting options, are available in the Paragraph dialog box or from the Line Spacing button ⁝≣ ▾.

1. Position the insertion point in the first paragraph.

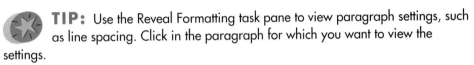

2. Display the Reveal Formatting task pane. (Choose Reveal Formatting from the Format menu).

TIP: Use the Reveal Formatting task pane to view paragraph settings, such as line spacing. Click in the paragraph for which you want to view the settings.

3. Press Ctrl+2 to double-space the paragraph. Notice that the line spacing is listed under Spacing in the Reveal Formatting task pane.

4. With the insertion point in the same paragraph, press Ctrl+5 to change the spacing to 1.5 lines. Press Ctrl+1 to restore the paragraph to single spacing.

5. With the insertion point in the same paragraph, click the down arrow to the right of the Line Spacing button on the Formatting toolbar and choose 2.0 to change the line spacing to double. Choose 1.0 to restore the paragraph to single spacing.

6. Right-click the first paragraph and choose Paragraph from the shortcut menu. (You can also open the Paragraph dialog box by choosing Paragraph from the Format menu.)

7. Click the down arrow to open the Line spacing drop-down list and choose Double. The change is reflected in the Preview box.

FIGURE 5-3
Line spacing options in the Paragraph dialog box

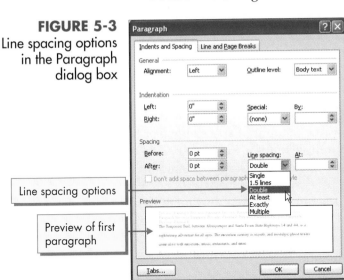

Line spacing options

Preview of first paragraph

8. With the dialog box still open, choose Single from the Line spacing drop-down list. The Preview box shows the change.

9. Choose Multiple from the Line spacing drop-down list. In the At box, key **1.25**. (Select the text that appears in the box and key over it.) Press Tab to see the change displayed in the Preview box.

10. Click OK. Word adds an extra quarter-line of space between lines in the paragraph. (See Figure 5-4 on the next page.)

NOTE: The At Least option applies minimum line spacing that Word can adjust to accommodate larger font sizes. The Exactly option applies fixed line spacing that Word does not adjust. This option makes all lines evenly spaced. The Multiple option increases or decreases line spacing by the percentage you specify. For example, setting line spacing to a multiple of 1.25 increases the space by 25 percent, and setting line spacing to a multiple of 0.8 decreases the space by 20 percent.

FIGURE 5-4
Examples of
line spacing

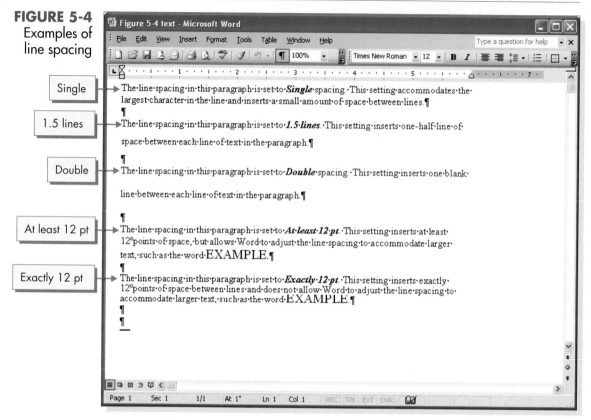

Paragraph Spacing

In addition to changing spacing between lines of text, you can change *paragraph space*. Paragraph space is the amount of space above or below a paragraph. Instead of pressing Enter multiple times to increase space between paragraphs, you can use the Paragraph dialog box to set a specific amount of space before or after paragraphs.

Paragraph spacing is set in points. If a document has 12-point text, one line space equals 12 points. Likewise, one-half line space equals 6 points and two line spaces equal 24 points.

EXERCISE 5-3 **Change the Space Between Paragraphs**

1. Select the whole document by pressing Ctrl+A. Press Ctrl+L to left-align all paragraphs.

2. Move the insertion point to the beginning of the document (Ctrl+Home) and use the keyboard shortcut Ctrl+1 to change the first paragraph back to single spacing.

3. Click the Bold button **B** to turn on bold, key **THE TURQUOISE TRAIL** in all capitals, and press Enter.

4. Move the insertion point into the heading you just keyed. Although this heading includes only three words, it is also a paragraph. Any text followed by a paragraph mark is considered a paragraph.

5. Open the Paragraph dialog box. You use the text boxes labeled Before and After to choose an amount of space for Word to insert before or after a paragraph.

6. Set the Before text box to 72 points (select the "0" and key **72**). Because 72 points equal one inch, this adds to the existing one-inch top margin and places the title two inches from the top of the page.

> **NOTE:** Most business documents start two inches from the top of the page. You can set this standard by using paragraph formatting, as done here, or by changing margin settings. Margins are discussed in Lesson 6.

7. Press Tab, set the After text box to 24 points, and click OK. The heading now starts at two inches (check the position of the insertion point on the Status bar and the notation under Spacing in the Reveal Formatting task pane) and is followed by two line spaces.

8. Click the Center button ≣ to center the heading.

> **TIP:** Word provides these keyboard shortcuts for paragraph spacing: Ctrl+0 adds 12 points of space before a paragraph; Ctrl+Shift+0 removes space before a paragraph; Ctrl+Shift+N removes all paragraph and character formatting, restoring the text to default formatting.

Paragraph Indents

An *indent* increases the distance between the sides of a paragraph and the two side margins (left and right). Indented paragraphs appear to have different margin settings. Word provides a variety of indents to emphasize paragraphs in a document, as shown in Figure 5-5.

To set paragraph indents, you can use one of these methods:

- Indent buttons on the Formatting toolbar
- Paragraph dialog box
- Keyboard
- Ruler

FIGURE 5-5
Types of
paragraph indents

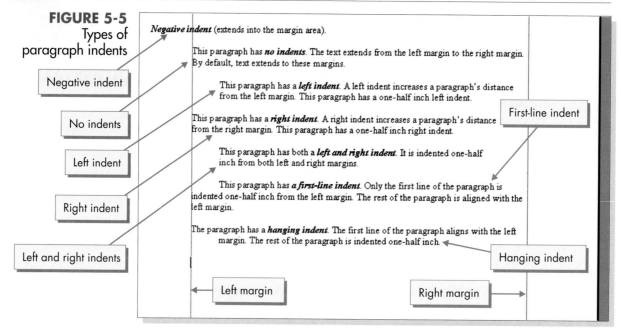

E X E R C I S E   **5-4**   **Set Indents by Using Indent Buttons and the Paragraph Dialog Box**

1. Select the paragraph that begins "From Albuquerque" through the end of the document.

 2. Click the Increase Indent button 📰 on the Formatting toolbar. The selected text is indented 0.5 inch from the left side.

3. Click the Increase Indent button 📰 again. Now the text is indented 1 inch.

 4. Click the Decrease Indent button 📰 twice to return the text to the left margin.

5. With the text still selected, open the Paragraph dialog box.

6. Under Indentation, change the Left setting to 0.75 inch and the Right setting to 0.75 inch.

> **NOTE:** To set a *negative indent*, which extends a paragraph into the left or right margin areas, enter a negative number in the Left or Right text boxes. Any indent that occurs between the left and right margins is known as a *positive indent*.

7. Click to open the Special drop-down list in the Paragraph dialog box and choose First line. Word sets the By box to 0.5" by default. Notice the change in the Preview box.

FIGURE 5-6
Setting indents

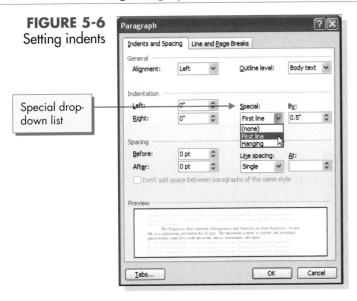

Special drop-down list

8. Click OK. Now each paragraph is indented from the left and right margins by 0.75 inch and the first line of each paragraph is indented another 0.5 inch. Notice that these settings are listed under Indentation in the Reveal Formatting task pane.

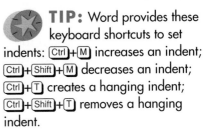

TIP: Word provides these keyboard shortcuts to set indents: Ctrl+M increases an indent; Ctrl+Shift+M decreases an indent; Ctrl+T creates a hanging indent; Ctrl+Shift+T removes a hanging indent.

EXERCISE **5-5** **Set Indents by Using the Ruler**

You can set indents by dragging the *indent markers* that appear at the left and right of the horizontal ruler. There are four indent markers:

- The *first-line indent marker* is the top triangle on the left side of the ruler. Drag it to the right to indent the first line of a paragraph.
- The *hanging indent marker* is the bottom triangle. Drag it to the right to indent the remaining lines in a paragraph.
- The *left indent marker* is the small rectangle. Drag it to move the first-line indent marker and hanging indent marker at the same time.
- The *right indent marker* is the triangle at the right side of the ruler, at the right margin. Drag it to the left to create a right indent.

Hanging indent marker   First-line indent marker

FIGURE 5-7
Indent markers
on the ruler

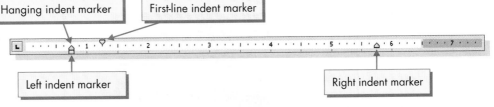

Left indent marker   Right indent marker

1. Make sure the ruler is displayed. If it is not, choose Ruler from the View menu.
2. Position the insertion point in the first paragraph below the title.
3. Point to the first-line indent marker on the ruler. A ScreenTip appears when you are pointing to the correct marker.
4. Drag the first-line indent marker 0.5 inch to the right. The first line of the paragraph is indented by 0.5 inch.

5. Drag the first-line indent marker back to the zero position. Point to the hanging indent marker and drag it 0.5 inch to the right. The lines below the first line are indented 0.5 inch, creating a hanging indent.

 NOTE: To make sure you're pointing to the correct indent marker, check the ScreenTip identifier before you drag the marker.

6. Drag the hanging indent marker back to the zero position. Drag the Left Indent marker (the small rectangle) 1 inch to the right. The entire paragraph is indented by 1 inch.

7. Select the first two paragraphs below the title and press Ctrl+Shift+N to remove all formatting from the paragraphs.

8. Close the Reveal Formatting task pane.

9. Position the insertion point in the second paragraph, which begins "From Albuquerque," and re-create the indents by using the ruler:

   ● Drag the Left Indent marker 0.75 inch to the right to indent the entire paragraph.

   ● Drag the first-line indent marker to the 1.25-inch mark on the ruler.

   ● Drag the Right Indent marker 0.75 to the left (to the 5.75-inch mark on the ruler). Now the paragraph is indented like the paragraphs below it.

10. Select all the indented paragraphs and drag the first-line indent marker to the 1-inch mark on the ruler. Now the opening line of each paragraph is indented only 0.25 inch. Display the Reveal Formatting task pane to see the indent settings.

FIGURE 5-8
Document with
indented text

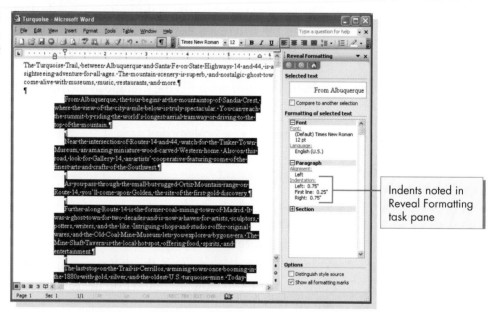

Indents noted in Reveal Formatting task pane

11. Save the document as *[your initials]*5-5 in a new folder for Lesson 5.

12. Print the document, but do not close it.

EXERCISE   **5-6**   **Use Click and Type to Insert Text**

You can use *Click and Type* to insert text or graphics in any blank area of a document. This feature enables you to position the insertion point anywhere in the document without pressing Enter repeatedly. Word automatically inserts the paragraph marks before that point and also inserts a tab.

1. Open the file **DukeCity**.
2. Choose Options from the Tools menu and click the Edit tab. Click Enable click and type if it is not already selected. Click OK.
3. Switch to Print Layout view by choosing Print Layout from the View menu.
4. Click anywhere in the last line of text.
5. Position the I-beam about five lines below the last line of text, in the center of the page. The I-beam is now the Click and Type pointer, which includes tiny lines that show right or center alignment.

FIGURE 5-9
Using Click
and Type

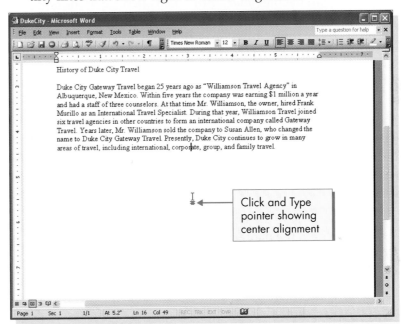

6. Move the I-beam back and forth until it shows center alignment. Double-click and key **Visit us at www.gwaytravel.com.** The text is centered and paragraph marks are inserted before it.
7. Save the document as *[your initials]***5-6** in your Lesson 5 folder.
8. Print and close the document.

Borders and Shading

To add visual interest to paragraphs or to an entire page, you can add a *border*—a line, box, or pattern—around text, a graphic, or a page. In addition, you can use *shading* to fill in the background behind the text of a paragraph. Shading can appear as a shade of gray, as a pattern, or as a color. Borders can appear in a variety of line styles and colors.

This lesson explains how to use the Borders and Shading dialog box to set border and shading options, and how to use the Borders button on the Formatting toolbar (which applies the most recently selected border style).

EXERCISE **5-7** **Add Borders to Paragraphs**

1. With the file *[your initials]*5-5 open, go to the end of the document. Press Enter twice to start a new paragraph and press Ctrl+Q to remove the formatting carried over from the previous paragraph.
2. Key the text shown in Figure 5-10.

FIGURE 5-10

Let Duke City Gateway Travel be your trail guide! We can provide you with a colorful printed guide of a trail you can follow at your leisure and in your own vehicle. Or we can lead the way with an escorted tour. Call us today at 555-1234 for more information.

3. Make sure the insertion point is to the left of the current paragraph mark or within the paragraph.
4. Choose Borders and Shading from the Format menu. The Borders and Shading dialog box appears. Click the Borders tab if it is not displayed.
5. Under Setting, click the Box option. The Preview box shows the Box setting. Each button around the Preview box indicates a selected border.
6. Scroll to see the options in the Style box. Choose the first border style (the solid line).
7. Open the Color drop-down list and choose Green. (ScreenTips identify colors by name.)
8. Open the Width drop-down list and choose 2 1/4 pt.

9. Click the top line of the box border in the Preview box. The top line is deleted and the corresponding button is no longer selected. Click the Top Border button or the top border area in the diagram to restore the top line border.

FIGURE 5-11
Borders and
Shading dialog box

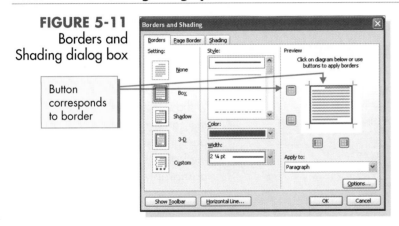

Button
corresponds
to border

10. Click the Options button. In the Border and Shading Options dialog box, change the top, bottom, left, and right settings to 5 points to increase the space between the text and the border. Click OK.

11. Change the setting from Box to Shadow. This setting applies a black shadow to the green border. Notice that the Apply to box is set to Paragraph.

Click OK. The shadow border is applied to the paragraph. The border settings appear in the task pane.

12. Click anywhere within the title "THE TURQUOISE TRAIL."

13. Click the down arrow next to the Borders button ▣▾ on the Formatting toolbar. A drop-down menu of border options appears.

FIGURE 5-12
Border options on
the Formatting
toolbar

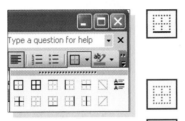

14. Click the Bottom Border button ▦ . A bottom border with the options previously set in the Borders and Shading dialog box is applied to the title.

15. Click the No Border button ▦ to delete the border.

16. Reapply the bottom border and click the Top Border button ▦ to add a top border as well.

NOTE: Borders and shading, when applied to a paragraph, extend from the left margin to the right margin or, if indents are set, from the left indent to the right indent.

EXERCISE 5-8 **Apply Borders to Selected Text and a Page**

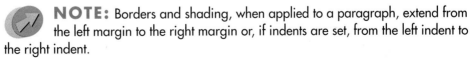

In addition to paragraphs, you can apply borders to selected text or to an entire page. When you apply a border to a page, you can choose whether to place the border on every page, the current page, the first page, or all but the first page in a document.

1. In the third paragraph below the title (which begins "Near the"), select the text "Gallery 14." Open the Borders and Shading dialog box.

2. From the Style box, scroll to the fifth line style from the bottom. Word automatically applies this style as the Box setting.

3. Change the <u>C</u>olor to Blue. Notice that the Apply to box indicates Text.

FIGURE 5-13
Applying borders
to selected text

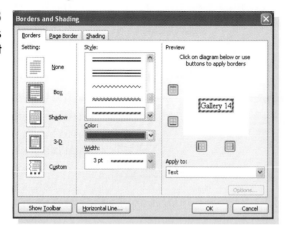

NOTE: When the Apply to box indicates Text, the borders are applied only to the selected text and not to the paragraph. If you include a paragraph mark in your selection, the borders are applied to the paragraph unless you change the Apply to setting to Text. It's important to notice the Apply to setting when applying borders and shading, or you might not get the results you intended.

4. Click the <u>P</u>age Border tab. Choose the third-to-last line style (a band of three shades of gray) and click the 3-<u>D</u> setting. The width should be 3 pt.

5. Click OK. Notice the text border added to "Gallery 14" and the page border. Deselect the text so you can see the border color.

6. In the Reveal Formatting task pane, click the down arrow, which is the Other Task Panes button ▼. Choose Styles and Formatting. This task pane provides a more visual display of the formatting used in the document.

7. At the bottom of the task pane, open the Show drop-down list and choose Formatting in use. The task pane now displays the formatting currently used in the document. Notice the different border formatting. Point to any format in the task pane (without clicking) for a descriptive ScreenTip.

FIGURE 5-14
Document with
border formatting

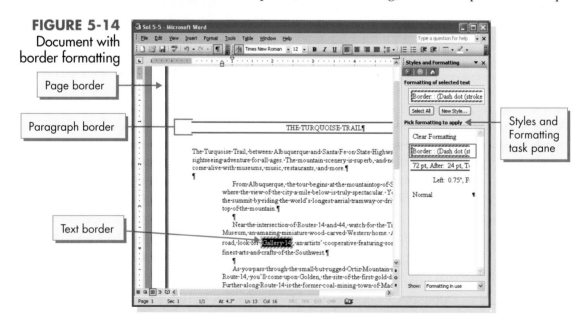

8. Click the Normal View button located at the bottom left of the Status bar to return to Normal view. The page border is not visible in this view.

9. Save the document as *[your initials]*5-8 in your Lesson 5 folder. Print the document. Leave it open.

EXERCISE **5-9** **Add a Horizontal Line**

Word provides special horizontal lines to divide or decorate a page. These lines are actually picture files (or "clips") in the shape of horizontal lines that are normally used when creating Web pages.

1. Position the insertion point anywhere in the last paragraph and open the Borders and Shading dialog box. Click the Borders tab.

2. Click None to remove the shadowed border. Click OK.

3. Position the insertion point at the paragraph mark directly above the last paragraph.

4. Press Ctrl+Q to remove the indent settings. (This ensures that the horizontal line you insert will extend from margin to margin.)

FIGURE 5-15
Inserting a
horizontal line

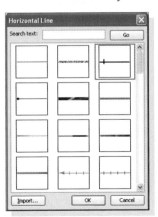

5. Open the Borders and Shading dialog box. Click the Horizontal Line button at the bottom of the dialog box.

6. In the Horizontal Line dialog box, click the third box in the first row. Click OK. The line is inserted in the document.

 NOTE: Available horizontal line clips might vary, depending on which files are installed on your computer. Check with your instructor if the specified line is not available.

EXERCISE **5-10** **Apply Shading to a Paragraph**

1. Click anywhere in the last paragraph, and open the Borders and Shading dialog box.

2. Click the Shading tab.

3. Under Fill, choose the third box in the first row of the color palette, which is 10 percent gray.

FIGURE 5-16
Shading options in
the Borders and
Shading dialog box

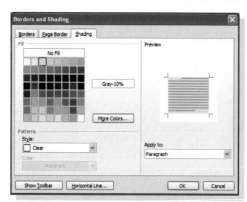

4. Open the Style drop-down list and scroll to view other shading options. Notice that you can apply a pattern, such as vertical lines or a grid. Close the Style drop-down list without choosing a style.

5. Click OK to apply the gray shading to the paragraph. Notice that the shading formatting is now listed in the task pane.

NOTE: Shading can affect the readability of text, especially when you use dark colors or patterns. It's a good idea to choose a larger type size and bold text when you use shading.

6. With the insertion point still in the last paragraph, remove the gray shading by choosing Clear Formatting from the Styles and Formatting task pane. This option clears all formatting from a paragraph (including borders, indents, and character formatting).

7. Click the Undo button 🔄 to restore the shading.

EXERCISE 5-11 Apply Borders Automatically

Word provides an AutoFormat feature to apply bottom borders. Instead of using the Borders button 🔲▾ or the Borders and Shading dialog box, you can key a series of characters and Word automatically applies a border.

1. Select the title, and use the Borders and Shading dialog box to remove the borders. (Choose None on the Borders tab.)

2. Place the insertion point at the end of the title (after "TRAIL"). Press Enter.

3. Key --- (three consecutive hyphens) and press Enter. Word applies a bottom border to the title. Delete the blank paragraph mark below the title.

TABLE 5-1

AutoFormatting Borders	
YOU KEY	**WORD APPLIES**
Three or more hyphens (–) and press Enter	A thin bottom border
Three or more underscores (_) and press Enter	A thick bottom border
Three or more equal signs (=) and press Enter	A double-line bottom border

TIP: If you do not want to format borders automatically, click the AutoCorrect Options button displayed after you key a series of characters and choose <u>S</u>top Automatically Creating Border Lines.

Repeating and Copying Formats

You can quickly repeat, copy, or remove paragraph formatting, using many of the techniques you learned in Lesson 3 for character formatting. For example, use F4 or Ctrl + Y to repeat paragraph formatting and the Format Painter button to copy paragraph formatting. You can also use the Reveal Formatting task pane to copy formatting to surrounding text and the Styles and Formatting task pane to apply formatting used elsewhere in a document.

EXERCISE 5-12 Repeat, Copy, and Remove Paragraph Formats

1. Click anywhere in the first paragraph under the title (which begins "The Turquoise Trail") and change the paragraph alignment to justified.

NOTE: You can click in a paragraph when repeating, copying, or removing formatting. You do not have to select the entire paragraph.

2. Click within the next paragraph and press F4 to repeat the formatting. Notice that the Styles and Formatting task pane lists two types of Justified formatting currently in use, one for regular text and one for the indented text.

3. Select the rest of the indented paragraphs, starting with the paragraph that begins, "Near the intersection of" through the paragraph that begins "The last stop on."

4. Point to the second Justified formatting listed in the task pane. A ScreenTip shows this is the Justified formatting for the indented text. Click this Justified formatting to apply it to the selected paragraphs.

5. Click anywhere in the last paragraph (with the shading).

6. Click the Format Painter button ; then click within the paragraph above the shaded paragraph to copy the formatting.

7. Click the Undo button to undo the paragraph formatting.

8. Click anywhere in the shaded paragraph. Choose Clear Formatting from the task pane to remove the formatting.

9. Click the Undo button to restore the formatting.

10. Close the task pane. Click just before the paragraph mark for the horizontal line you inserted above the last paragraph. Open the Paragraph dialog box, add 24 points of spacing before the paragraph, and click OK.

11. Save the document as *[your initials]*5-12 in your Lesson 5 folder.

12. Print and close the document.

NOTE: If you are using an inkjet printer, the bottom border might not print.

Bulleted and Numbered Lists

Bulleted lists and *numbered lists* are types of hanging indents you can use to organize important details in a document. In a bulleted list, a bullet (•) precedes each paragraph. In a numbered list, a sequential number or letter precedes each paragraph. When you add or delete an item in a numbered list, Word automatically renumbers the list.

To create bulleted lists or numbered lists, you can use the Bullets and Numbering dialog box or the Bullets button 📋 or Numbering button 📋 on the Formatting toolbar (which apply the most recently selected bullet or numbering style).

EXERCISE **5-13** **Create a Bulleted List**

---

1. Open the file **Memo2**. This document is a two-page memo. Key the current date in the memo date line, and then display the Reveal Formatting task pane.

2. Scroll toward the end of the document and select the list of items from "Airline tickets" to "Flight bag."

3. Click the Bullets button 📋 on the Formatting toolbar. Word applies the bullet style that was most recently chosen in the Bullets and Numbering dialog box. In the Reveal Formatting task pane, scroll down until you see <u>List</u>. Notice the description of the bullets.

FIGURE 5-17
Bulleted list

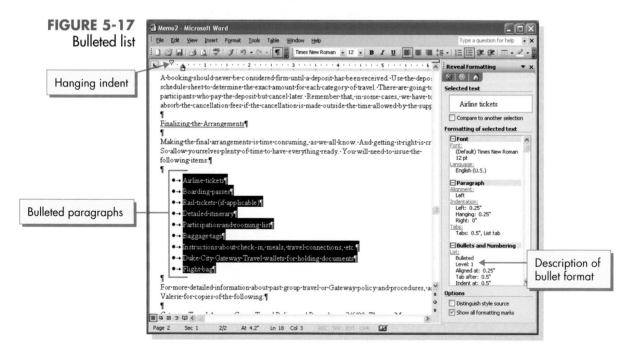

NOTE: When you create bulleted or numbered lists, Word automatically sets a 0.25-inch hanging indent.

4. With the list still selected, choose Bullets and Numbering from the Format menu. The Bullets and Numbering dialog box appears.

FIGURE 5-18
Bullet options in the Bullets and Numbering dialog box

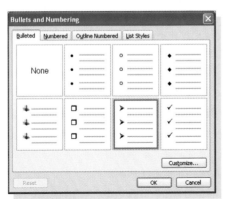

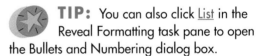 **TIP:** You can also click List in the Reveal Formatting task pane to open the Bullets and Numbering dialog box.

5. Click the Bulleted tab if it is not displayed. Choose the arrow bullet or another bullet shape.

6. Click OK. Word applies the new bullet shape to the list.

EXERCISE 5-14 Create a Numbered List

1. Select the last three paragraphs in the document, from "Gateway Travel Agency" through "IATA."

 2. Click the Numbering button to format the list with the style that was most recently chosen from the Bullets and Numbering dialog box.

3. With the list still selected, click List in the Reveal Formatting task pane. The Bullets and Numbering dialog box appears.

4. Choose the Roman numeral format and click OK. Word reformats the list with Roman numerals.

EXERCISE 5-15 Change a Bulleted or Numbered List

Word's bulleting and numbering feature is very flexible. When a list is bulleted or numbered, you can change it in several ways. You can

- Convert bullets to numbers or numbers to bullets in a list.
- Add or remove items in a bulleted or numbered list and Word renumbers the list automatically.
- Interrupt a bulleted or numbered list to create several shorter lists.
- Customize the list formatting by changing the symbol used for bullets or changing the alignment and spacing of the bullets and numbers.
- Turn off bullets or numbering for part of a list or the entire list.

1. Select the bulleted list that starts with "Airline tickets."

NOTE: When you select a bulleted or numbered list, the list is highlighted but the bullets or numbers are not.

2. Open the Bullets and Numbering dialog box and display the Numbered tab.

3. Choose a numbered format that starts with "1" and click OK to convert the bullets to numbers.

4. Select and delete the sixth item in the list, "Baggage tags." Word renumbers the list automatically.

5. Place the insertion point at the end of the last item in the numbered list, between "Flight bag" and the paragraph mark.

6. Press Enter and key **Baggage tags**. The formatting is carried to the new line.

7. Place the insertion point at the end of the sixth item (after "etc.") and press Enter.

8. Key in italic *These items are optional:*.

 9. Click within the italic text and click the Numbering button 🔢 on the Formatting toolbar to turn off numbering for this item. The list continues with the following paragraph.

10. Select and right-click the numbered text below the italic text and choose Bullets and Numbering from the shortcut menu.

11. Click the Restart numbering option so the list doesn't continue numbering from the previous list. Click OK. The new list starts with "1."

12. Insert a blank line above the italic text (click to the left of *These* and press Enter).

TIP: To change the size, color, and alignment of a bullet, display the Bulleted tab in the Bullets and Numbering dialog box, choose a bullet style, click Customize, and choose new options. If you click Picture in the Customize Bulleted List dialog box, you can insert a picture bullet—a decorative bullet often used in Web pages.

EXERCISE 5-16 **Create Lists Automatically**

Word provides an AutoFormat feature to create bulleted and numbered lists as you type. When this feature is selected, you can enter a few keystrokes, key your list, and Word inserts the numbers and bullets automatically.

1. Scroll to the paragraph under the heading "Record Keeping." At the end of the paragraph, change the period after "master file" to a colon.

2. Press Enter twice. Key **1.** and press Spacebar.

3. Key **Name** and press Enter. Word automatically formats your text as a numbered list.

4. Key the following text to complete the list, pressing Enter at the end of each line except the last line:

Itinerary
Accommodations
Options bought
Money paid

5. Position the insertion point at the end of the paragraph under "<u>Deposits and Cancellations</u>." (To display the end of the paragraph, you can use the horizontal scrollbar or close the task pane.)

6. Press Enter twice. Key an asterisk (*) and press Spacebar. Key **Deposit paid**.

7. Press Enter to start a new line and key **Cancellation deadline**. Word formats your text as a bulleted list.

TABLE 5-2 | **AutoFormatting Numbered and Bulleted Lists**

YOU KEY	WORD CHANGES THE FORMAT TO
A number and a period, closing parenthesis, or hyphen, followed by a space or tab, followed by text. Example, **1.**, **1)**, or **1-**. Press Enter.	A numbered list
An asterisk (*) or hyphen (-), followed by a space or tab, followed by text. Press Enter.	A bulleted list

EXERCISE 5-17 **Create an Outline Numbered List**

An *outline numbered list* has indented subparagraphs. For example, your list can start with item number "1)," followed by another level of indented items numbered "a)," "b)," and "c)." An outline numbered list can have up to nine levels and is often used for technical or legal documents.

1. Select the numbered list that begins with "1. Name" and open the Bullets and Numbering dialog box.

2. Click the Outline Numbered tab. Notice the outline numbering styles available. (See Figure 5-19 on the next page.)

3. Click the outline numbering style that begins with 1) and click OK. The list has the outline number format with different indent settings from the previous numbered list.

4. Position the insertion point at the end of "1) Name" and press Enter.

FIGURE 5-19
Outline numbered
options in the Bullets
and Numbering
dialog box

5. Click the Increase Indent button
   (or press Tab) and key **Group name**.

6. Press Enter and key **Participant
   name**. The numbered list now has
   two indented subparagraphs.

7. With the insertion point at the end of
   "b) Participant name," press Enter
   and click the Increase Indent button
   (or press Tab). You can now add a
   third level to your list.

8. Click the Decrease Indent button twice (or press Shift+Tab twice). You
   can now add a new item numbered "2)."

9. Delete the paragraph mark in the new line numbered "2)."

10. Under "2) Itinerary," add the following three indented subparagraphs:
    Air
    Land
    Sea

NOTE: You can create an outline numbered list from scratch. Key the first
paragraph, apply the desired outline numbered formatting, and then continue
keying the list, using the indent buttons or (Tab and Shift+Tab)) to change outline levels.

Symbols and Special Characters

The fonts you use with Word include *special characters* that do not appear on your
keyboard, such as those used in foreign languages (for example, ç, Ö, and Ω).
There are additional fonts, such as *Wingdings* and *Symbol,* that consist entirely of
special characters.

To insert symbols and special characters in your documents, you can use the
Symbol command on the Insert menu or use keyboard shortcuts.

EXERCISE 5-18 Insert Symbols

1. Scroll toward the beginning of the document. Position the insertion point to
   the immediate left of the title "Deadlines."

2. Choose Symbol from the Insert menu. The Symbol dialog box appears.

3. Make sure the Symbols tab is displayed and choose (normal text) from the
   Font drop-down list box.

4. Scroll through the grid of available symbol characters for normal text.

5. At the bottom of the dialog box, click the arrow to open the from drop-down list box and choose Unicode (hex). At the top of the dialog box, open the Subset drop-down list box and scroll the list of categories available for normal text.

6. Close the list without changing the subset.

7. Click the arrow to open the Font drop-down list box and choose Symbol. Review the available symbol characters.

8. Change the font to Wingdings. The characters included in the Wingdings font appear in the grid.

FIGURE 5-20
Symbol dialog box

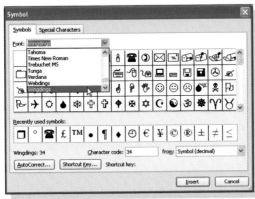

9. Scroll down several rows until you see the clock symbols. Click one of the clock symbols.

TIP: Notice the recently used symbols shown at the bottom of the Symbol dialog box. Word displays the 16 most recently used symbols.

10. Click Insert, and then click Close. The symbol appears in the document.

11. Press Tab. Select the clock and the space between it and the text "Deadlines." Remove the underlining. Use the Font Size drop-down list on the Formatting toolbar to increase the clock size to 18 points.

TIP: You can assign shortcut keys or AutoText to a symbol by clicking Shortcut Key or AutoCorrect in the Symbol dialog box. You can also press Alt and key the numeric code (using the numeric keypad, if you have one) for a character. For example, if you change the font to Wingdings and press Alt+0040, you will insert the character for a Wingdings telephone. Remember to change the Wingdings font back to your normal font after inserting a special character.

12. Select the list with Roman numerals, at the end of the document. You're going to change the Roman numerals to a symbol character.

13. Open the Bullets and Numbering dialog box. Display the Bulleted tab, click Customize, and then click Character.

14. Choose a symbol from either the Symbol or Wingdings font that would be appropriate for this bulleted list. Click OK, and then click OK again. The Roman numerals are replaced with your chosen bullet symbol.

EXERCISE 5-19 **Insert Special Characters**

You can use the Symbol dialog box and shortcut keys to insert characters such as an en dash, an em dash, or SmartQuotes. An *en dash* is a dash slightly wider than a hyphen. An *em dash*, which is twice as wide as an en dash, is used in sentences where you would normally insert two hyphens. *SmartQuotes* are quotation marks that open a quote curled in one direction (") and close a quote curled in the opposite direction (").

 NOTE: By default, Word inserts SmartQuotes automatically.

1. Make sure nonprinting characters are displayed in the document. If they are not, click the Show/Hide ¶ button ¶ .

2. On page 1, locate the paragraph that begins "July 1." Position the insertion point to the immediate right of the "1," hold down [Shift], and press [→] three times to select the space, hyphen, and space. (This technique is much easier than trying to drag to select the three characters.)

3. Choose Symbol from the Insert menu and click the Special Characters tab.

4. Choose Em Dash from the list of characters. (Notice the keyboard shortcut listed for the character.) Click Insert, and then click Close. The em dash replaces the three characters.

5. Select the three characters immediately following "July 15." Press [Alt]+[Ctrl]+ − (the minus sign on the numeric keypad). An em dash is inserted. (If you don't have a numeric keypad, press [F4] to repeat the character.)

6. Insert em dashes after "July 17," "August 1," and "August 15."

EXERCISE 5-20 **Create Symbols Automatically**

You can use Word's AutoCorrect feature to create symbols as you type. Just enter a few keystrokes and Word converts them into a symbol.

1. Scroll to the numbered lists under "Finalizing the Arrangements." In the second numbered list, edit "Flight bag" to read **Easy Carry flight bag**.

2. Position the insertion point to the immediate right of "Carry" and key **(tm)**. Word automatically creates a trademark symbol ™.

3. Position the insertion point at the end of the first paragraph below the memo heading (after "destinations!"). Press [Spacebar] and key **:)** with no spaces. Word creates a happy-face character.

 NOTE: To review the symbols AutoCorrect can enter automatically, choose AutoCorrect Options from the Tools menu and click the AutoCorrect tab.

4. Delete the happy face.

5. Format the first line of the memo with 72 points of paragraph spacing before it. This starts the first line two inches from the top of the page. (See Appendix B: "Standard Forms for Business Documents.")

6. Decrease the length of the memo to two pages by selecting the entire document and changing the font size to 11 points.

REVIEW: Use the keyboard shortcuts Ctrl+A to select all and Ctrl+Shift+< to reduce the font size.

7. Save the document as *[your initials]*5-20 in your Lesson 5 folder.

8. Print and close the document.

USING ONLINE HELP

In this lesson, you learned about special characters. In addition to being able to insert special characters, such as an em dash, you can use keyboard shortcuts to insert foreign characters, such as è.

Learn more about special characters:

1. Click the Ask a Question box and key **special characters**.

2. Choose the topic Keyboard shortcuts for international characters.

3. In the Help window, note the keyboard shortcuts for the French characters á and è.

FIGURE 5-21
Keyboard shortcuts
for international
characters

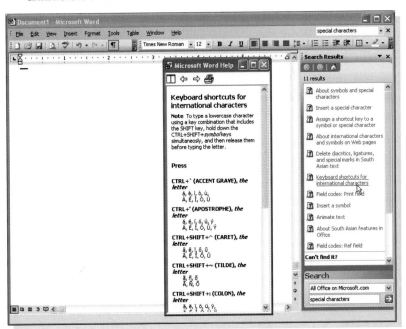

TIP: You might want to print this Help topic for future reference. Click the Print button 🖨 in the Help window; then choose Print the selected topic.

4. Close Help. Open a new document.

5. Key the following text, using the keyboard shortcuts for the characters à and é:

   Duke City Gateway Travel can help you plan a spectacular vacation in Montréal. Just call us and you can be on your way. Voilà!

6. Close the document without saving it.

LESSON 5 Summary

➤ A paragraph is any amount of text followed by a paragraph mark.

➤ Paragraph alignment determines how the edges of a paragraph appear horizontally. Paragraphs can be left-aligned, centered, right-aligned, or justified.

➤ Line space is the amount of vertical space between lines of text in a paragraph. Lines can be single-spaced, 1.5-line spaced, double-spaced, or set to a specific value.

➤ Paragraph space is the amount of space above or below a paragraph. Paragraph space is set in points—12 points of space equals one line space for 12-point text. Change the space between paragraphs by using the Before and After options in the Paragraph dialog box or by using the Ctrl+0, Ctrl+Shift+0 keyboard shortcuts to add or remove 12 points before a paragraph.

➤ A left indent or right indent increases a paragraph's distance from the left or right margin. A first-line indent indents only the first line of a paragraph. A hanging indent indents the second and subsequent lines of a paragraph.

➤ To set indents by using the horizontal ruler, drag the left indent marker (small rectangle), the first-line indent marker (top triangle), or the hanging indent marker (bottom triangle), which are all on the left end of the ruler, or drag the right indent marker (triangle) on the right end of the ruler.

➤ The Click and Type feature enables you to insert text in any blank area of a document by simply positioning the insertion point and double-clicking. (The document must be in Print Layout view.)

➤ A border is a line or box added to selected text, a paragraph, or a page. Shading fills in the background of selected text or paragraphs. Borders and shading can appear in a variety of styles and colors.

➤ In addition to regular borders, Word provides special decorative horizontal lines that are available from the Borders and Shading dialog box.

➤ The AutoFormat feature enables you to create a border automatically. Key three or more hyphens ⊟, underscores ⊡, or equal signs ⊑, and press ⏎. See Table 5-1.

➤ Repeat paragraph formats by pressing F4 or Ctrl+Y. Copy paragraph formats by using the Format Painter button 🖌 . Apply existing formatting to selected text by using the Styles and Formatting task pane. Remove paragraph formats by pressing Ctrl+Q or choosing Clear Formatting from the task pane.

➤ Format a list of items as a bulleted or numbered list. In a bulleted list, each item is indented and preceded by a bullet character or other symbol. In a numbered list, each item is indented and preceded by a sequential number or letter.

➤ Remove a bullet or number from an item in a list by clicking the Bullets button ▤ or the Numbering button ▤ on the toolbar. Press ⏎ mid-list to add another bulleted or numbered item automatically. Press ⏎ twice in a list to turn off bullets or numbering. Change the bullet symbol or the numbering type by using the Bulleted or Numbered tabs in the Bullets and Numbering dialog box.

➤ The AutoFormat feature enables you to create a bulleted or numbered list automatically. See Table 5-2.

➤ Create a list with outline numbering by using the Outline Numbered tab in the Bullets and Numbering dialog box. An outline numbered list has indented subparagraphs, such as paragraph "1)" followed by indented paragraph "a)" followed by indented paragraph "i)." To increase the level of numbering for each line item, click the Increase Indent button ▤ or press Tab. To decrease the level of numbering, click the Decrease Indent button ▤ or press Shift+Tab.

➤ Insert symbols, such as foreign characters, by using the Symbols tab in the Symbol dialog box. Wingdings is an example of a font that contains all symbols.

➤ Insert special characters, such as an em dash (—), by using the Special Characters tab in the Symbol dialog box.

➤ Create symbols automatically as you type by keying AutoCorrect shortcuts, such as keying :) to produce the ☺ symbol.

LESSON 5   Command Summary

FEATURE	BUTTON	MENU	KEYBOARD
Left-align text	▤	Format, Paragraph	Ctrl+L
Center text	▤	Format, Paragraph	Ctrl+E
Right-align text	▤	Format, Paragraph	Ctrl+R
Justify text	▤	Format, Paragraph	Ctrl+J

continues

LESSON 5 Command Summary *continued*

FEATURE	BUTTON	MENU	KEYBOARD
Single space		Fo̲rmat, P̲aragraph	Ctrl + 1
Double space		Fo̲rmat, P̲aragraph	Ctrl + 2
1.5-line space		Fo̲rmat, P̲aragraph	Ctrl + 5
Borders and shading		Format, Borders and Shading	
Remove paragraph formatting		Fo̲rmat, P̲aragraph	Ctrl + Q
Restore text to Normal formatting			Ctrl + Shift + N
Increase indent		Fo̲rmat, P̲aragraph	Ctrl + M
Decrease indent		Fo̲rmat, P̲aragraph	Ctrl + Shift + M
Hanging indent		Fo̲rmat, P̲aragraph	Ctrl + T
Bulleted list		Fo̲rmat, Bullets and N̲umbering	
Numbered list		Fo̲rmat, Bullets and N̲umbering	
Symbols and special characters		I̲nsert, S̲ymbol	

Concepts Review

Each of the following statements is either true or false. Indicate your choice by circling T or F.

T F 1. You can use the Formatting toolbar to right-align paragraphs.

T F 2. Text that is left-aligned has a ragged left edge.

T F 3. You can open the Paragraph dialog box from the shortcut menu.

T F 4. The keyboard shortcut Ctrl+5 changes line spacing to 1.5 lines.

T F 5. You can use Word's AutoCorrect feature to create symbols as you type by using certain keyboard combinations.

T F 6. You can use keyboard shortcuts to insert some foreign characters.

T F 7. Ctrl+Q removes all paragraph formatting.

T F 8. A hanging indent indents all lines in a paragraph except the first line.

Write the correct answer in the space provided.

1. Which type of paragraph alignment adjusts spacing between words?

2. What is the keyboard shortcut for centering text?

3. Single, 1.5, and double are examples of what type of spacing?

4. If you click 🔳 once, what happens to selected text?

5. With an outline numbered list, instead of clicking 🔳, what key can you press to achieve the same result?

6. Which keystrokes apply a double-line border automatically?

_____

7. What can you click in the Reveal Formatting task pane to open the Bullets and Numbering dialog box?

_____

8. Which indent marker is the top triangle on the left side of the ruler?

_____

CRITICAL THINKING

Answer these questions on a separate page. There are no right or wrong answers. Support your answers with examples from your own experience, if possible.

1. You can use keyboard shortcuts to change paragraph alignment, or you can use the alignment buttons on the Formatting toolbar. Which method do you prefer? Why? When might you use the other method?

2. Many people use bulleted lists and numbered lists interchangeably. Are there times when it would be more appropriate to use a bulleted list than a numbered list and vice versa? Explain your answer.

Skills Review

EXERCISE 5-21

Change paragraph alignment and line spacing.

1. Start a new document.

2. Change the character formatting and paragraph alignment for the first paragraph by following these steps:

   a. Select the paragraph mark and set the font to 14-point Arial.

   b. Click the Center button ▤ on the Formatting toolbar to center the text you are about to key.

   c. Key the first two lines shown in Figure 5-22 in all capitals. Use a line break (press Shift+Enter) after the first line to make the two lines one paragraph.

3. At the end of this paragraph, press Enter twice, turn off Caps Lock, and key the four names shown in Figure 5-22. Again, use a line break after each name to make the names all one paragraph.

FIGURE 5-22

Duke City Gateway

Senior Travel Counselors

Frank Murillo

Nina Chavez

Tom Carey

Alice Fung

4. Change the alignment of each paragraph by following these steps:

    a. Move the insertion point into the first paragraph and click the Align Right button 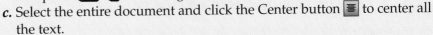 on the Formatting toolbar to right-align the first two lines.

    b. Move the insertion point into the paragraph containing the four names and press Ctrl+L to left-align the names.

    c. Select the entire document and click the Center button ≡ to center all the text.

5. Change line spacing by following these steps:

    a. Click within the second paragraph.

    b. Press Ctrl+5 to change the line spacing to 1.5 lines.

6. Save the document as *[your initials]***5-21** in your Lesson 5 folder.

7. Print and close the document.

EXERCISE 5-22

Change paragraph spacing and set indents.

1. Open the file **Prices2**. (Make sure the Show/Hide ¶ button ¶ is turned on.)

2. Change spacing between paragraphs by following these steps:

    a. Select the entire document, click the right mouse button, and choose Paragraph from the shortcut menu.

    b. Click the Indents and Spacing tab if it is not displayed.

    c. Click the up arrow to the right of After to set the spacing after paragraphs to 12 points. Click OK.

    d. Delete the blank lines between the three paragraphs of text.

3. Press Ctrl+Home to move to the top of the document. Key in bold uppercase letters **TOUR INFORMATION**.

4. Press Enter once. Click within the new title and open the Paragraph dialog box. Change spacing to 72 points before and 24 points after. Change the Alignment to Centered and click OK.

5. Apply a first-line indent to the paragraphs by following these steps:
   a. Select the three paragraphs below the title.
   b. Make sure the horizontal ruler is displayed. Point to the first-line indent marker on the ruler. When a ScreenTip identifies it, drag it 0.5 inch to the right.

6. Change the indentation of the last paragraph by following these steps:
   a. Left-click to deselect the paragraphs.
   b. Right-click the last paragraph and open the Paragraph dialog box.
   c. Set the Left and Right indentation text boxes to 1 inch.
   d. Remove the first-line indent by choosing (none) from the Special drop-down list. Click OK.

7. Key **Note:** at the beginning of the newly indented paragraph.

8. Justify the three paragraphs below the title.

9. Save the document as *[your initials]*5-22 in your Lesson 5 folder.

10. Print and close the document.

EXERCISE 5-23

Apply borders and shading; repeat and copy formatting.

1. Open the file **Rockies2**.

2. At the end of the document, key the following text as a separate paragraph. Use bold text and be sure to insert a blank line before the new paragraph.

   Now is the time to schedule this exciting tour. Call Duke City Gateway at 555-1234 for reservations and more information. Ask for Steve Ross.

3. Apply borders and shading to the new paragraph by following these steps:
   a. Place the insertion point in the paragraph.
   b. Choose Borders and Shading from the Format menu. Click the Borders tab if it is not displayed.
   c. Use the first line style and change the Width to 1 1/2 pt.
   d. Change the line color to blue.
   e. In the Preview box, click the Top Border button and the Bottom Border button.
   f. Click the Shading tab. From the Fill palette, choose the fourth gray box in the first row (12.5 percent gray). Click OK.

4. Repeat the formatting by clicking within the first paragraph and pressing F4.

5. Click the Undo button to undo the formatting in the first paragraph.

6. Apply a border to text automatically by following these steps:

    a. In the first paragraph, place the insertion point before "Experience" in the second sentence and press (Enter) to split the paragraph.

    b. Key === and press (Enter) to automatically insert a double-line border under the first paragraph.

7. Copy formatting from one paragraph to another by following these steps:

    a. Click in the last paragraph.

    b. Click the Format Painter button 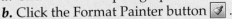 .

    c. Click in the first paragraph.

8. Change the first paragraph to bold and all capitals, and delete the period. Add 72 points of spacing before and 24 points after the paragraph and center-align it.

9. Save the document as *[your initials]*5-23 in your Lesson 5 folder.

10. Print and close the document.

EXERCISE 5-24

Align paragraphs, change paragraph spacing, create bulleted lists, and insert symbols and special characters.

1. Open the file **Currency**.

2. Format the title as bold, all caps, centered. Set the paragraph spacing to 72 points before and 12 points after.

3. Create a bulleted list by following these steps:

    a. Select the text under the heading "In England" (from "$2" through "$10").

    b. Choose Bullets and Numbering from the Format menu. Click the Bulleted tab if it is not displayed.

    c. Choose a bullet option and click OK.

4. Repeat the bullet formatting for the text under the heading "Information about Eurocurrency" (from "$1" through "dollars.") by selecting the text, and then clicking the Bullets button [≡] .

5. Insert symbols by following these steps:

    a. In the line of text below the heading "In England," delete the word "pound" and position the insertion point immediately in front of the "1."

    b. Choose Symbol from the Insert menu. Click the Symbols tab if it is not displayed. Make sure the Font box is set to (normal text).

    c. Scroll down to the seventh row in the grid of symbols. Double-click the British pound symbol (£) in the third box from the right.

    d. Click Close.

6. Two lines down, repeat the pound symbol before "5" and delete the word "pounds."

7. In the line that begins "$1 equals," delete the text "in Eurocurrency" without deleting the period.

8. Position the insertion point immediately in front of "0.87." Press [Alt]+[Ctrl]+[E] to insert the Euro symbol (€).

9. In the last paragraph of the document, insert an em dash by following these steps:

   a. Select the comma and the space after the word "change" and before the word "but."

   b. Press [Alt]+[Ctrl]+− (the minus key on the numeric keypad). Or choose Symbol from the Insert menu and use the Special Characters tab to select the em dash.

10. Create a symbol automatically by following these steps:

   a. Position the insertion point to the immediate left of "In England."

   b. Key these three characters with no spaces between them: = = > (remember to press [Shift] to key the ">").

   c. Repeat the symbol before the heading that begins "Information."

11. Save the document as *[your initials]*5-24 in your Lesson 5 folder.

12. Print and close the document.

Lesson Applications

EXERCISE 5-25

Change alignment, line spacing, and paragraph spacing and apply shading.

1. Start a new document.

2. Key the text shown in Figure 5-23, including the corrections. Use Times New Roman and single spacing. Use line breaks (⟨Shift⟩+⟨Enter⟩) to format the text as two paragraphs.

FIGURE 5-23

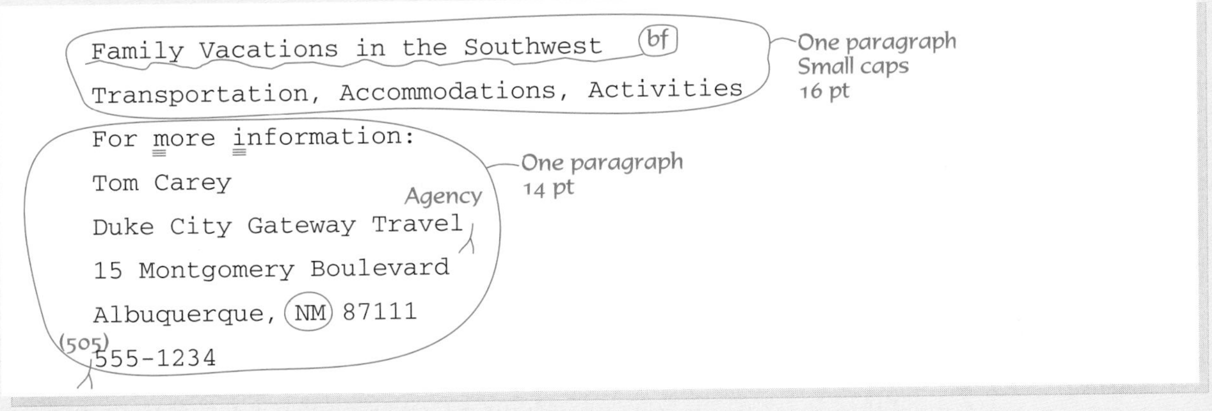

Family Vacations in the Southwest (bf)   ⎤ One paragraph
                                                  Small caps
Transportation, Accommodations, Activities ⎦   16 pt

For more information: ⎤ One paragraph
                                            14 pt
Tom Carey   Agency

Duke City Gateway Travel

15 Montgomery Boulevard

Albuquerque, (NM) 87111

(505)
    555-1234

3. Center all the text.

4. Change the line spacing in only the second paragraph to 1.5 lines.

5. Change the paragraph spacing for only the first paragraph to 72 points before and 36 points after.

6. Add 10 percent gray shading to the first paragraph. Add a page border, using the box setting and the double wavy line style.

7. Save the document as *[your initials]*5-25 in your Lesson 5 folder.

8. Print and close the document.

EXERCISE 5-26

Change alignment, line spacing, and paragraph spacing; add a border; repeat formatting; and add symbols.

1. Open the file **Offices1**.

2. Change the title to "East Coast Gateway Travel Offices" and delete all office locations except New York, Boston, and Washington. There should be one

blank line below the title and a blank line after each of the remaining office locations.

3. Format all lines of the first office location with 1.5-line spacing.

4. Repeat the formatting for the remaining two locations.

5. Change the title to 24-point bold with paragraph spacing of 72 points before and 24 points after. Delete the blank paragraph mark after the title.

6. Add a 3/4-point double-line bottom border to the title.

7. Center the second office location text and right-align the last office location text.

8. Replace the text "Tel." throughout the document with the telephone symbol from the Wingdings font (first row, eighth symbol). Leave the space between the symbol and the telephone number.

9. Save the document as *[your initials]*5-26 in your Lesson 5 folder.

10. Print and close the document.

EXERCISE 5-27

Indent paragraphs, create bulleted and numbered lists, and change paragraph spacing.

1. Start a new document.

2. Key the text shown in Figure 5-24, including the corrections. Use single spacing and 12-point Arial. Remember to insert a blank line between paragraphs (but not between list items). Use the standard round bullet for the bulleted list and the standard number format (1., 2.) for the numbered list.

3. Set 0.25-inch first-line indents for all paragraphs except the title, the bulleted and numbered lists, and the final paragraph.

4. Format the final paragraph with 0.75-inch left and right indents.

5. Add 5 percent gray shading to the final paragraph.

6. Format the title as 14-point, bold, shadowed text, and centered. Add 72 points of paragraph spacing before the title.

7. Save the document as *[your initials]*5-27 in your Lesson 5 folder.

8. Print and close the document.

FIGURE 5-24

2 blank lines   center-align

Gateway Travel Amazon Trip (bf)

#date

Congratulations on choosing our most exciting special tour to. For ten days, join our expert guide on a tour of the Amazone rainforest. The beauty and diversity of the rainforest ecosystem are unequalled. You'll see plant, bird, and animal life that live nowhere else on t he planet.

You'll also visit with native peoples whose ancestors have inhabited the region for centuries. Native culture and ceremonies ~~with~~ will be yours to participate in and admire. Keep ing in mind that the Amazon climate is very different from our own. We recommend that you pack the following:

Mosquito netting
Insect repellent
Light cotton clothing   bulleted list
Camera
Extra film
Head scarf or soft hat

You'll need the following documentation:

Vacinnation certificate   numbered list
Complete physical exam

We also recommend, because of the time difference, And concerns about water quality, that you consider the following: ital

Pack some remedy for jet lag, such as an over-the-counter or homeopathic remedy, and planning to drink only boiled or bottled water.

EXERCISE 5-28 *Challenge Yourself*

Indent and align paragraphs, change paragraph spacing, apply a border, copy formatting, create bulleted lists, and insert a special character.

1. Open the file **Layout**.

2. Format the heading "**The Layout of a Cruise Ship**" in uppercase with a negative left indent of –0.25 inch and 12 points of spacing before the paragraph.

3. Use the same formatting for the heading "**The Staterooms.**"

4. Format the paragraphs from "**Decks**" through "**Cruise Ship Features**" as a bulleted list, using the bullet of your choice, with six points of spacing after paragraphs.

5. Format the paragraphs from "**Amenities**" through "**Price**" with a different bullet and with six points of spacing after paragraphs.

6. Add two blank paragraph marks above "**THE STATEROOMS.**" At the middle blank paragraph mark, add a horizontal line of your choice.

 NOTE: If horizontal line clips aren't available, use a border.

7. In the first paragraph below the title, replace the comma and space after the words "floating city" with an em dash.

8. Justify all text in the document.

9. Add a page border to the document, using the third-to-last line style, 3-point width, and the 3-D setting.

10. If the document is two pages long, delete the blank paragraph mark at the top of page 2.

11. Save the document as *[your initials]*5-28 in your Lesson 5 folder.

12. Print and close the document.

On Your Own

In these exercises you work on your own, as you would in a real-life work environment. Use the skills you've learned to accomplish the task—and be creative.

EXERCISE 5-29
Create a flyer for an event, such as a meeting or concert. Use a variety of paragraph alignment settings, line spacing, and paragraph spacing. Add shading to one or more paragraphs and a page border. Include a bulleted list. Save the document as *[your initials]*5-29. Print the document.

EXERCISE 5-30
Create a set of instructions for how to make something. Use a bulleted list for the materials needed. Use a numbered list to describe the step-by-step instructions. Use borders or shading for emphasis or for paragraphs containing special notes or tips. Save the document as *[your initials]*5-30 and print it.

EXERCISE 5-31
Review the foreign-language characters in the Symbols dialog box (normal text). Use a foreign-language dictionary (such as an online dictionary on the Internet) to find a few foreign words you can key in a document by inserting the appropriate foreign-language characters. Include the English translations. Display the Research task pane to help you define and translate vocabulary. Save the document as *[your initials]*5-31 and print it.

LESSON 6

Margins and Printing Options

OBJECTIVES

After completing this lesson, you will be able to:

1. Change margins.
2. Preview a document.
3. Change paper size and orientation.
4. Use hyphenation.
5. Insert the date and time as a field.
6. Print envelopes and labels.
7. Choose print options.

MICROSOFT OFFICE SPECIALIST ACTIVITIES

In this lesson:
    WW03S-1-2
    WW03S-3-5
    WW03S-5-5
    WW03S-5-6
    WW03S-5-7

See Appendix C.

 Estimated Time: 1½ hours

In a Word document, text is keyed and printed within the boundaries of the document's margins. *Margins* are the spaces between the edges of the text and the edges of the paper. Adjusting the margins can significantly change the appearance of a document.

Word offers many useful printing features: changing the orientation—the direction (horizontal or vertical) in which a document is printed, selecting paper size, and printing envelopes and labels.

Changing Margins in Normal View

By default, a document's margin settings are:

- Top margin   1 inch
- Bottom margin   1 inch
- Left margin   1.25 inches
- Right margin   1.25 inches

FIGURE 6-1
Default margin settings

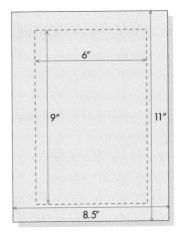

FIGURE 6-2
Actual workspace using
default margin settings and
standard-size paper

Using standard-size paper (8.5 × 11 inches) and Word's default margin settings, you have 6 × 9 inches on the page for your text. To increase or decrease this workspace, you can change margins by using the Page Setup dialog box or the rulers in Print Layout view or Print Preview.

EXERCISE   **6-1**   **Change Margins for a Document Using the Page Setup Dialog Box**

When working in Normal view, you use the Page Setup dialog box to change margin settings for a document.

1. Open the file **Rockies2**.
2. Choose Page Setup from the File menu. (Make sure no text is selected.)

3. Click the Margins tab, if it is not active. The dialog box shows the default margin settings.

4. Edit the margin text settings so they have the following values (or click the arrow boxes to change the settings). As you do so, notice the changes in the Preview box.

Top	**1.5**
Bottom	**1.5**
Left	**2**
Right	**2**

FIGURE 6-3
Changing margins
in the Page Setup
dialog box

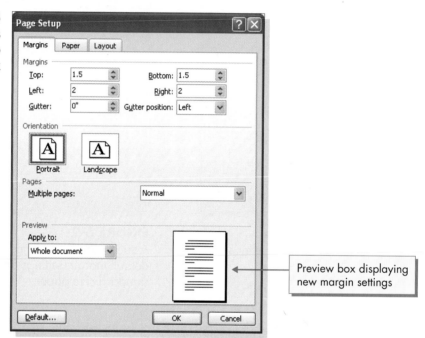

Preview box displaying
new margin settings

TIP: Press [Tab] to move from one margin text box to the next and to see the new settings in the Preview box. Press [Shift]+[Tab] to move to the previous margin text box.

5. Click the down arrow to open the Apply to drop-down list. Notice that you can choose either Whole document or This point forward (from the insertion point forward). Choose Whole document and click OK to change the margins of the entire document.

6. Display the Reveal Formatting task pane and scroll down to Section. If Section is preceded by a plus sign (+), click the plus sign to display the current margin settings for the document.

FIGURE 6-4
Reveal Formatting
task pane showing
margin settings

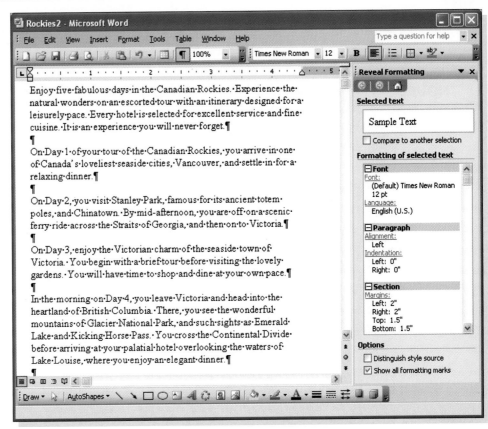

E X E R C I S E   **6-2**   **Change Margins for Selected Text by Using the Page Setup Dialog Box**

When you change margins for selected text, you create a new section. A *section* is a portion of a document that has its own formatting. When a document contains more than one section, you see double-dotted lines, or *section breaks,* between sections to indicate the beginning and end of a section.

1. Select the text from the second paragraph to the end of the document.

2. Click <u>Margins</u> under Section in the Reveal Formatting task pane.

3. Change the margins to the following settings:

<u>Top</u>	**2**
Bottom	**2**
<u>L</u>eft	**1.5**
<u>R</u>ight	**1.5**

4. Choose Selected text from the Apply to box. Click OK.

5. Deselect the text and scroll to the beginning of the selection. Word applied the margin changes to the selected text and created a new section, as indicated by the section break lines. The section appears on a new page.

FIGURE 6-5
Changing margins
for selected text
creates a
new section.

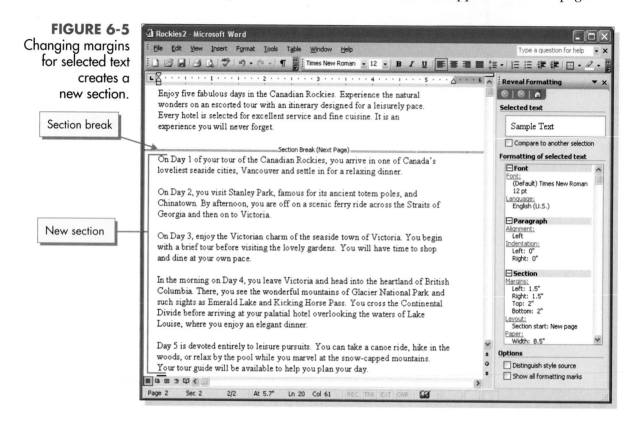

Section break

New section

EXERCISE   **6-3**   **Change Margins for a Section by Using the Page Setup Dialog Box**

After a section is created, you can change the margins for just the section (not the entire document) by using the Page Setup dialog box.

1. Move the insertion point anywhere in the new section (section 2) and open the Page Setup dialog box.

2. Change the left and right margin settings to 1.25 inches.

3. Open the Apply to drop-down list to view the options. Notice that you can apply the new margin settings to the current section, to the whole document, or from the insertion point forward.

4. Choose This section and click OK to apply the settings to the new section.

EXERCISE   **6-4**   Change Margins in Print Layout View

In Print Layout view, you can see how text will be positioned on the printed page. Print Layout view displays headers, footers, and other page elements not visible in Normal view. You can change margins quickly and easily in Print Layout view by using the ruler.

There are two ways to switch to Print Layout view:

● Click the Print Layout View button 🔲 at the left of the horizontal scroll bar.

● Choose Print Layout from the View menu.

FIGURE 6-6
View buttons

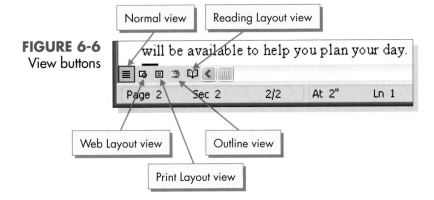

1. Place the insertion point at the beginning of the document (Ctrl+Home). In bold uppercase letters, key **CANADIAN ROCKIES TOUR**, and then press Enter. Center the title and add two blank line spaces below it.

 2. Switch to Print Layout view by clicking the Print Layout View button 🔲 at the left of the horizontal scroll bar (or choose Print Layout from the View menu). You see only page 1 of the document. If the rulers are not displayed, choose View, Ruler.

 3. Drag the scroll box down or click the Next Page button 🔻 on the vertical scroll bar. Notice that the new section starts on a new page.

  NOTE: If the Next Page button 🔻 is blue, click the Select Browse Object button 🔘 and choose the page icon to browse by page.

4. Click in the new section. The Status bar shows that the document contains two pages and two sections. Notice the extra space at the top of the page. The new section has a larger top margin (two inches), and the Reveal Formatting task pane displays the margin settings.

5. Move the insertion point to the end of the document, press Enter twice, and key the two paragraphs shown in Figure 6-7. Use single spacing and insert one blank line between paragraphs.

FIGURE 6-7

Tour prices include all transportation and meals as listed in the itinerary, as well as taxes and gratuities, except for the customary gratuities to your tour guide.

All tours include enough free time for travelers to do some exploring on their own. Gateway Tour Hosts can provide information specific to individual interests, such as golfing, tennis, swimming, or other activities that may be available during free time.

6. To see more of the page, including the margin areas, choose Zoom from the View menu, and then choose Page width. Click OK.

7. Move the insertion point to the top of the document (the first section). The blue area on the vertical ruler shows the 1.5-inch top margin. The blue areas on the horizontal ruler show the 2-inch left and right margins. The white area in the horizontal ruler shows the text area, which is a line length of 4.5 inches. (See Figure 6-8.)

TIP: You can close the task pane for a clearer view of rulers and how they affect your document. Open the task pane when you are through working with rulers in Print Layout view, so you can continue to see all settings for the formatting in your document.

FIGURE 6-8
Rulers in Print Layout view

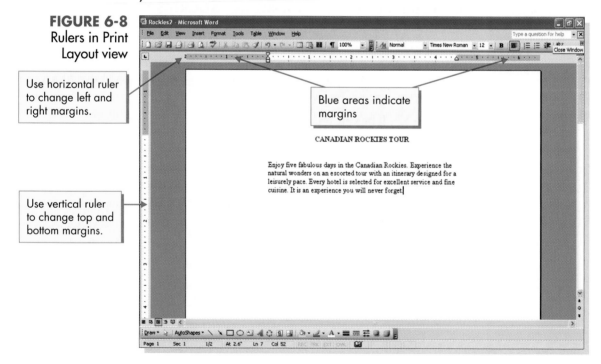

Use horizontal ruler to change left and right margins.

Use vertical ruler to change top and bottom margins.

8. To change the top margin, position the pointer over the top margin boundary on the vertical ruler. The top margin boundary is between the blue area and the white area on the ruler. The pointer changes to a two-headed arrow and a ScreenTip displays the words "Top Margin."

9. Press and hold down the left mouse button. The margin boundary appears as a dotted horizontal line.

10. Drag the margin boundary slightly up and release the mouse button. The text at the top of the document moves up to align with the new top margin.

11. Click the Undo button [↻ ▾] to restore the 1.5-inch top margin.

12. Hold down the [Alt] key and drag the top margin boundary down until it is at 2 inches on the ruler. Release [Alt] and the mouse button. Holding down the [Alt] key as you drag shows the exact margin and text-area measurements.

13. To change the left margin, position the pointer over the left margin boundary on the horizontal ruler. The left margin boundary is between the blue area and the white area on the ruler. The pointer changes to a two-headed arrow and a ScreenTip displays the words "Left Margin."

NOTE: You might have to fine-tune the pointer position to place it directly on the left margin boundary. Move the pointer slowly until you see the two-headed arrow and the "Left Margin" ScreenTip.

14. Hold down the [Alt] key and drag the margin boundary to the left to create a 1.75-inch left margin. (See Figure 6-9.)

FIGURE 6-9
Adjusting the
left margin

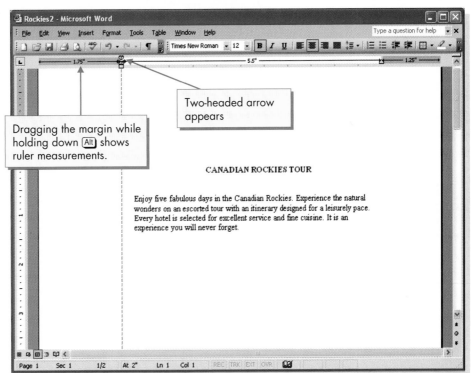

15. Using the same procedure, drag the right margin boundary until it is located 1.75 inches from the right. Be sure to watch for the two-headed arrow before dragging. The first section now has 1.75-inch left and right margins and a 2-inch top margin.

16. Scroll to the next page (section 2). Click within the text to activate this section's ruler. Change the top margin to 1.75 inches.

17. Save the document as *[your initials]*6-4 in a new Lesson 6 folder and leave it open.

NOTE: If you move the pointer to the top of the page in Print Layout view, you'll see the Hide White Space button . Click the button to hide the white space (the margin area) at the top and bottom of each page, and the gray space between pages so you can see more document text. Point to the top of the page and click the Show White Space button to restore the space.

EXERCISE **6-5** **Set Facing Pages with Gutter Margins**

If your document is going to be *bound*—put together like a book, with printing on both sides of the paper—you'll want to use mirror margins and gutter margins. *Mirror margins* are inside and outside margins on facing pages that mirror one another. *Gutter margins* add extra space to the inside margins to allow for binding.

FIGURE 6-10
Mirror margins on facing pages

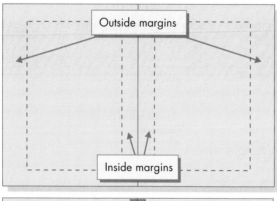

FIGURE 6-11
Facing pages with gutter margins

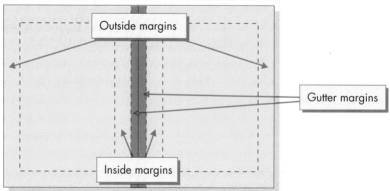

1. Open the file **Mexico1**.

2. Open the Page Setup dialog box and open the Multiple pages drop-down list. Choose Mirror margins. Notice that the Preview box now displays two pages. The left and right margins are called the *inside* and *outside margins*.

3. Change the inside margins to 1.25 inches and the outside margins to 1 inch.

4. Set the gutter margin to 1 inch and press Tab to reflect the change in the Preview box. Click OK. A 1-inch gutter margin is added to the document. (Make sure you use at least 1-inch gutter margins to allow room for binding.)

5. Switch to Print Layout view and scroll through the document.

6. Switch back to Normal view and save the document as *[your initials]*6-5 in your Lesson 6 folder.

7. Print and close the document.

TIP: Visualize the document as double-sided, facing pages in a book by placing the back of page 2 against the back of page 1 and placing page 3 beside page 2. The gutter margin of page 1 is on the left. The gutter margin on the right of page 2 and on the left of page 3 allows space for the binding and represents facing pages. *Facing pages* appear as a two-page spread with odd-numbered right pages and even-numbered left pages.

Using Print Preview

Viewing a document in Print Preview is the best way to check how a document will look when you print it. You can view multiple pages at a time, adjust margins and tabs, and edit text.

 To display a document in Print Preview, click the Print Preview button on the Standard toolbar or choose Print Preview from the File menu.

EXERCISE 6-6 **View a Multiple-Page Document in Print Preview**

Print Preview displays entire pages of a document in reduced size. You can view one page at a time or multiple pages at a time.

1. Move the insertion point to the beginning of *[your initials]*6-4.

 2. Click the Print Preview button to open the Print Preview window. Click the One Page button on the Print Preview toolbar to display the first page of the document.

 3. Click the Multiple Pages button. You use the grid that appears to choose the number of pages you want to view and how they are configured in the window. Move the pointer across the top row of the grid to highlight the first two pages. Click the second page. You now see the entire document.

TIP: If you drag the pointer as you move across the grid, you can expand the grid to display additional rows and pages, which is useful in a long document.

FIGURE 6-12
Viewing multiple pages in Print Preview

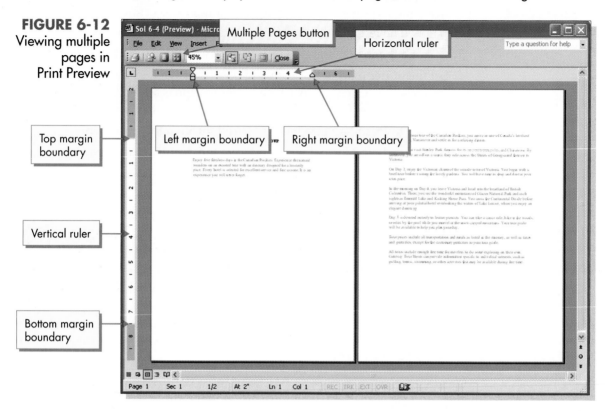

4. To zoom into page 2, click once on the page with the arrow pointer to make page 2 active, and then click again with the magnifier pointer.

5. Click again to zoom out. Notice that the horizontal ruler shows the settings for page 2.

6. Click Close on the Print Preview toolbar to close the Print Preview window and return to Print Layout view. (See Table 6-1 on the next page.)

EXERCISE **6-7** **Change Margins in Print Preview**

When you view a page in Print Preview, you can see all four margins and adjust margin settings using the horizontal and vertical rulers or by opening the Page Setup dialog box within the Print Preview window.

1. Move the insertion point to the beginning of the document (page 1, section 1).

 2. Click the Print Preview button on the Standard toolbar.

TABLE 6-1   Print Preview Toolbar Buttons

BUTTON	DESCRIPTION	FUNCTION
	Print	Prints the document in the Print Preview window.
	Magnifier	Changes the I-beam pointer to a magnifying glass and vice versa. With the magnifying glass pointer, click in the document to zoom in and out. With the I-beam pointer, edit document text.
	One Page	Displays one page at a time.
	Multiple Pages	Displays the number of pages you select from a grid.
23%	Zoom	Use to choose a magnification to reduce or enlarge the page or pages displayed.
	View Ruler	Displays or hides the Print Preview ruler, which you can use to change margins, tabs, and indents.
	Shrink to Fit	Shrinks a document to fit on one less page when the last page contains only a few lines of text.
	Full Screen	Hides most screen elements to show more of the document. Click ▣ again to restore the previous view.
Close	Close Preview	Closes the Print Preview window and returns to the previous view.

3. Click the One Page button ▣ on the Print Preview toolbar to display only page 1, and click the View Ruler button 🔲 to display the rulers if they are not showing.

4. Move the pointer to the top margin boundary on the vertical ruler. The pointer changes to the two-headed arrow and the top margin is identified in a ScreenTip.

NOTE: Changing margins in Print Preview is similar to changing margins in Print Layout view. You use the vertical and horizontal rulers to drag the margins to the desired positions.

5. Hold down [Alt] as you drag the margin boundary down to 1.5 inches on the blue area of the vertical ruler. Word adjusts the top margin to 1.5 inches.

6. Use the File menu to open the Page Setup dialog box.

7. Change the top margin to 2 inches and click OK.

8. Use the horizontal ruler to change the left and right margins to 1.5 inches.

TIP: You can check the exact measurement of a margin in Print Preview or Print Layout view by moving the pointer over the margin boundary, holding down the left mouse button without dragging it, and holding down [Alt].

EXERCISE   6-8   **Edit a Document in Print Preview**

To edit text in Print Preview, you magnify a page to the desired size, and then switch to Edit mode. You would not, however, want to make extensive changes in Print Preview.

1. Click on page 1 to zoom in. The view of the document is enlarged to 100 percent.
2. Click the Magnifier button 🔍 to change the pointer to the I-beam.
3. Select "Canadian Rockies" in the first paragraph, and press Ctrl+I to make the text italic.
4. Click the Magnifier button 🔍 to cancel edit mode.
5. Click Close to close the Print Preview window, and notice the italic text.
6. Save the document as *[your initials]*6-8.
7. Print and close the document.

Paper Size and Orientation

When you open a new document, the default paper size is 8.5 × 11 inches. Using the Page Setup dialog box, you can change the paper size to print a document on legal paper or a custom-size paper.

The page setup dialog box also gives you a choice between two page orientation settings: Portrait and Landscape. A *portrait* page is taller than it is wide. This orientation is the default in new Word documents. A *landscape* page is wider than it is tall. You can apply page-orientation changes to sections of a document or to the entire document.

EXERCISE   6-9   **Change Paper Size and Page Orientation**

1. Open the file **Mexico1**. Switch to Print Preview. Display all pages of the document by clicking the Multiple Pages button 🔲.
2. Choose Page Setup from the File menu.
3. On the Margins tab, click Landscape.
4. Click the Paper tab. Notice the default paper size for Letter paper.
5. Open the Paper size drop-down list and choose Legal. Click OK. Notice how the orientation and paper size changed.
6. Press Ctrl+Z to undo the changes to paper size and orientation.
7. Click Close to close the Print Preview window.
8. Position the insertion point to the left of the bold heading **Stateroom Locations and Rates** on page 2.

FIGURE 6-13
Changing page
orientation

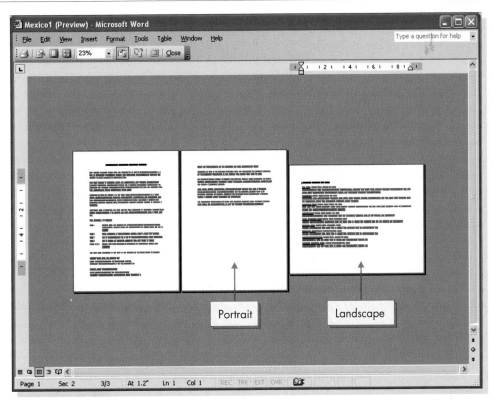

9. Open the Page Setup dialog box again and choose Landscape. Open the Apply to drop-down list and choose This point forward. Click OK. Word applies landscape orientation from just the insertion point to the end of the document, creating a separate section on a new page.

10. View all pages of the document in Print Preview; then close the Print Preview Window.

11. Save the document as *[your initials]*6-9.

> **NOTE:** You can change orientation for an entire document or from the insertion point forward by choosing an option from the Apply to drop-down list. When you choose This point forward, a new section is created with the orientation you choose.

Using Hyphenation

Hyphenation is used to divide words at the end of a line. You can hyphenate text manually or Word can hyphenate text for you automatically. Word uses three types of hyphens:

- A normal hyphen is used for words that should always be hyphenated, such as "twenty-three" or "mother-in-law."

- A *nonbreaking hyphen* is used when a hyphenated word should not be divided at a line break. For example, you could use a nonbreaking hyphen in a hyphenated proper name, such as "Minneapolis-St. Paul." Nonbreaking hyphens are similar in purpose to nonbreaking spaces, which were discussed in Lesson 2.

- An *optional hyphen* indicates where a word should be divided if the word falls at the end of a line. If the word does not fall at the end of a line, the optional hyphen disappears from the screen and is not printed.

EXERCISE **6-10** Insert Normal, Nonbreaking, and Optional Hyphens

It is a good idea to use hyphenation when a document contains lines that seem to break very irregularly. You can also use hyphenation to potentially shorten a document's length.

1. In the current document, scroll to the paragraph that begins "For reservation information" on page 2. Delete the hyphen between **555** and **1234**. Insert a nonbreaking hyphen by pressing Ctrl+Shift+- (the Hyphen key). The nonbreaking hyphen will prevent the telephone number from being divided between two lines.

2. Locate the paragraph that begins "All Mexican Riviera." Place the insertion point between the "k" and the "i" of "kayaking."

3. Insert an optional hyphen by pressing Ctrl+-. The optional hyphen indicates where kayaking should be divided.

4. Locate the phrase "seven-day cruises" in the paragraph that begins "This cruise." The phrase contains a normal hyphen. Notice the difference in shape and size among the three types of hyphens. All hyphens, however, have the same appearance when printed. (See Figure 6-14 on the next page.)

EXERCISE **6-11** Hyphenate a Document

Word offers automatic hyphenation to control ragged edges at the ends of paragraph lines and to reduce the amount of space Word inserts in justified text. In the Hyphenation dialog box, you can:

- Hyphenate the document automatically as you key text.
- Hyphenate words written in all-capital letters.
- Set a Hyphenation Zone measurement to control the amount of raggedness in the right margin. The default setting is 0.25 inches. A lower number reduces the raggedness and a higher number reduces the amount of hyphenation.

FIGURE 6-14
Normal,
nonbreaking, and
optional hyphens

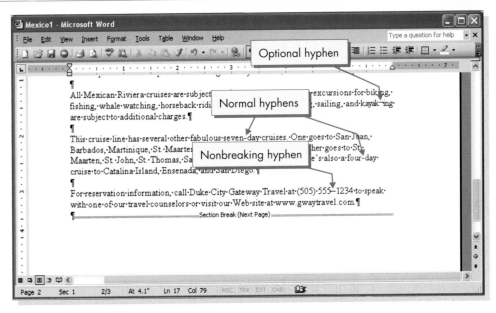

- Limit the number of consecutive hyphens in lines of text by a number you enter.
- Manually hyphenate a document, which lets you confirm each hyphen.

FIGURE 6-15
Hyphenation
dialog box

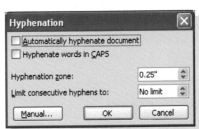

1. Move the insertion point to the beginning of the document. Choose Language from the Tools menu; then choose Hyphenation. The Hyphenation dialog box appears.

2. Set the Limit consecutive hyphens box to **2**, so no more than two consecutive lines end with hyphens.

3. Click Manual. Word begins the hyphenation process by changing the document to Print Layout view. The program asks for confirmation for each word to be hyphenated.

4. Click Yes to accept the hyphenation, but avoid hyphenating proper names or words containing fewer than six letters, such as "ha-ven."

 TIP: You can move the hyphenation point for a word in the Hyphenate at box by clicking another point in the word or by using ⊟ or ⊞.

5. When the dialog box appears to tell you hyphenation is complete, click OK to return to the document.

NOTE: The Gregg Reference Manual contains several guidelines for preferred hyphenation practices. For example, you should not hyphenate abbreviations or contractions (such as "shouldn't"), or hyphenate a word to create a one-letter syllable (such as "a-ware"). For more information about hyphenation, see Section 9: "Word Division" in The Gregg Reference Manual.

Inserting the Date and Time

You've seen that when you begin keying a month, AutoComplete displays the suggested date, and you press [Enter] to insert the date as regular text. You can also insert the date or time in a document as a field. A *field* is a hidden code that tells Word to insert specific text that might need to be updated automatically, such as a date or page number. If you insert the date or time in a document as a field, Word automatically updates it each time you print the document.

There are two ways to insert the date or time as a field:

- Choose Date and Time from the Insert menu and choose the desired format.
- Press [Alt]+[Shift]+[D] to insert the date and [Alt]+[Shift]+[T] to insert the time.

EXERCISE **6-12** **Insert the Date and Time**

You can enter date and time fields that can be updated automatically. You can also choose not to update these fields automatically.

1. Move the insertion point to the end of the current document and press [Enter] twice.

2. Press [Alt]+[Shift]+[D] to enter the default date field.

3. Click the Undo button .

4. Choose Date and Time from the Insert menu.

FIGURE 6-16
Date and Time
dialog box

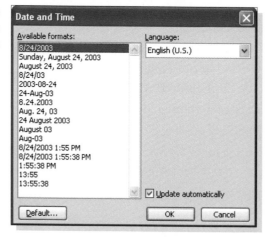

5. Scroll the list of available time and date formats, and choose the third format in the list (the standard date format for business documents).

6. Check the Update automatically check box, so the date is automatically updated each time you print the document. Click OK.

> **NOTE:** You can also use this dialog box to insert the date and time in a particular text format without inserting it as an updatable field.

7. Move the insertion point after the date field and press [Spacebar] twice.

8. Press [Alt]+[Shift]+[T] to insert the time as a field.

TIP: Although printing updates a field, you can also update a field on-screen by clicking the field and pressing F9.

9. Save the document as *[your initials]***6-12** in your Lesson 6 folder.

10. Print and close the document.

TIP: Remember that the Update Automatically option will change the date in your document. If you are sending correspondence, do not choose this option because the date in the letter will then always reflect the current date, not the date on which you wrote the letter.

Printing Envelopes and Labels

Word provides a tool to print different size envelopes and labels. Using the Envelopes and Labels command, you can:

- Print a single envelope without saving it or attach an envelope to a document for future printing. The attached envelope is added to the beginning of the document as a separate section.

- Print labels without saving them, or create a new document that contains the labels. You can print a single label or a full page of the same label.

EXERCISE **6-13** **Print an Envelope**

Printing envelopes often requires that you manually feed the envelope to your printer. If you print labels that are on other than 8½ × 11 inch sheets, you might need to feed the labels manually. Your printer will display a code and not print until you feed an envelope or label sheet manually.

1. Open the file **Caliente**. This document is a one-page letter.

2. Open the Tools menu. Choose Letters and Mailings, Envelopes and Labels.

3. Click the Envelopes tab if it is not active. Notice that Word detected the address in the document and placed this text in the Delivery address text box. You can edit this text as needed. (See Figure 6-17 on the next page.)

4. In the Delivery address text box, enter the full ZIP+4 Code by keying **–1129** after "87110."

5. Make sure the Omit box is not checked. Select and delete any text in the Return address text box, and then key the following return address, starting with your name:

[your name]
Duke City Gateway Travel
15 Montgomery Boulevard
Albuquerque, NM 87111-2307

FIGURE 6-17
Envelopes and
Labels dialog box

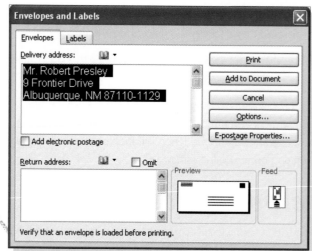

6. Place a standard business-size envelope in your printer. The Feed box illustrates the feeding method accepted by your printer.

NOTE: If you don't have an envelope, you can use a blank sheet of paper to test the placement of the addresses. Ask your instructor how to proceed. You might have to feed the envelope or blank sheet manually.

7. Click Print. When Word asks if you want to save the return address as the default return address, click No. Word prints the envelope with the default font and text placement settings.

NOTE: Check your printer to see what you need to do to complete a manual envelope feed. If the printer is flashing or displaying a message, you might have to press a button.

EXERCISE 6-14 Choose Envelope Options

Before printing an envelope, you can choose additional envelope options. For example, you can add the envelope content to the document for future use. You can also click the Options button in the Envelopes and Labels dialog box to:

- Change the envelope size. The default size is Size 10, which is a standard business envelope.
- Print codes for automated handling by the U.S. Postal Service. For example, you can print a bar code that represents the ZIP Code for the delivery address or print a FIM-A code that identifies the front of an envelope (used on preprinted courtesy reply envelopes). To print these codes, your printer must be able to print graphics.
- Change the font and other character formatting of the delivery address or return address.

NOTE: If your computer is equipped with electronic postage software, you can choose options to apply postage to the envelope.

1. Open the Envelopes and Labels dialog box again.

2. Type your name and address in the Return address box.

3. Click the Options button in the Envelopes and Labels dialog box to open the Envelope Options dialog box. Click the Envelope Options tab if it is not active.

4. Under Envelope size, click the down arrow to look at the different size options. Click the arrow again to close the list.

FIGURE 6-18
Envelope Options
dialog box

5. Click the Delivery point barcode check box to select it. Notice the change in the Preview box.

6. Click the Font button for the Delivery address. The Envelope Address dialog box for the delivery address opens.

7. Make the text bold and all caps and change the font size to 10. Click OK. Click OK to close the Envelope Options dialog box.

8. Delete the punctuation from the delivery address and add **-1129** to the ZIP Code.

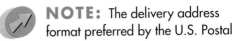 **NOTE:** The delivery address format preferred by the U.S. Postal Service is all caps with no punctuation.

9. Click Add to Document to add the envelope information to the top of the document as a separate section. Don't save the return address as the default address.

 NOTE: Once the envelope is added to the document, you can also format or edit the envelope text just as you would any document text. The default font for envelope addresses is Arial.

10. Place the insertion point at the beginning of the letter to Mr. Robert Presley. Add the date at the top of the letter and press (Enter) four times. Place 72-point spacing before the date. Correct any spacing between the elements of the letter. Add your reference initials followed by **Enclosures**. To make sure the letter follows the correct format, see Appendix B: "Standard Forms for Business Documents."

11. View the letter and envelope in Print Preview.

12. Close the Print Preview window and save the document as *[your initials]*6-14 in your Lesson 6 folder.

13. Click the Print button on the Standard toolbar to print the envelope and letter.

> **NOTE:** If you are asked to manually feed the envelope, you might be asked to manually feed the letter as well.

14. Leave the document open for use in the next exercise.

EXERCISE 6-15 Print Labels

The Labels tab in the Envelopes and Labels dialog box makes it easy to print different size labels for either a return address or a delivery address.

1. Position the insertion point in the envelope section of the document. Open the Envelopes and Labels dialog box and click the Labels tab.

2. Click the Use return address check box to create labels for the letter sender.

3. Select the address text and press Ctrl+Shift+A to turn on all caps. Delete the comma after the city.

4. Click the option Full page of the same label, if it is not active, to create an entire page of return address labels.

5. Click the Options button to choose a label size.

FIGURE 6-19
Printing labels

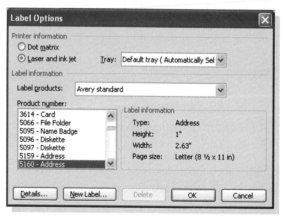

6. Scroll the Product number list to see the various label options and choose 5160, the product number for a standard Avery address label.

7. Click OK, and then click New Document to save the labels as a separate document. (If you click Print, you can print the labels without saving them.) Do not save the return address.

8. Select all text in the new document and reduce the font size to 11 points and Arial font.

9. Save the document as *[your initials]*6-15 in your Lesson 6 folder.

10. Switch to Print Preview to view the labels on the page.

11. Prepare the printer for a sheet of 5160-size labels or feed a blank sheet of paper into the printer, and then print the labels.

12. Close the document. Save and close *[your initials]*6-14.

Setting Print Options

When you click the Print button ![print icon] to print a document, Word prints the entire document. If you open the Print dialog box, however, you can choose to print only part of a document. You can also select other print options from the dialog box, including collating copies of a multipage document, printing selected text, or printing multiple document pages on one sheet of paper.

EXERCISE 6-16   **Choose Print Options from the Print Dialog Box**

1. Open the file **Duke2**.

2. Choose Print from the File menu to open the Print dialog box.

3. Click Current Page and click OK. Word prints page 1 of the document.

4. Open the Print dialog box again. Key **1-2** in the Pages text box. You can also enter specific page numbers or page ranges.

5. In the Number of copies text box, use the up arrow to change the number of copies to 2.

6. If the Collate check box is not active, click it to select it. Notice the change in the preview of the number of copies. Click the Collate check box to uncheck it (make sure it is unchecked). With this box not checked, Word will print two copies of page 1, and then two copies of page 2.

7. Change the number of copies back to 1.

8. Click the down arrow to open the Print what drop-down list. It shows the various elements you can print, in addition to the entire document. Click again to close the list.

9. Click the down arrow to open the Print drop-down list, which gives you the option to print even or odd pages. Click again to close the list.

10. Click the down arrow to open the Pages per sheet drop-down list, which gives you the option to print your selection over a specified number of sheets. Choose the 2 pages setting. This lets you print two pages on one sheet of 8 ½ × 11 inch paper, with each page reduced to fit on the sheet. (See Figure 6-20 on the next page.)

11. Open the Scale to paper size drop-down list, which gives you the option to print on a different paper or envelope size (Word adjusts the scaling of the fonts, tables, and other elements to fit the new size). Close the drop-down list.

12. Click OK. Word prints reduced versions of pages 1 and 2 on one sheet of paper.

13. On page 3 of the document, select the "Corporate Travel" heading and the rest of the text on page 3. Open the Print dialog box again.

FIGURE 6-20
Print dialog box

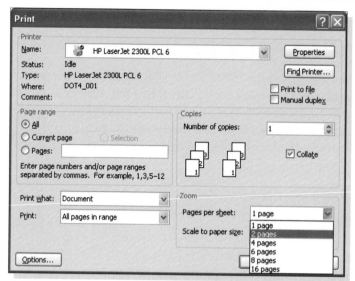

14. Click Selection, and then click OK. Word prints just the selected text.
15. Close the document without saving.

USING ONLINE HELP

In addition to using the Ask a Question box to access Help, you can also use the Office Assistant, an animated figure that displays alerts and tips as you work in Word. Now that you've worked in Print Layout view and Print Preview to change margins, learn about another feature in the Page Setup dialog box called book fold.

Use Help to learn about book fold:

1. Choose Show the Office Assistant from the Help menu.
2. Click the animated figure, and key **page margins** in the Office Assistant text box, and then press (Enter).
3. Click the topic, About page margins.
4. Read the information on using the book-fold feature.
5. Hide the Office Assistant and close the Help window when you finish.

LESSON 6 Summary

➤ In a Word document, text is keyed and printed within the boundaries of the document's margins. Margins are the spaces between the edges of the text and the edges of the paper.

➤ Change the actual space for text on a page by changing margins (left, right, top, and bottom). You can key new margin settings in the Page Setup dialog box.

➤ Changing margins for selected text results in a new section for the selected text. A section is a portion of a document that has its own formatting. When a document contains more than one section, you see double-dotted lines, or section breaks, between sections to indicate the beginning and end of a section.

➤ Print Layout view shows how text is positioned on the printed page. Use the View buttons to the left of the horizontal scroll bar to switch between Print Layout view and Normal view.

➤ Print Preview shows how an entire document looks before printing. Use the Multiple Pages button ⊞, the One Page button ▣, and the scroll bar to view all or part of the document. Change the Zoom as needed.

➤ Change margins in Print Layout view or in Print Preview by positioning the pointer over a margin boundary on a ruler and dragging. Press Alt to see the exact ruler measurement as you drag.

➤ Edit a document in Print Preview by clicking the Magnifier button 🔍 to change the magnifier pointer to the I-beam pointer.

➤ For bound documents, use mirror margins and gutter margins. Mirror margins are inside and outside margins on facing pages that mirror one another. Gutter margins add extra space to the inside margins to allow for binding

➤ A document can print in either portrait (8 1/2" × 11") or landscape (11" × 8 1/2") orientation. Choose an orientation in the Page Setup dialog box, Margins tab.

➤ A document can be scaled to fit a particular paper size. Choose paper size options in the Page Setup dialog box, Paper tab.

➤ Hyphenation is the division of words that cannot fit at the end of a line. A normal hyphen is used for words that should always be hyphenated. A nonbreaking hyphen is used when a hyphenated word should not be divided at a line break. An optional hyphen indicates where a word should be divided if the word falls at the end of a line.

➤ Insert the date and time in a document as an automatically updated field, which is a hidden code that tells Word to insert specific information—in this case, the date and/or time. Use the Date and Time dialog box to choose different date and time formats.

➤ Use Word to print different size envelopes. You can print delivery codes, change address formatting, and make the envelope part of the document for future printing. Use Word to print different size address labels—either a single label or a sheet of the same label.

➤ Choose print options, such as printing only the current page, specified pages, selected text, collated copies of pages, and reduced pages by opening the Print dialog box.

LESSON 6   Command Summary

FEATURE	BUTTON	MENU	KEYBOARD
Print Preview		File, Print Preview	Ctrl + F2
Print Layout view		View, Print Layout	Alt + Ctrl + P
Normal view		View, Normal	Alt + Ctrl + N
Choose print options		File, Print	Ctrl + P
Print envelopes or labels		Tools, Letters and Mailings, Envelopes and Labels	
Insert date		Insert, Date and Time	Alt + Shift + D
Insert time		Insert, Date and Time	Alt + Shift + T

Concepts Review

Each of the following statements is either true or false. Indicate your choice by circling T or F.

T F   **1.** Word has default settings for margins that are automatically set for each new document.

T F   **2.** You can change margins in Normal view by using the ruler.

T F   **3.** You can edit a document in Print Preview.

T F   **4.** Both Print Preview and the Page Setup dialog box use horizontal and vertical rulers to drag margins to desired positions.

T F   **5.** You can change margins for an entire document only by using the Page Setup dialog box.

T F   **6.** Gutter margins are outside margins on a bound document.

T F   **7.** Words containing nonbreaking hyphens cannot be broken across two lines.

T F   **8.** Landscape is the default page orientation.

SHORT ANSWER QUESTIONS

Write the correct answer in the space provided.

1. Which view is displayed when you click ▣?

_____

2. What does the pointer look like when it is located over the margin boundary on the ruler in Print Layout view?

_____

3. What is created when you change margins for selected text?

_____

4. Which kind of document needs gutter margins?

_____

5. Which tab, in which dialog box, do you use to change page orientation?

_____

6. How do you switch to edit mode in Print Preview?

_____

7. Which kind of hyphen do you use if you don't want a word to break at the end of the line?

_____

8. Which keyboard combination inserts the date?

_____

CRITICAL THINKING

Answer these questions on a separate page. There are no right or wrong answers. Support your answers with examples from your own experience, if possible.

1. Collect samples of printed documents with interesting treatments of margins (such as books, advertisements, or reports). Pay particular attention to mirror margins and gutter margins. How does the margin treatment contribute to the overall feeling of the document?

2. What might you notice about a document in Print Preview that you wouldn't notice in Normal view?

Skills Review

EXERCISE 6-17

Set margins for an entire document and for selected text by using the Page Setup dialog box.

1. Open the file **Baggage**.

2. Change the margins for the entire document by following these steps:
   a. Display the Reveal Formatting task pane. Under Section, click Margins. (Click the plus sign (+) to display the current margin settings if necessary.)
   b. Set the Top and Bottom margins to 2" and the Left and Right margins to 1". Click OK.

3. Change the margins and create a new section for selected text by following these steps:
   a. Select the paragraphs that begin "A." and "B."
   b. Open the Page Setup dialog box.
   c. Set the left and right margins to 2 inches.
   d. Choose Selected text from the Apply to drop-down list and click OK.

 4. Click the Undo button to undo the new section.

5. Format the paragraphs that begin "A." and "B." as a numbered list using the A. B. C. style.

6. Save the document as *[your initials]*6-17 in your Lesson 6 folder.

7. Print and close the document.

EXERCISE 6-18

Set margins in Print Layout view and Print Preview, and change orientation.

1. Open the file **OldTown**.
2. Change the left and right margins in Print Layout view by following these steps:

   a. Click the Print Layout View button  to the left of the horizontal scroll bar. Click the left arrow on the horizontal scroll bar to see the left edge of the document.
   b. If the rulers are not displayed, choose View, Ruler.
   c. Using the horizontal ruler, position the pointer on the left margin boundary until it becomes a two-headed arrow (and the ScreenTip "Left Margin" appears).
   d. Hold down [Alt] and drag the margin boundary until the left margin measures 1.5 inches.
   e. Position the pointer on the right margin boundary and use the same method to drag it to 1.5 inches.

3. Click the Normal View button  to switch back to Normal view.

4. Change the top margin in Print Preview by following these steps:

   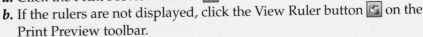

   a. Click the Print Preview button.
   b. If the rulers are not displayed, click the View Ruler button on the Print Preview toolbar.
   c. Using the vertical ruler, position the pointer on the top margin boundary until it becomes a two-headed arrow.
   d. Hold down [Alt] and drag the top margin boundary to 2 inches.
   e. Click Close or click the Normal View button to return to Normal view.
5. Add a bold, centered, uppercase title **OLD TOWN** to the top of the document, with two blank line spaces below it (or 24 points of paragraph spacing after it).
6. Change the orientation to Landscape by following these steps:
   a. Open the Page Setup dialog box.
   b. Click the Margins tab and choose Landscape.
   c. Change the top and bottom margins to 2 inches, and change the left and right margins to 1.5 inches.
   d. Click OK to close the Page Setup dialog box.
   e. Position the pointer on the right margin boundary and use the same method to drag it to 1.5 inches.
7. Save the document as *[your initials]*6-18 in your Lesson 6 folder.
8. Print and close the document.

EXERCISE 6-19

Set mirror and gutter margins, and use hyphenation.

1. Open the file **Duke2**.
2. Switch to Print Layout view and scroll through the entire document.

3. Use the Page Setup dialog box to set mirror and gutter margins by following these steps:
   a. Open the Page Setup dialog box.
   b. Choose Mirror margins from the Multiple pages drop-down list.
   c. Set the Gutter margin to 0.5 inches.
   d. Click OK.
4. Scroll through the document in Print Layout view to see the new margin settings.
5. Switch back to Normal view.
6. Delete the text from the heading "**Information**" through the end of the document.
7. Insert a nonbreaking hyphen by following these steps:
   a. Locate the word "full-service" in the paragraph on page 2 that begins "Welcome to."
   b. Replace the normal hyphen with a nonbreaking hyphen by pressing Ctrl + Shift + -.
8. In the paragraph under "**Group Travel**" on page 3, replace the hyphen in "wine-tasting" with a nonbreaking hyphen.
9. Hyphenate the document manually by following these steps:
   a. Move the insertion point to the beginning of the document.
   b. From the Tools menu, choose Language, Hyphenation.
   c. Clear both check boxes, if they are active, and set the Limit consecutive hyphens to box to 2.
   d. Click Manual. Click Yes to confirm each hyphenation, but avoid hyphenating proper names. In addition, avoid hyphenating the end of the last full line in a paragraph.
   e. Click OK when hyphenation is complete.
10. Save the document as *[your initials]*6-19 in your Lesson 6 folder.
11. Print and close the document.

EXERCISE 6-20

Set print options, insert the date, print an envelope, and print labels.

1. Open the file **Offices2**.
2. Print a portion of the document by following these steps:
   a. Select the text from "Ontario Gateway Travel" through the end of the document.
   b. Press Ctrl + P to open the Print dialog box.
   c. Choose Selection and click OK.
3. Insert the date by following these steps:
   a. On a new line at the top of the document, turn off the bold formatting. Key the text **Updated** and press Spacebar.
   b. Open the Insert menu and choose Date and Time.
   c. Choose the fourth date format, make sure Update automatically is not checked, and click OK.

d. Insert a blank line after the date.

4. Select the page break and press `Delete`.

5. Prepare an envelope addressed to the Chicago Gateway Travel office by following these steps:

   a. Select the name and address lines for the Chicago office. Do not select the telephone and fax numbers.

   TIP: Because this document contains many addresses, you need to select the text you want to appear in the Delivery address box of the Envelopes and Labels dialog box.

   b. Open the Tools menu. Choose Letters and Mailings, Envelopes and Labels. Click the Envelopes tab.

   c. In the Delivery address box, delete the comma in the address and key –3301 at the end of the ZIP Code.

   d. In the Return address box, key your name, followed by the office address:
   Duke City Gateway Travel
   15 Montgomery Boulevard
   Albuquerque, NM 87111-2307

6. Choose additional envelope options and add the envelope to the document by following these steps:

   a. Click the Options button and choose the Envelope Options tab, if it is not already displayed.

   b. Make sure the envelope size is 10 and check Delivery point barcode.

   c. Click the Font button for the Delivery Address. Change the style to bold all caps and click OK.

   d. Click OK in the Envelope Options dialog box. Click Add to Document in the Envelopes and Labels dialog box. Do not save the return address as the default.

7. Prepare the printer for a standard business envelope (or feed a blank sheet of paper into the printer). Print the document (envelope included).

8. Save the document as *[your initials]*6-20a in your Lesson 6 folder.

9. Create and print a page of return address labels by following these steps:

   a. With the insertion point in section 1 (the envelope), open the Envelopes and Labels dialog box and click the Labels tab.

   b. Check the Use return address box. (Word recognizes the return address you previously entered for the envelope.)

   c. Add **Agency** after "Travel."

   d. Choose the option Full page of the same label.

   e. Click Options. Set the Product number to 5160 and click OK.

   f. Click New Document to save the labels as a separate document. Do not save the return address as the default.

   g. Save the labels as *[your initials]*6-20b in your Lesson 6 folder.

   h. Print the labels on a blank sheet of paper.

10. Close both documents.

Lesson Applications

EXERCISE 6-21

Set margins for a document and for selected text.

1. Open the file **Offices1**.

2. At the beginning of the document, key the text shown in Figure 6-21, including the corrections. Use single spacing and insert one blank line between paragraphs.

FIGURE 6-21

Than^k you for selecting Duke City Gateway Ta^rvel as your full-service corporate travel agency. We began as ^a small down_town office 25 yea^ars ago. Today we have offices all^over the world^# and we provide quality service to all our customers.

All of our offices work together (ot) meet ~~all of~~ your travel needs. We offer expertise in the following four specialty areas:

Corporate travel
International travel   *Create as one paragraph*
Group travel   *using line breaks*
Family travel

For your convenience as a corporate customer, Gateway offers complete service from any one of our locations. Our ^*domestic* office locations and phone numbers are listed on the following page.^ *You can also reach us at www.gwaytravel.com on the Web.*

3. At the top of the document, key the title **Corporate Travel at Gateway** in 14-point bold small caps as a separate line. Center the title and add 24 points of paragraph spacing after it.

4. Using the Page Setup dialog box, set the top margin to 2 inches, and then set the left and right margins to 1.25 inches.

5. Select the text beginning with "U.S. Gateway Travel Offices" through the end of the document. Use the Page Setup dialog box to change the left and right margins for *only* the selected text to 3 inches.

6. Format the first line of the new section (U. S. Gateway Travel Offices) as bold small caps. Add 12 additional points of spacing after the first line.

7. Save the document as *[your initials]*6-21 in your Lesson 6 folder.

8. Print the document using the 2 Pages per sheet option in the Print dialog box. Close the document.

EXERCISE 6-22

Set margins for a document and for selected text, change page orientation, update a date field, insert an optional hyphen, and set print options.

1. Open the file **Memo2**.
2. Insert today's date in the date line.
3. In the opening paragraph of the memo, replace the text "items below" with **schedule**.
4. At the bottom of page 1, delete the text from "Record Keeping" through the end of the document.
5. Change the top margin to 2 inches and the left and right margins to 1.25 inches.
6. Select the text from "July 1" through the end of the document. Change the orientation to landscape, change the top margin to 2 inches, change the bottom margin to 1 inch, and change the left and right margins to 2.25 inches.
7. Above the section break, in the paragraph directly below "Deadlines," insert an optional hyphen between the "e" and the "t" in "sometimes."
8. Save the document as *[your initials]*6-22 in your Lesson 6 folder.
9. Print the document 2 pages per sheet. Close the document.

EXERCISE 6-23

Set margins, set mirror and gutter margins, and address an envelope.

1. Open the file **ElMorro1**.
2. Change the left and right margins to 1.25 inches.
3. Set mirror margins and a 0.5-inch gutter margin.
4. Format the title as 14-point bold with a shadow effect. Add 72 points of paragraph spacing before the title.
5. Justify all text below the title.
6. At the end of the document, format the text from "For more information" through the telephone number with 1.5-inch left and right indents (not margins), a 1-point box border, and 10 percent gray shading.
7. Add an envelope to the document, using the El Morro National Monument address. Use your name and address as the return address.
8. Save the document as *[your initials]*6-23 in your Lesson 6 folder.
9. Print and close the document.

EXERCISE 6-24 *Challenge Yourself*

Set margins, insert the date, insert hyphens, and create labels.

1. Open the file **Balloon**.
2. At the beginning of the second paragraph, key the following text using a nonbreaking hyphen: **If you love hot-air balloons, you should know that**
3. Make sure "hot-air balloon" appears throughout the paragraph with a nonbreaking hyphen.
4. In the second sentence of the same paragraph, replace the comma and space after "Fiesta" with an em dash. Following the em dash, change "a nine-day event" to **nine days**.
5. In the same paragraph, change the sentence that begins "A minimum of 1000" to **More than 1000**.
6. Key the text shown in Figure 6-22 as a closing paragraph.

FIGURE 6-22

As you look through the brochures, pay particular attention to the New Mexico State Fair, the Great Rio Grande Raft Race, the Cowboy Classic Western Art Show, and the San Felipe Festival. Please feel free to contact me or visit our Web site, www.gwaytravel.com, for more information.

7. Format the document as a standard business letter. (Refer to Appendix B: "Standard Forms for Business Documents" for margin and spacing requirements.) Enter the date as a field. The letter is from you with the title **Associate Travel Counselor** and to the following person:

Mr. Raymond Haas
12 Evonne Avenue
Rohnert Park, CA 94928

8. Add an enclosure notation to the letter.
9. Switch to Print Preview for a final view of the document.
10. Save the document as *[your initials]*6-24a in your Lesson 6 folder.
11. Create a sheet of labels (Avery Standard 5160) of Mr. Haas's address as a new document. Use all caps and no punctuation in the address.
12. Save the labels as *[your initials]*6-24b in your Lesson 6 folder.
13. Print both documents, and then close them.

On Your Own

In these exercises, you work on your own, as you would in a real-life work environment. Use the skills you've learned to accomplish the task—and be creative.

EXERCISE 6-25
Write a summary about a book you have recently read. Change the margins for the document and change the margins of one of the sections of the summary that you want to emphasize. Save the document as *[your initials]*6-25 and print it.

EXERCISE 6-26
Write a letter to a friend about your progress in school. Use the standard business letter format. Insert an automatically updatable date field in the date line. Hyphenate the entire letter, using manual hyphenation. Add an envelope to the document. Save the document as *[your initials]*6-26 and print it.

EXERCISE 6-27
Log on to the Internet and find five Web sites about today's hottest political topic. Create a document summarizing the topic with a pro and con approach. Format the document using landscape orientation and manually hyphenate the text. Save the document as *[your initials]*6-27 and print it.

LESSON 7

Tabs and Tabbed Columns

OBJECTIVES

After completing this lesson, you will be able to:

1. Set tabs.
2. Set leader tabs.
3. Clear tabs.
4. Adjust tab settings.
5. Create tabbed columns.
6. Sort paragraphs and tabbed columns.

MICROSOFT OFFICE
SPECIALIST
ACTIVITIES
In this lesson:
WW03S-3-2

See Appendix C.

 Estimated Time: 1 hour

A *tab* is a paragraph-formatting feature used to align text. When you press Tab, Word inserts a tab character and moves the insertion point to the position of the tab setting, called the *tab stop*. You can set custom tabs or use Word's default tab settings.

As with other paragraph-formatting features, tab settings are stored in the paragraph mark at the end of a paragraph. Each time you press Enter, the tab settings are copied to the next paragraph. You can set tabs before you key text or for existing text.

Setting Tabs

Word's default tabs are left-aligned and set every half inch from the left margin. These tabs are indicated at the bottom of the horizontal ruler by tiny tick marks.

FIGURE 7-1
Default tabs

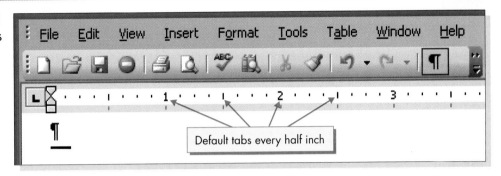

If you don't want to use the half-inch default tab settings, you have two choices:

- Change the distance between the default tab stops.
- Create custom tabs.

The four most common types of custom tabs are left-aligned, centered, right-aligned, and decimal-aligned. Custom tab settings are indicated by *tab markers* on the horizontal ruler. Additional custom tab options, such as leader tabs and bar tabs, are discussed in the section "Setting Leader Tabs" and in the exercise "Insert Bar Tabs."

FIGURE 7-2
Types of tabs

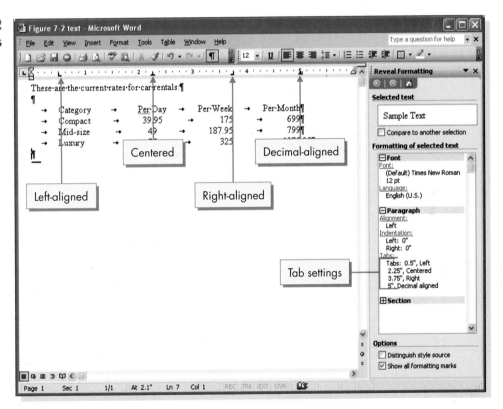

TABLE 7-1 Types of Tabs

RULER SYMBOL	TYPE OF TAB	DESCRIPTION
L	Left-aligned	The left edge of the text aligns with the tab stop.
⊥	Centered	The text is centered at the tab stop.
⌐	Right-aligned	The right edge of the text aligns with the tab stop.
⊥·	Decimal-aligned	The decimal point aligns with the tab stop. Use this option for columns of numbers.
I	Bar	Inserts a vertical line at the tab stop. Use to create a divider line between columns.

There are two ways to set tabs:

- Use the Tabs dialog box.
- Use the ruler.

EXERCISE **7-1** **Set Tabs by Using the Tabs Dialog Box**

1. Open the file **Europe1**.
2. Display the Reveal Formatting task pane (Format menu, Reveal Formatting).

FIGURE 7-3
Tabs dialog box

3. Select the paragraph that begins "Duke City."

   ⭐ **TIP:** Instead of selecting a paragraph, you can place the insertion point within a paragraph when setting tabs or applying other paragraph formatting.

4. Choose Tabs from the Format menu. The Tabs dialog box appears. Notice that the Default tab stops text box is set to 0.5 inch.

5. Key **.25** in the Tab stop position text box. The alignment is already set to Left, by default.

6. Click OK. The ruler shows a left tab marker, the symbol used to indicate the type and location of a tab stop on the ruler. The Reveal Formatting task pane displays the tab setting.

 NOTE: When you set tabs, the settings are listed in the Reveal Formatting task pane under Paragraph. You can click the Tabs link to open the Tabs dialog box.

7. In the same paragraph, move the insertion point to the left of the first word, "Duke."

8. Press Tab. The first line of the paragraph is now indented 0.25 inch. This produces the same effect as creating a first-line indent.

 REVIEW: Tabs are nonprinting characters that can be displayed or hidden. Remember, to display or hide nonprinting characters, click the Show/ Hide ¶ button ¶ on the Standard toolbar or choose Show all formatting marks in the Reveal Formatting task pane.

9. Click within the tabbed text under "**$1 equals**." Notice that the second column of this text is aligned at a 2-inch decimal tab setting.

10. Select the text under the heading "European Weather in June" from "City" through "53" at the end of the list. This text contains tab characters, but no tab stops are set. The text is aligned at the default tab settings. You'll set two center tabs to align the numeric text.

11. Open the Tabs dialog box. Key **1.5** in the Tab stop position text box.

12. Under Alignment, choose Center. Click Set. Notice that the tab setting appears below the Tab stop position text box. The setting is automatically selected so that another tab setting can be keyed.

13. Key **3** in the Tab stop position box and click Set. (Center alignment is already set.) Click OK. The column headings "Average High" and "Average Low," along with the numeric text below the headings, are now centered at the 1.5-inch and 3-inch tab settings.

 NOTE: When you set a custom tab, Word clears all default tabs to the left of the new tab marker.

EXERCISE **7-2** **Set Tabs by Using the Ruler**

Setting tabs by using the ruler is an easy two-step process: Click the Tab Alignment button on the left of the ruler to choose the type of tab alignment, and then click the position on the ruler to set the tab.

1. At the top of the document, place the insertion point after the word "about" in the paragraph that begins "Duke City" and key **hotel prices,** (include the comma).

2. Key a comma after the next word, "weather."

3. Position the insertion point in the next line (which is blank) and press (Enter).

4. Key **Hotel Prices** in bold and press (Enter).

5. Turn off bold. Key **Economy** and press (Enter).

6. Place the insertion point to the left of "Economy."

7. Click the Tab Alignment button on the horizontal ruler until it shows right alignment . Each time you click the button, the alignment changes.

FIGURE 7-4
Tab alignment on
the ruler

TIP: When choosing tab settings for information in a document, keep in mind that left-aligned text and right- or decimal-aligned numbers are easier to read.

8. With the Tab Alignment button showing right alignment, click the ruler at 1.5 inches. Word inserts a right-aligned tab marker at 1.5 inches and deletes the default tab markers to the left of the new tab.

9. Click the Tab Alignment button until it changes to decimal alignment , and then click the ruler at 3 inches, 4 inches, and 5 inches. This line of text now has four custom tab settings.

10. With the insertion point to the left of "Economy," press (Tab). Word right-aligns "Economy" at the 1.5-inch tab stop.

11. Move the insertion point to the right of "Economy" and key the following figures, pressing (Tab) before each number:

    99.50   216.00   278.00

12. Press (Shift)+(Enter) to start a new line.

13. Key the following information, pressing (Tab) before the hotel type and before each new number:

    Luxury   245.75   389.00   519.50

14. Move the insertion point to the beginning of the heading "**Hotel Prices**."

15. Click the Tab Alignment button until it shows center alignment .

TIP: As you toggle through the Tab Alignment button symbols, notice the appearance of the first-line indent symbol and the hanging indent symbol. You can display one of these symbols, and then just click the ruler to the desired indent position instead of using the point and drag method described in Lesson 5.

16. Click the ruler at 3 inches. Press (Tab) to center the line at the tab stop. (With 1.25-inch left and right margins, 3.0 is the center point of a 6-inch line.)

FIGURE 7-5
Document with
tabbed text

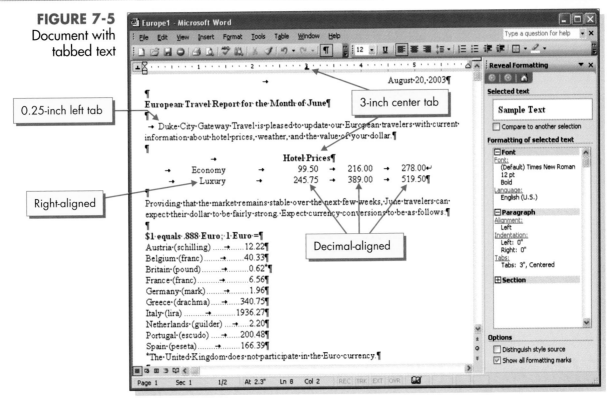

0.25-inch left tab

3-inch center tab

Right-aligned

Decimal-aligned

TIP: You can copy tab settings from one paragraph to another. Click in the paragraph whose tab settings you want to copy. Click the Format Painter button. Click in the paragraph to which you are copying the tab settings.

Setting Leader Tabs

You can set tabs with *leader characters,* patterns of dots or dashes that lead the reader's eye from one tabbed column to the next. Leaders are often found in a table of contents, in which dotted lines fill the space between the headings on the left and the page numbers on the right.

Word offers three leader patterns: dotted line, dashed line, and solid line.

FIGURE 7-6
Leader patterns

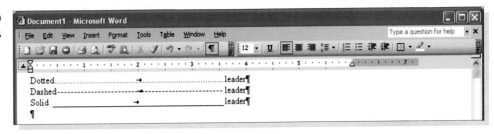

EXERCISE   **7-3**   **Set Leader Tabs**

1. Select the two columns of text under the heading "**$1 equals**." The figures are aligned at a 2-inch decimal tab.

2. Open the Tabs dialog box. The 2-inch tab is highlighted in the Tab stop position box.

3. Under Leader, click the second leader pattern (the dotted line).

4. Click Set and click OK. A dotted-line leader fills the space to the left of the 2-inch tab setting.

 NOTE: Leader patterns always fill the space to the left of a leader tab setting.

5. At the top of the document, delete the date and the blank line below it. Delete the last paragraph so the document prints on one page.

6. Save the document as *[your initials]*7-3 in your Lesson 7 folder.

7. Print the document, but don't close it.

Clearing Tabs

You can clear custom tabs all at once or individually. When you clear custom tabs, Word restores the default tab stops to the left of the custom tab stop.

There are three ways to clear a tab:

● Use the Tabs dialog box.

● Use the ruler.

● Press Ctrl+Q.

EXERCISE   **7-4**   **Clear a Tab by Using the Tabs Dialog Box and the Keyboard**

1. Open the file **Country**.

2. Position the insertion point to the left of "For" in the first sentence (after the tab character).

3. Open the Tabs dialog box. The 0.75-inch tab is highlighted in the Tab stop position box.

4. Click the Clear button and click OK. Word clears the 0.75-inch custom tab and restores the 0.5-inch default tab.

 5. Click the Undo button to restore the 0.75-inch custom tab.

6. Press Ctrl+Q. The tab is deleted.

7. Press Backspace to delete the tab character to the left of "For."

 NOTE: Remember, to remove tabs from text, you must delete the tab characters.

EXERCISE 7-5 Clear a Tab by Using the Ruler

1. Position the insertion point at the beginning of the line of text that starts with "Antigua."

2. Position the pointer on the 4.5-inch centered tab marker on the ruler.

3. When the ScreenTip "Center Tab" appears, drag the tab marker down and off the ruler. The custom tab is cleared and the word "Paraguay" moves to a default tab stop.

4. Undo the last action to restore the tab setting.

 NOTE: When clearing or adjusting tabs by using the ruler, watch for the ScreenTip to correctly identify the item to which you are pointing. If no ScreenTip appears, you might inadvertently add another tab marker.

Adjusting Tab Settings

You can adjust tabs inserted in a document by using either the Tabs dialog box or the ruler. Tabs can be adjusted only after you select the text to which they have been applied.

EXERCISE 7-6 Adjust Tab Settings

1. Select all the tabbed text from "Antigua" through "Zambia."

2. Open the Tabs dialog box.

3. Click the 4.5-inch tab setting in the Tab stop position box to select it. Adjust the alignment for the setting by clicking Right. Click OK. The last column is now right-aligned at 4.5 inches.

4. With the same text still selected, close the task pane, and then place the pointer on the 4.5-inch right tab marker.

5. When the ScreenTip indicates "Right Tab," drag the marker to the 6-inch position on the ruler. The column is now right-aligned at 6 inches.

FIGURE 7-7
Using the ruler to
adjust a tab setting

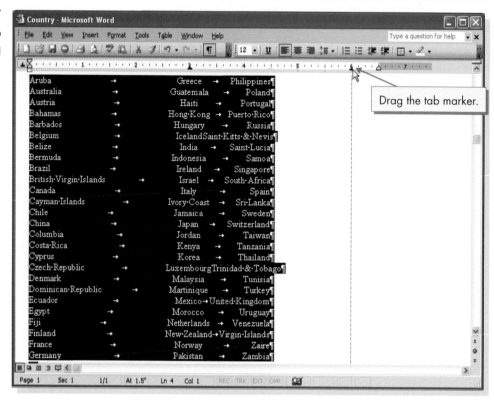

 NOTE: When you change tab settings with the ruler, be careful to drag the tab marker only to the right or to the left. If you drag the tab marker up or down, you might clear it from the ruler. If you inadvertently clear a tab marker, undo your action to restore the tab.

6. With the text still selected, drag the 3-inch tab marker to 3.25 inches on the ruler and apply a 0.5-inch left indent. The text is now centered horizontally between the margins.

7. Reopen the Reveal Formatting task pane to see the new tab settings.

TIP: If you press Alt while positioning the pointer on the tab marker, Word displays the tab stop measurements.

Creating Tabbed Columns

As you've seen in these practice documents, you can use tabs to present information in columns.

When you set up a table using tabbed columns, follow these general rules based on *The Gregg Reference Manual*:

● The table should be centered horizontally within the margins.

● Columns within the table should be between six and ten spaces apart.

- The width of the table should not exceed the width of the document's body text.
- At least one blank line should separate the top and bottom of the table from the body text of the document.

EXERCISE 7-7 Set Tabbed Columns

1. Position the insertion point at the end of the document and press Enter twice.
2. Press Ctrl+Q to remove the indent and tab settings from the paragraph mark; then key the text shown in Figure 7-8. Use single spacing.

FIGURE 7-8

Whenever your travel plans include international travel, contact our travel specialists to help you assemble all the information you need about your destination. Don't forget, Duke City Gateway Travel can help you arrange individual or group tours, family trips, or business travel anywhere in the world.

3. Press Enter twice and key **Please call us any time:**.
4. Press Enter twice. Create a guide line that contains the longest item in each column of Figure 7-10 by keying the following with ten spaces between each group of words:
   Frank Murillo   Senior Travel Counselor   International Travel

5. Click the Center button ≡ on the Formatting toolbar to center the line.
6. Scroll down until the guide line is directly below the ruler.

7. Change the Tab Alignment button to center alignment ⊥. Using the I-beam as a guide, click the ruler to set a center-aligned tab in the middle of each group of words. (See Figure 7-9 on the next page.)
8. Delete the text in the guide line up to the paragraph mark. Do not delete the paragraph mark, which is now storing your centered tab settings.

9. Click the Align Left button ≣ to left-align the insertion point.
10. Key the table text as shown in Figure 7-10 (on the next page), pressing Tab before each item and single-spacing each line. Underline each column heading.

FIGURE 7-9
Guide line for
centering tabbed
columns

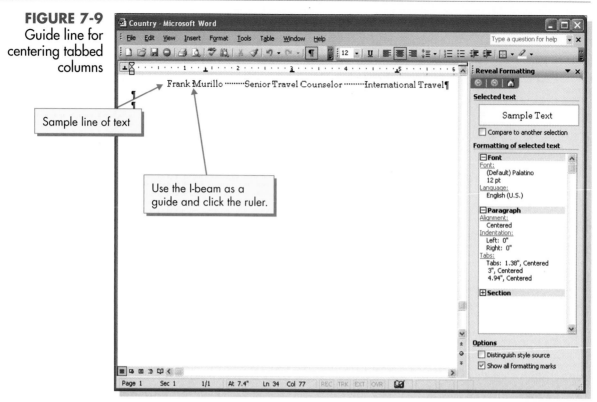

FIGURE 7-10

Name	Title	Specialty
Susan Allen	President/Owner	Group Travel
Frank Murillo	Senior Travel Counselor	International Travel
Nina Chavez	Senior Travel Counselor	Corporate Travel

EXERCISE   **7-8**   **Select a Tabbed Column**

After text is formatted in tabbed columns, you can select columns individually by selecting a vertical block of text. Selecting tabbed text can be helpful for formatting or deleting text. You use [Alt] to select a vertical block of text.

NOTE: If you do not press [Alt] when trying to select text vertically, you will select the entire first line of text, rather than just the column header for the column you are selecting.

1. Hold down [Alt] and position the I-beam before the tab character in the line that contains the column heading "Name."

2. Drag across the heading, and then down until all three names are selected. Don't select the tab characters to the right of the column.

FIGURE 7-11
Selecting text
vertically

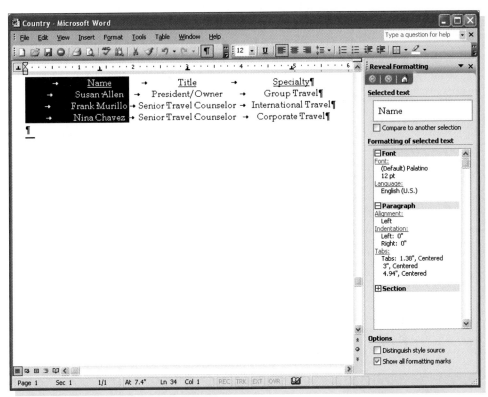

3. Press Delete to delete the column.

4. Undo the deletion.

5. Select the column again, this time selecting only the names under the column head "Name."

6. Click the Italic button I to make the names italic.

EXERCISE 7-9 **Insert Bar Tabs**

Bar tabs are used to make tabbed columns look more like a table with gridlines. A bar tab inserts a vertical line at a fixed position, creating a border between columns. You can set bar tabs by using the ruler or the Tabs dialog box.

1. At the bottom of the document, select the three lines of tabbed text below the headings "Name," "Title," and "Specialty."

2. Open the Tabs dialog box. To set a 0.6-inch bar tab, key **0.6** in the text box, click Bar, and click OK. Do not deselect the tabbed text.

3. To set bar tabs by using the ruler, click the Tab Alignment button until it changes to a bar tab . Click the ruler at 2 inches, 4 inches, and 5.75 inches. The bar tab markers appear as short vertical lines on the ruler.

> **NOTE:** You might need to close the task pane if it is open to accomplish this step. If you do, reopen it after you have completed the step.

4. Adjust the bar tab markers on the ruler to make them more evenly spaced, as needed.

5. Deselect the tabbed text. Click the Show/Hide ¶ button ¶ to view the document without nonprinting characters. The bar tabs act as dividing borders between the columns.

FIGURE 7-12
Tabbed text with
bar tabs

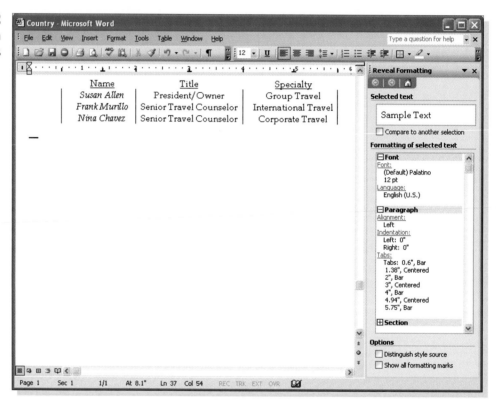

6. Position the insertion point in the line that contains "Frank Murillo."

7. Point to the 4-inch bar tab on the ruler and drag it off. The vertical line in the table before "International Travel" disappears.

8. Undo the deletion to restore the bar tab.

9. Save the document as *[your initials]*7-9 in your Lesson 7 folder.

10. Print and close the document.

Sorting Paragraphs and Tabbed Columns

Sorting is the process of reordering text alphabetically or numerically. You can sort to rearrange text in ascending order (from lowest to highest, such as 0–9 or A–Z) or descending order (from highest to lowest, such as 9–0 or Z–A).

You can sort any group of paragraphs, from a single-column list to a multiple-column table, such as one created by tabbed columns. When sorting a tabbed table, you can sort by any of the columns.

FIGURE 7-13
Sorting paragraphs and tables

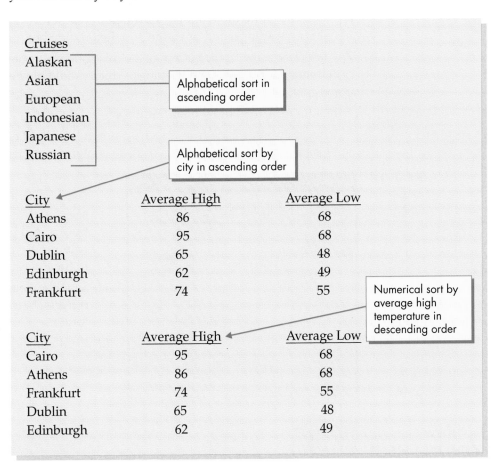

Cruises
Alaskan
Asian   Alphabetical sort in ascending order
European
Indonesian
Japanese
Russian   Alphabetical sort by city in ascending order

City	Average High	Average Low
Athens	86	68
Cairo	95	68
Dublin	65	48
Edinburgh	62	49
Frankfurt	74	55

Numerical sort by average high temperature in descending order

City	Average High	Average Low
Cairo	95	68
Athens	86	68
Frankfurt	74	55
Dublin	65	48
Edinburgh	62	49

EXERCISE 7-10 Sort Tabbed Tables

1. With the file *[your initials]***7-3** open, select the table under the text "**$1 equals**." Notice that each paragraph of this table is sorted alphabetically by country.

2. Choose Sort from the Table menu to open the Sort Text dialog box. Notice that the sort settings for the selected lines indicate the paragraphs are sorted by text in ascending order.

3. Open the Sort by drop-down list to view the other sort options. Fields 1 and 2 indicate the first and second columns. Open the Type drop-down list. Notice that the type options include Text, Number, and Date.

4. Click Descending to change the sort order and click OK. The text is sorted alphabetically in descending order.

5. Press [Ctrl]+[Z] to undo the sort.

6. Select the entire table under "**European Weather in June**," including the column headings, and open the Sort Text dialog box.

7. Click Header row at the bottom of the dialog box. This option indicates that the selection includes column headings, which should not be sorted with the text.

8. Open the Sort by drop-down list. Now you can sort by the table's column headings instead of by field numbers.

9. Choose Average High from the drop-down list. Word changes the Type to Number, recognizing that this column contains numbers.

FIGURE 7-14
Sorting options in the
Sort Text dialog box

10. To the right of Number, click Descending if it is not already selected. Click OK. The table is now sorted by the Average High column, from the highest number to the lowest.

11. Save the document as *[your initials]*7-10 in your Lesson 7 folder.

12. Print and close the document.

USING ONLINE HELP

Instead of setting individual tabs in a document, you might want to change only the spacing between Word's default tab stops. For example, you can change the 0.5-inch default tabs to 1-inch tabs. Word's Help can tell you how.

Use Help to find out how to change default tab stops:

1. Key **change default tabs** in the Ask a Question text box.

2. Press [Enter].

3. Click Change the spacing between default tab stops.

4. Review the information in the Help window that tells you how to change default tab stops.

5. Close the Help window.

LESSON 7 Summary

➤ Tabs are a paragraph-formatting feature used to align text. When you press `Tab`, Word inserts a tab character and moves the insertion position to the tab setting, called the tab stop.

➤ Word's default tabs are left-aligned and set every half-inch from the left margin, as indicated at the bottom of the horizontal ruler.

➤ The four most common types of custom tabs are left-aligned, centered, right-aligned, and decimal-aligned. Custom tab settings are indicated on the horizontal ruler by tab markers.

➤ Set tabs by using the Tabs dialog box or the ruler. To use the ruler, click the Tab Alignment button on the left of the ruler to choose the type of tab alignment, and then click the position on the ruler to set the tab. See Table 7-1.

➤ A leader tab uses a series of dots, dashes, or solid underlines to fill the empty space to the left of a tab stop. Use the Tabs dialog box to set a leader tab.

➤ Clear custom tabs all at once or individually. When you clear custom tabs, Word restores the default tab stops to the left of the custom tab stop. To clear a tab, use the Tabs dialog box, or the ruler, or press `Ctrl`+`Q`.

➤ To adjust tab settings, position the insertion point in the tabbed text (or select the text), and then either open the Tabs dialog box or drag the tab markers on the ruler.

➤ Use tabs to present information in columns. Tabbed columns are a side-by-side vertical list of information.

➤ To select a tabbed column (for formatting or deleting the text), hold down `Alt` and drag the I-beam over the text.

➤ Use bar tabs to make tabbed columns look more like a table with gridlines. A bar tab inserts a vertical line at a fixed position, creating a border between columns. You can set bar tabs by using the ruler or the Tabs dialog box.

➤ Sorting is the process of reordering text alphabetically or numerically. You can sort to rearrange text in ascending order (from lowest to highest, such as 0–9 or A–Z) or descending order (from highest to lowest, such as 9–0 or Z–A).

Concepts Review

Each of the following statements is either true or false. Indicate your choice by circling
T or F.

T F **1.** In Word, you can use ⌨Spacebar to precisely align text.

T F **2.** When you set a custom tab, Word clears all default tabs to the
left of the tab stop.

T F **3.** You cannot set tabs for existing text.

T F **4.** You can either select a paragraph or place the insertion point
within the paragraph when setting tabs for the paragraph.

T F **5.** When custom tabs are cleared, you must reestablish default tabs
or you will have no tabs at all.

T F **6.** Tabs inserted in a document cannot be adjusted after they are set.

T F **7.** The symbol for a bar tab marker is a short vertical line.

T F **8.** You can use the ruler to set a leader tab.

Write the correct answer in the space provided.

1. What are the two ways to set tabs?

_____

2. What are the five types of tabs you can set when using the Tab Alignment
button?

_____

3. Which dialog box is used to set a leader tab?

_____

4. What type of tab do you use ⬚ to create?

_____

5. What interval does Word use for default tabs?

_____

6. How do you clear a tab by using the ruler?

_____

7. Which key is used to select a tabbed column?

_____

8. Which menu contains the Sort feature?

_____

CRITICAL THINKING

Answer these questions on a separate page. There are no right or wrong answers. Support your answers with examples from your own experience, if possible.

1. Some books and magazines use leader tabs in the table of contents and index. Based on representative books and magazines, create a few general guidelines on when to use leader tabs. Support your view with examples.

2. Do you find it easier to read text that is centered in a column or text that is left-aligned? What about numbers that are centered or right-aligned? Create samples to support your position.

Skills Review

EXERCISE 7-11

Set tabs and create a business memo.

1. Start a new document with a 2-inch top margin and 1.25-inch left and right margins.

2. Set a 1-inch left-aligned tab by following these steps:

    a. Make sure the Tab Alignment button on the horizontal ruler shows left alignment [L].

    b. Click the ruler at the 1-inch mark.

3. Key the text in Figure 7-15 (shown on the next page), using the spacing shown. Press (Tab) after each colon. Refer to Appendix B: "Standard Forms for Business Documents."

4. In the date line, key today's date.

5. Add your reference initials at the end of the document.

6. Spell-check the document.

7. Save the document as *[your initials]*7-11 in your Lesson 7 folder.

8. Print and close the document.

FIGURE 7-15

```
            MEMO TO:     Susan Allen
  1 blank    FROM:       Alexis Johnson
  line
            DATE:
  2 blank    SUBJECT:    Ontario Gateway Travel          Single space
  lines

            The Ontario Gateway Travel office in Toronto has added a second phone
            line. The number is (416) 555-0015. Please make a note of it.
```

EXERCISE 7-12

Set leader tabs.

1. Start a new document. Format the paragraph mark as 14 points Arial Narrow.
2. Key the first paragraph in Figure 7-16. Press Enter twice.

FIGURE 7-16

```
    Enter a drawing for a free trip to Hawaii, all expenses paid. Fill in
    the form below.

    Name_____

    Address_____

    City/State/ZIP_____

    Telephone_____
```

3. Before keying the remaining text, set solid leader tabs that extend to the right margin by following these steps:
   a. Choose Tabs from the Format menu to open the Tabs dialog box.
   b. In the Tab stop position box, key **6**.
   c. Under Alignment, click Right.
   d. Choose the fourth leader option, click Set, and click OK.
4. Key the remaining information, beginning with **Name**. Press Tab to move to the 6-inch right-aligned tab setting, and then press Enter. Continue keying the text in the figure.
5. Format the text with the leaders as double-spaced and small caps.
6. Change the top margin to 2 inches.
7. Save the document as *[your initials]*7-12 in your Lesson 7 folder.
8. Print and close the document.

EXERCISE 7-13

Adjust and clear tab settings.

1. Open the file **Duke1**.

2. Position the insertion point in the tabbed text that begins "Businesspeople." (The tabbed text is a single paragraph.)

3. Set a 1-inch left-aligned tab by using the ruler.

4. Use the ruler to adjust the tab setting by following these steps:

   a. Point to the 1-inch left tab marker.
   b. When you see the ScreenTip identifying the tab marker, drag the marker to 2.5 inches on the ruler.

5. Click the Tab Alignment button until it shows center alignment . Click the ruler at 2 inches to set a tab.

6. Use the Tabs dialog box to clear both tabs by following these steps:

   a. Open the Tabs dialog box.
   b. Click Clear All and click OK.

7. Click the Tab Alignment button until it shows right alignment . Click the ruler at 1.5 inches.

8. Drag the 1.5-inch tab marker on the ruler to adjust it to 2 inches.

9. Save the document as *[your initials]*7-13 in your Lesson 7 folder.

10. Print and close the document.

EXERCISE 7-14

Create tabbed columns and sort text.

1. Start a new document. Key **ALBUQUERQUE'S MOST POPULAR RESTAURANTS** in uppercase bold. Center the text and press Enter three times.

2. Left-align the paragraph mark and turn off bold and uppercase.

3. Create a table with single-spaced, tabbed columns that are horizontally centered between the left and right margins by following these steps:

   a. Key a guide line containing the longest text from each column in Figure 7-17, with ten spaces between columns. (Include the column headings when determining the longest item in each column.)
   b. Center the text.
   c. Scroll until the guide line is directly under the ruler.

 NOTE: If the task pane is open, you might want to close it to get a better view of the ruler.

d. Using the I-beam for guidance, set a left-aligned tab for the first and second columns and a centered tab for the third column.

e. Delete the guide line up to the paragraph mark. Left-align the paragraph mark.

f. Key the text shown in Figure 7-17, pressing `Tab` before each item in each column.

FIGURE 7-17

Name	Cuisine	Average Dinner Price
Tuscany	Italian	$15
El Greco	Spanish	$18
Desert Star	Southwestern	$16

4. Bold the column headings.

5. Change the paragraph spacing for the table (including the column headings) to double-spacing.

6. Set a 2-inch top margin for the document.

7. Sort the table alphabetically by cuisine by following these steps:

a. Select the entire table, including the column headings.

b. Choose Sort from the Table menu.

c. Click Header row to display the column headings in the Sort by drop-down list.

d. Choose Cuisine from the Sort by drop-down list.

e. Choose Text from the Type list and choose Ascending. Click OK.

8. Sort the table by price, from highest to lowest, by following these steps:

a. With the table still selected, open the Sort Table dialog box.

b. Make sure Header row is selected. Sort by Average Dinner Price, by Number, and in Descending order. Click OK.

9. Save the document as *[your initials]*7-14 in your Lesson 7 folder.

10. Print and close the document.

Lesson Applications

EXERCISE 7-15

Set tabs for a memo, and then adjust the tab settings.

1. Start a new document. Set a 2-inch top margin and a 1.25-inch left-aligned tab. Create a memo to Frank Youngblood from Nina Chavez. The subject is La Fonda Hotel. Remember to include today's date and insert a blank line between lines in the memo heading.

2. Press Enter three times after the subject line and key the text shown in Figure 7-18, including the corrections. Use single-spacing and insert a blank line between paragraphs.

FIGURE 7-18

Renovation work is complete at La Fonda, ~~the~~ oldest hotel ~~in~~ [Santa Fe's] ~~Santa Fe~~. The rooms have been upgraded and New Mexican décor is standard throughout.

La fonda is still the only hotel actually right on the plaza, at the southeast corner. The hotel has been the favorite resting place of some famous people, including Kit Carson, Ulysses S. Grant, and [President] Rutherford B. Hayes. Billy the Kid is rumored to have washed dishes in the hotel kitchen.

The hotel is included in our walking tour of Santa Fe.

3. Change the font for the entire document to Arial.

4. Change the font for the memo-heading guidewords (MEMO TO:, FROM:, and so on) to Arial Black.

5. Spell-check the document.

6. Add your reference initials to the document.

7. Save the document as *[your initials]*7-15 in your Lesson 7 folder.

8. Print and close the document.

EXERCISE 7-16

Set leader tabs and sort paragraphs.

1. Open the file **Glossary**.

2. Change the top margin to 1.5 inches and the left and right margins to 1 inch.

3. Increase the title size to 14 points.

4. Position the insertion point in the blank line under the title, press Enter twice, and key the paragraph shown in Figure 7-19. Use single-spacing. Include one blank line below the paragraph.

FIGURE 7-19

```
Cruises are among the most popular vacation choices, and for good
reason. Travelers enjoy luxurious accommodations and the ability to
visit numerous destinations while on the open seas. To help you enjoy
your adventure, our travel specialists prepared the following glossary
of nautical terms. The seafaring world has a language all its own!
```

5. Select the list of terms and definitions. Remove the hanging indents in the list by using the Paragraph dialog box or the ruler. Clear the 1.5 tab setting. Set a dotted leader tab that right-aligns the glossary definitions at the right margin.

6. Sort each paragraph in the glossary alphabetically, from A to Z.

7. Change the spacing for each glossary definition to 6 points after paragraphs.

8. If the document is now two pages long, delete the blank paragraph mark at the end of the document.

9. Apply a page border, using the box setting and the double wavy line style.

10. Spell-check the document.

11. Save the document as *[your initials]*7-16 in your Lesson 7 folder.

12. Print and close the document.

EXERCISE 7-17

Set and adjust tab settings.

1. Start a new document. Set the top margin to 2 inches.

2. Key the title in Figure 7-20 (shown on the next page). Format it as 12-point Arial bold, uppercase, and centered.

3. Insert two blank lines below the title and key the paragraph that begins "Below." Use single-spacing and Arial.

4. Key the remaining text in the figure, beginning with "New York," using a 1.5-inch left indent and an appropriate right tab setting for the dollar amounts. The text should be evenly spaced between the left and right margins, as shown in the figure.

5. Format the paragraph that begins with "Below" with justified alignment and a dropped capital letter (use the default drop cap settings). There should be one blank line between this paragraph and the table below it.

FIGURE 7-20

THE MOST EXPENSIVE CITIES TO VISIT IN AMERICA

Below are the five most expensive cities for business travelers in this country, according to Fuller-Reilly International, a Minnesota-based consulting firm. Figures are based on an average daily total for three meals, lodging, gratuities, and taxes.

New York (Manhattan)	$405
Washington, D.C.	365
Chicago	338
San Francisco	332
Boston	325

6. Adjust the tab setting for the table, so it includes a dotted leader.

7. Change the line spacing for the table text to 1.5 lines.

8. Increase the font size of the title to 16 points.

9. Spell-check the document.

10. Save the document as *[your initials]***7-17** in your Lesson 7 folder.

11. Print and close the document.

EXERCISE 7-18 ✚ *Challenge Yourself*

Create a memo with tabbed columns, sort the text, and add bar tabs.

1. Start a new document. Using the proper line spacing, margin settings, and a 1-inch left tab setting, create a memo to Nina Chavez from Steve Ross. The subject is **Weekend Hours for U.S. Gateway Flagship Offices**.

2. For the body of the memo, key the text in Figure 7-21 (shown on the next page). Use single-spacing. For the tabbed columns, create a guide line to set the tabs. Align the tabbed columns as indicated. Insert a blank line above and below the tabbed columns.

3. Apply a ¾-point box border around the entire memo heading (excluding the blank lines below "SUBJECT:"). Add 10 percent gray shading to the memo heading.

4. Adjust the tab setting for the memo heading to 1.25 inches and set a 1-inch bar tab to create a vertical dividing line.

5. Select the first column of the memo heading (which begins "MEMO TO") and make it bold.

FIGURE 7-21

```
As you requested, the following is a list of the time zones and
weekend hours for the U.S. Gateway flagship offices:

Office                      Time Zone        Days              Hours
Golden Gateway Travel       Pacific          Sat.             10-5:30
Duke City Gateway Travel    Mountain         Sat., Sun.          11-4
Windy City Gateway Travel   Central          Sat., Sun.          9-12
Big Apple Gateway Travel    Eastern          Sat.                 9-4

Let me know if you need additional information.
```

Left-align these columns 　　　　　　　　　　　　　　　　 *Right-align this column*

6. Sort the tabbed table of Gateway offices by office in ascending order.
7. Add your reference initials to the bottom of the memo.
8. Spell-check the document.
9. Save the document as *[your initials]*7-18 in your Lesson 7 folder.
10. Print and close the document.

On Your Own

In these exercises, you work on your own, as you would in a real-life work environment. Use the skills you've learned to accomplish the task—and be creative.

EXERCISE 7-19
Write a short business memo. The memo is from you, to a person and about a subject of your choosing. Use the correct spacing and tab settings for the memo heading. Save the document as *[your initials]*7-19 and print it.

EXERCISE 7-20
Create a monthly budget in the form of a tabbed table. Set and adjust the tabs, using the ruler. Sort the table. Save the document as *[your initials]*7-20 and print it.

EXERCISE 7-21
Log on to the Internet and find airlines and ticket prices for a place you would like to visit. Create a tabbed table containing the airline names and ticket prices. Use leaders and sort the information. Save the document as *[your initials]*7-21 and print it.

Unit 2 Applications

UNIT APPLICATION 2-1

Work with paragraph spacing, tabs, and margins; add a horizontal line; create symbols automatically; and sort text.

1. Open the file **Toronto**.
2. Display the Reveal Formatting task pane.
3. Set a 2-inch top margin and 1.25-inch left and right margins.
4. Remove the shading and borders from the document.
5. Replace the first paragraph with a title that reads **Accommodation Recommendations of the Bed and Breakfast Registry of Toronto**.
6. Format the title as follows:
   - 14-point bold
   - Small caps
   - Centered
   - Line break before the word "Bed"
7. Under the title, apply a single-line border automatically, without using a dialog box.
8. Format the next three paragraphs as follows:
   - 12 points of spacing before
   - 1/2-inch first-line indent (or 1/2-inch left tab at the start of each paragraph)
   - Left alignment
9. Sort the three paragraphs alphabetically in ascending order.
10. Replace the last paragraph with text that reads **Some of these accommodations do not accept credit cards as payment.**
11. Make sure two blank lines appear above the last paragraph, and format the paragraph as follows:
    - Bold italic
    - 12.5 percent shading
    - 0.25-inch left and right indent
12. Change the names "The Fountaines," "The St. Clair," and "Lowter House" to bold italic without quotation marks.
13. Add the following text approximately five lines below the last line of text using Click and Type: **For more information and availability, call (800) 555-1234 or visit www.gwaytravel.com.** Center the text and make sure the Internet address is a hyperlink.

14. To the left of the sentence that begins "Some of these accommodations," key ==> to insert a thick, right-pointing arrow and press (Spacebar). At the end of the sentence, press (Spacebar) and key <== to insert a thick, left-pointing arrow. Center the paragraph.

15. Save the document as *[your initials]*u2-1 in a new folder for Unit 2 Applications.

16. Print and close the document.

UNIT APPLICATION 2-2

Work with paragraph spacing, tabs, and margins; create a bulleted list; insert symbols; add a horizontal line; add paragraph borders; change page orientation; and sort text.

1. Open the file **Itin2**.

2. Format the title in the first line of the document as 14-point bold uppercase, centered.

3. Insert a new paragraph below the title "Travel in the Southwest" by keying the text shown in Figure U2-1. Include the corrections. Use single spacing and insert the degree symbol (°) where indicated. Two blank lines should appear above the new paragraph and one blank line below it.

FIGURE U2-1

All of us at
Duke City Gateway Travel—a full-service travel agency, pride ourselves on creating the perfect travel experience for our clients. We can provide interested travelers with custom-designed itineraries to hlep them make the most of their travel time and to help them be sure they'll see everything they want to see. Following are some sample itineraries for travel in the Southwest and in other pars of the U.S. and Canada. These itinerearies were designed with specific clients in mind. We also like to let travelers know aobut the range of weather conditions they'll be facing—for example 50-60° at night, and 75-80° during the day.

4. Center the heading "Travel in the Southwest" and change it to small caps.

5. Sort the paragraphs beginning "In New Mexico" through "In Nevada" in ascending order. Format the paragraphs as a bulleted list, using the solid diamond-shaped bullet. (Hint: You might have to choose a regular bullet first, and then click Customize to choose the solid diamond bullet.)

6. Indent the list so that the bullets are 0.5 inch from the left margin.

7. Format the paragraphs under the heading "Sample Itinerary—The Great Southwest" from "Day 1" through "Day 9" with a 0.45-inch hanging indent.

8. In the same paragraphs, set a bar tab at 0.63 inches. Set a left-justified tab at 0.75 inches. Adjust the hanging indent to 0.75 inches in each paragraph. In each paragraph, delete the space and press Tab between the day number and the description of the day's events.

9. On the line below Day 9, automatically apply a single-line border.

10. Format "Sample Itinerary—The Great Southwest," with a single-line bottom border. There should not be any blank space between the border and the Day 1 text.

11. In Print Layout view, under "Sample Itinerary—The Great Southwest," select all the noncontiguous day numbers (not "Day" or the description of the day's events). Change the font color of the numbers to white, make the numbers bold, and apply 95 percent shading to the text.

12. For the same text, change the paragraph spacing to 6 points before.

13. Copy all of this formatting to the text under the heading "Sample Itinerary—The California Coast." Don't forget to copy the day number shading and to replace the space after each number with a tab character.

14. Center the heading "Travel to Other Parts of the United States and Canada" and change it to small caps.

15. Format the document with 1.25-inch left and right margins.

16. Apply mirror margins and a 0.5-inch gutter margin to the entire document.

17. Select the text from "Sample Itinerary—The Great Southwest" through the horizontal line after Day 9. Change the selected text's page orientation to Landscape. Repeat this formatting for the sample itinerary for the California coast.

18. Format the first section with a 2-inch top margin.

19. Spell-check the document.

20. Save the document as [your initials]u2-2 in your Unit 2 Applications folder.

21. Print and close the document.

UNIT APPLICATION 2-3

Apply and change bulleted lists, create tabbed columns, insert the current date and insert special characters, print an envelope, print labels, and choose print options.

1. Start a new document with a 2-inch top margin and 1.5-inch left and right margins.

2. Key the text shown in Figure U2-2, using 12-point Arial. Use leader tabs to create the lines under "Date Completed." The leaders should extend to the right margin.

FIGURE U2-2

> *Before you go on vacation, use this handy checklist to make sure you don't forget anything. Have a safe, happy trip and remember—don't drink and drive.*
>
> Date completed
>
> Stop the newspaper _____
> Ask neighbor to take in the mail _____
> Arrange for pet care _____
> Leave a small light on inside _____
> Check faucets inside and outside _____
> Turn off appliances _____
> Pack clothes _____
> Confirm reservations _____
> Have airline tickets on hand _____
> Take list of important phone numbers _____

 TIP: You need to set two tabs—one for the leader and one to begin the second column.

3. Center the text "Date completed" above the leader characters and insert two blank lines above it.

4. Format the list with 18-point spacing before paragraphs. Apply bullets to the list, using the checkmark bullet.

5. Format the opening paragraph as bold italic, justified.

6. Select the list with the checkmark bullets and customize the bullet as the 3-D box (❑) Wingding character.

7. Customize the 3-D box bullet format as follows:
   - Increase the font size of the bullet to 14 points.
   - Adjust the tab for the bullet to 0.63 inches.

8. Apply a 3-D page border, using the fourth-to-last line style.

9. Insert a blank line at the beginning of the document. Select the blank line paragraph mark and clear all formatting. Restore the font to 12-point Times New Roman.

10. With the paragraph mark still selected, apply 1.25-inch left and right margins to the selection (not to the whole document) to create a new section.

11. Remove the page border from the new section.

12. In the new section, key the letter shown in Figure U2-3. Insert an automatically updating date field for the date. Use the correct spacing between elements and add your reference initials and an enclosure notation.

FIGURE U2-3

[Date]

Dr. Edward Wells

12 Simms Avenue

Larchmont, NY 10573

Dear Dr. Wells:

Enclosed please find a checklist to help you plan your upcoming vacation. Please let me know if you have any questions or if I may be of further assistance.

Sincerely,

Nina Chavez

Senior Travel Counselor

13. Add to the document an envelope addressed to Dr. Wells in all caps, without punctuation, but with a delivery point barcode. Use your name and address as the return address.

14. Using the same all-caps formatting, create a sheet of Avery Standard 5160 address labels, using the doctor's address. Change the font for the sheet of labels to Arial. Save the label file as *[your initials]*u2-3labels in your Unit 2 Applications folder. Print and close the labels document.

15. Save the document as *[your initials]*u2-3 in your Unit 2 Applications folder.

16. Print the entire document, and then close it.

UNIT APPLICATION 2-4 *Using the Internet*

Work with a variety of paragraph-formatting features, include hyperlinks in a document, work with e-mail, and sort text.

Choose any city in the world as the destination for your dream vacation. Using the Internet, research the following:

- Attractions to see
- A list of everything you need to take with you on your trip
- A list of what you need to take care of before leaving for and upon returning from your vacation

In a new document, list a minimum of five attractions you want to see. Include a short paragraph describing each attraction. Include Web page addresses for further information about each of the attractions. Apply various formats to this document. For example, each attraction's name could be bold and set at the left margin. Descriptions might be indented from the right and left margins, with a first-line indent.

In the same document, create tabbed columns for the other two lists you need to create. Separate the lists by using shading and borders. Format the lists with numbers or bullets. Sort the lists alphabetically. Save the document as *[your initials]*u2-4 in your Unit 2 Applications folder. Print, and then close the document.

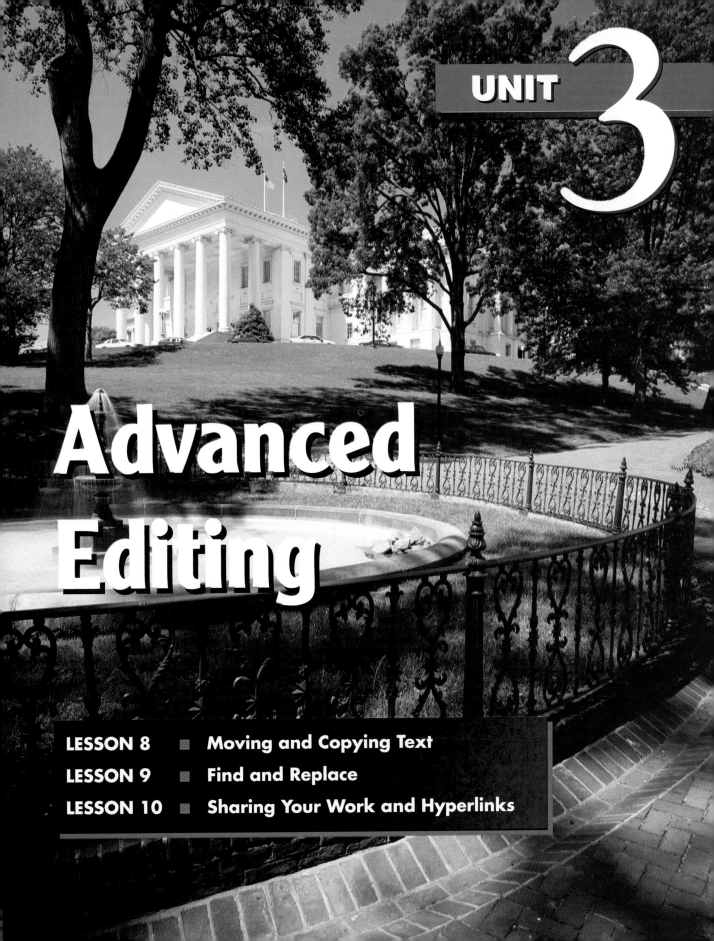

LESSON 8

Moving and Copying Text

MICROSOFT OFFICE SPECIALIST ACTIVITIES

In this lesson:
WW03S-1-1
WW03S-5-7

See Appendix C.

OBJECTIVES

After completing this lesson, you will be able to:

1. Use the Office Clipboard.
2. Move text by using cut and paste.
3. Move text by dragging.
4. Copy text by using copy and paste.
5. Copy text by dragging.
6. Use Smart Cut and Paste.
7. Work with multiple document windows.
8. Move and copy text among windows.

 Estimated Time: 1 hour

One of the most useful features of word processing is the capability to move or copy a block of text from one part of a document to another or from one document window to another, without rekeying the text. In Word, you can move and copy text quickly by using the Cut, Copy, and Paste commands or the drag-and-drop editing feature.

Using the Office Clipboard

Perhaps the most important tool for moving and copying text is the *Clipboard,* which is a temporary storage area. Here's how it works: cut or copy text from your document and store it on the Clipboard. Then move to a different location in your document and insert the Clipboard's contents.

There are two types of Clipboards:

- The system Clipboard stores one item at a time. Each time you store a new item on this Clipboard, it replaces the previous item. This Clipboard is available to many software applications on your system.
- The Office Clipboard can store 24 items, which are displayed on the Clipboard task pane. The Office Clipboard lets you collect multiple items without erasing previous items. You can store items from all Office applications.

EXERCISE   **8-1**   **Display the Clipboard Task Pane**

1. From the Edit menu, choose Office Clipboard. The Clipboard task pane opens. At the top of the task pane, notice the Paste All [Paste All] and Clear All [Clear All] buttons. At the bottom of the screen, at the right end of the taskbar, notice the Clipboard icon [icon], indicating that the Office Clipboard is in use.

FIGURE 8-1
Clipboard
task pane

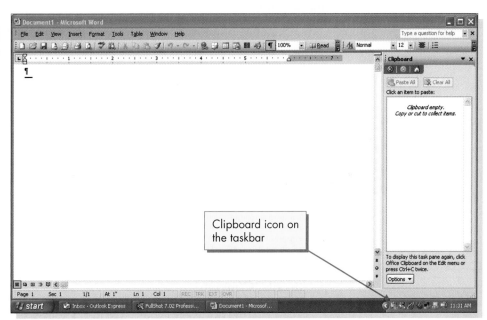

 NOTE: You can also press Ctrl+C twice to open the Office Clipboard on the task pane. Hold down Ctrl and press C twice.

2. If the Office Clipboard contains items from previous use, click the Clear All button to empty the Clipboard.

3. Click the Options button Options ▾ at the bottom of the task pane. Notice the options available for using the Office Clipboard.

> **NOTE:** If the option Show Office Clipboard Automatically is selected, the Clipboard task pane will open automatically when you copy twice in a row without pasting. You can choose to turn off this option.

4. Click outside the task pane, making sure not to choose any of the options in the list.

Moving Text by Using Cut and Paste

To move text by using the *cut-and-paste* method, start by highlighting the text you want to move and using the Cut command. Then move to the location where you want to place the text and use the Paste command. When you use cut and paste to move paragraphs, you can preserve the correct spacing between paragraphs by following these rules:

- Include the blank line below the paragraph you're moving as part of the selection.
- When you paste the selection, click to the left of the first line of text following the place where your paragraph will go—not on the blank line above it.

There are multiple ways to cut and paste text. The most commonly used methods are

- Use the Cut ✂ and Paste buttons 🖺 on the Standard toolbar.
- Use the shortcut menu.
- Use the keyboard shortcuts Ctrl+X to cut and Ctrl+V to paste.
- Use the Clipboard task pane.

EXERCISE 8-2 **Use the Standard Toolbar to Cut and Paste**

1. Open the file **NYmemo**.

> **NOTE:** The documents you create in this course relate to the Case Study about Duke City Gateway Travel, a fictional travel agency (see pages 1 through 4).

2. Drag over the date to select it and key the current date.

3. Select the text "*New York*" in the subject line of the memo.

4. Click the Cut button on the Standard toolbar to remove the text from the document and place it on the Clipboard. Notice the Clipboard item in the task pane.

NOTE: You may have to use the Toolbar Options button to locate the Cut, Copy, and Paste buttons.

5. Position the insertion point to the left of "Night on the Town" in the Subject line to indicate where you want to insert the text.

6. Click the Paste button to insert *"New York"* in its new location. The Paste Options button appears below the pasted text and the Clipboard item remains in the task pane.

7. Move the I-beam over the Paste Options button . (The I-beam will change to an arrow when it passes over the Paste Options button.) When you see the button's drop-down arrow, click to view the list of options.

8. Choose Keep Text Only. This option lets you paste just the text, without the italic formatting.

NOTE: The Paste Options button is available to make sure the text you paste has the type of formatting you want. You could also choose Match Destination Formatting to remove the italic formatting and match the text around the pasted text.

E X E R C I S E   **8-3**   **Use the Shortcut Menu to Cut and Paste**

---

1. Select the paragraph near the bottom of the document that begins "All the hotels are." Include the paragraph mark on the blank line following the paragraph.

FIGURE 8-2
Using the shortcut menu to cut

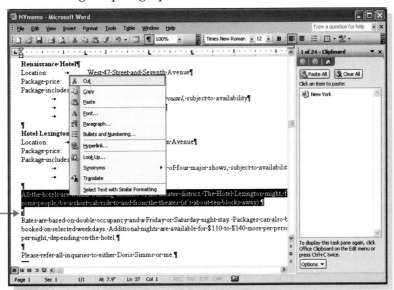

Include this blank line when moving an entire paragraph.

2. Point to the selected text and right-click to display the shortcut menu.

3. Click Cut. The item is added to the Clipboard task pane.

 NOTE: Each new item you cut (or copy) is added to the top of the Clipboard task pane.

4. Position the I-beam to the left of "Please refer" on the last line. Right-click and choose Paste from the shortcut menu. The paragraph moves to its new location and the Paste Options button 🗐 appears below the pasted text.

FIGURE 8-3
Using the shortcut menu to paste

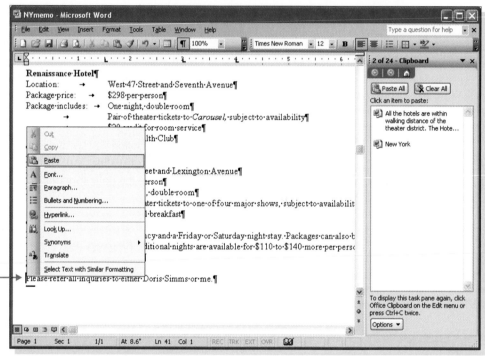

Click here and paste the paragraph.

E X E R C I S E   8-4   Use Keyboard Shortcuts to Cut and Paste

---

If you prefer using the keyboard, you can press Ctrl+X to cut text and Ctrl+V to paste text. You can also use Ctrl+Z to undo an action. The location of these shortcut keys is designed to make it easy for you to move your mouse with your right hand while you press command keys with your left hand.

1. Select the first sentence in the paragraph that begins "All the hotels." Don't forget to include the period. Press Ctrl+X to cut the text. A new item appears in the task pane.

2. In the same paragraph, position the insertion point just before the paragraph mark. Press Ctrl+V to paste the text.

3. Press Ctrl+Z to undo the paste. Press Ctrl+Z again to undo the cut. (Remember, you can also click the Undo button ⟲⋅ to undo actions.) Notice that the Clipboard item remains in the task pane.

 NOTE: The Cut, Copy, Paste, and Undo commands are also available from the Edit menu.

EXERCISE 8-5 **Use the Office Clipboard to Paste**

Each time you cut text in the previous exercises, a new item was added to the Office Clipboard. You can paste that item directly from the task pane.

1. Select all the information that goes with the "Peabody Hotel," including the title "Peabody Hotel" and the blank line that follows the hotel information.

2. Cut this text, using the Cut button 🔪 on the Standard toolbar. The text is stored as a new item at the top of the Clipboard task pane.

3. Position the insertion point to the left of the paragraph that begins "Hotel Lexington."

4. Click the task pane item for the Peabody Hotel that you just cut. (Do not click the drop-down arrow.) This pastes the text at the location of the insertion point.

5. Press Ctrl+Z to undo the paste. Press Ctrl+Z again to undo the cut. The Clipboard item remains in the task pane.

6. Point to this Clipboard item in the task pane and click the drop-down arrow that appears to its right.

7. Choose Delete from the list to delete the item from the Clipboard.

NOTE: Choosing the Paste option from the drop-down list pastes that item, just like clicking directly on the item. The Paste All button 📋 Paste All on the Clipboard task pane is used to copy all Office Clipboard items to the location of the insertion point.

Moving Text by Dragging

You can also move selected text to a new location by using the *drag-and-drop* method. Text is not transferred to the Clipboard when you use drag-and-drop.

EXERCISE | 8-6 | **Use Drag-and-Drop to Move Text**

1. Select the paragraph beginning "All the hotels," including the blank line below the paragraph.

2. Point to the selected text. Notice that the I-beam changes to a left-pointing arrow.

3. Click and hold down the left mouse button. The pointer changes to the drag-and-drop pointer. Notice the dotted insertion point near the tip of the arrow and the dotted box at the base of the arrow.

4. Drag the pointer until the dotted insertion point is positioned to the left of the line beginning "Rates are based on." Release the mouse button. The paragraph moves to its new location and the Paste Options button appears.

FIGURE 8-4
Drag-and-drop
pointer

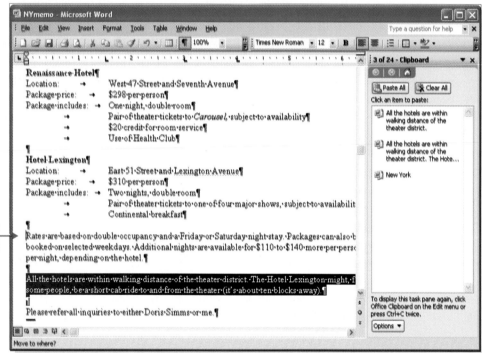

TIP: Use Cut and Paste to move text over long distances—for example, onto another page. Use drag-and-drop to move text short distances where you can see both the selected text and the destination on the screen at the same time.

Copying Text by Using Copy and Paste

Copying and pasting text is similar to cutting and pasting text. Instead of removing the text from the document and storing it on the Clipboard, you place a copy of it on the Clipboard.

There are several ways to copy and paste text. The most common methods are:

- Use the Copy and Paste buttons on the Standard toolbar.
- Use the shortcut menu.
- Use keyboard shortcut Ctrl+C to copy and Ctrl+V to paste.
- Use the Clipboard task pane.

EXERCISE   8-7   Use Copy and Paste

---

1. Under "Hotel Lexington," select the entire line that contains the text "Continental breakfast." Include the tab character to the left of the text and the paragraph mark to the right of the text. (Click the Show/Hide button ¶ to display formatting characters.)

2. Click the Copy button on the Standard toolbar to transfer a copy of the text to the Clipboard. Notice that the selected text remains in its original position in the document.

3. Position the insertion point to the left of the blank paragraph mark above "Hotel Lexington."

4. Right-click and choose Paste from the shortcut menu. A copy of the paragraph is added to the "Renaissance Hotel" package description and the Paste Options button appears.

5. Point to the Paste Options button. When you see the down arrow, click the button. Notice that the same options are available when you copy or paste text. Click in the document window to close the list of options and keep the source formatting.

6. Position the insertion point to the left of the blank paragraph mark above "Renaissance Hotel." Press Ctrl+V to paste the text into the "Peabody Hotel" package description.

EXERCISE **8-8** **Use the Office Clipboard to Paste Copied Text**

A new item is added to the Office Clipboard each time you copy text. You can click this item to paste the text into the document.

1. Under "Renaissance Hotel," select the entire line that contains the text "$20 credit for room service." Include the tab character to the left of the text and the paragraph mark to the right of the text.

2. Press Ctrl+C to copy this text.

3. Position the insertion point to the left of the blank paragraph mark above the paragraph that begins "All the hotels are within."

4. Click the Clipboard that contains the text, "$20 credit for room service." The Clipboard content is pasted into the document at the location of the insertion point.

NOTE: You can store up to 24 cut or copied items on the Office Clipboard. When the Clipboard is full and you cut or copy text, the bottom Clipboard item is deleted and the new item is added to the top of the task pane.

Copying Text by Dragging

To copy text by using the drag-and-drop method, press Ctrl while dragging the text. Remember, drag-and-drop does not store text on a Clipboard.

EXERCISE **8-9** **Use Drag-and-Drop to Copy Text**

1. Scroll until you can see the text under "Peabody Hotel" and "Renaissance Hotel."

2. Under "Package includes" for the Renaissance Hotel, select the text after "*Carousel*" from the comma through the word "availability." Do not include the paragraph mark.

3. While pressing Ctrl, drag the selected text to the immediate right of the word "*King*" in the "Peabody Hotel" section. The plus (+) sign attached to the drag-and-drop pointer indicates the text is being copied rather than moved (see Figure 8-5 on following page).

FIGURE 8-5
Copying with the
drag-and-drop
pointer

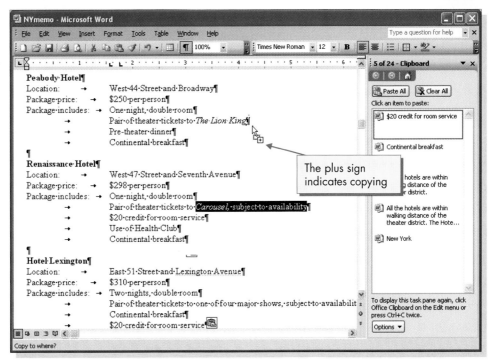

TIP: Dragging is not effective over long distances within a document. Try these alternative methods: To cut, select the text, hold down Ctrl, scroll as needed, and right-click where you want to paste the text. To copy, select the text, hold down Ctrl and Shift, scroll as needed, and right-click where you want to copy the text.

Using Smart Cut and Paste

You may already have noticed that when you delete, cut, move, or paste text, Word automatically adjusts the spacing between words. For example, if you cut a word at the end of a sentence, Word automatically deletes the leftover space. If you paste a word between two other words, Word automatically adds the needed space as part of its Smart Cut and Paste feature.

NOTE: The Smart Cut and Paste option is turned on by default. To check that your computer has it turned on, choose Options from the Tools menu and click the Edit tab. The Smart Cut and Paste box should be checked.

EXERCISE **8-10** **Observe Smart Cut and Paste**

1. In the last paragraph of the document, double-click the word "either." The space after the word is selected along with the word. Delete the word.

2. Select the text "or me" at the end of the sentence, not including the space before or the period after the text. Click the Cut button ⬚ . The selected text is removed and Word removes the extra space before the period.

3. With the insertion point positioned at the end of the sentence, before the period, click the Paste button ⬚ to reinsert the cut text. The needed space is inserted along with the text.

4. Before saving the document, change the top margin to 2 inches and insert your own reference initials at the bottom.

5. Switch to Print Preview and display both pages of the document.

6. Locate and click the Shrink to Fit button ⬚ . The document is reduced to a one-page document.

7. Close the Print Preview window.

8. Save the document as *[your initials]***8-10** in a new folder for Lesson 8.

9. Click the Clear All button ⬚ on the Office Clipboard to clear all items. Click the Close button ⬚ on the task pane to close the Office Clipboard.

10. Print and close the document.

Working with Multiple Document Windows

In Word, you can work with several open document windows. Working with multiple windows makes it easy to compare different parts of the same document or to move or copy text from one document to another.

EXERCISE   **8-11**   **Split a Document into Panes**

Splitting a document divides it into two areas separated by a horizontal line called the *split bar.* Each of the resulting two areas, called *panes,* has its own scroll bar.

To split a screen, choose <u>S</u>plit from the <u>W</u>indow menu or use the split box at the top of the vertical scroll bar.

1. Open the file **Temps**.

2. Open the <u>W</u>indow menu and choose <u>S</u>plit. A gray bar appears along with the split pointer ⬚ .

3. Move your mouse up or down (without clicking) until the gray bar is just below the paragraph that begins "To answer."

FIGURE 8-7
Splitting a
document into
two panes

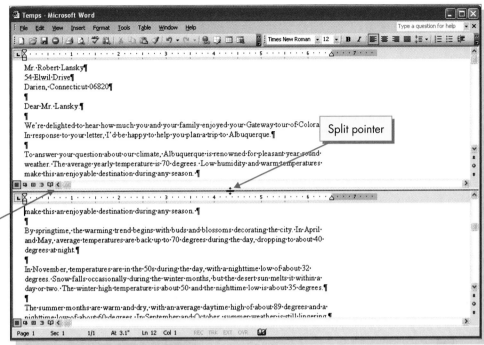

Split pointer

Split bar

4. Click the left mouse button to set the split. The document divides into two panes, each with its own ruler and scroll bar.

5. To change the split position, move the mouse pointer over the split bar (between the top and bottom panes) until you see the split pointer and a ScreenTip that says "Resize." Then, drag the bar up one paragraph.

TIP: To see more of each document, you can hide the rulers by clicking <u>R</u>uler on the <u>V</u>iew menu.

6. To remove the split bar, move the mouse pointer over it. When you see the split pointer, double-click. The split bar is removed.

7. Position the pointer over the *split box*—the thin gray rectangle at the top of the vertical scroll bar.

FIGURE 8-6
Double-click the
split box to create
two window panes.

Split box

8. When you see the split pointer ═, double-click. Once again, the document is split into two panes. (You can also remove the split bar by choosing Remove <u>S</u>plit from the <u>W</u>indow menu.)

EXERCISE **8-12** **Move Between Panes to Edit Text**

After you split a document, you can scroll each pane separately and easily move from pane to pane to edit separate areas of the document. To switch panes, press F6 or click the insertion point in the pane you want to edit.

1. Press F6. The insertion point appears in the top pane. If it doesn't, press F6 again.

2. With the insertion point in the top pane, press F6 again to move the insertion point to the bottom pane.

3. Use the scroll bar in the bottom pane to scroll to the top of the document. Both panes should now show the inside address.

4. In the bottom pane, change the street address to "55 Elwil Drive" and the state to "CT." Notice that the changes also appear in the top pane.

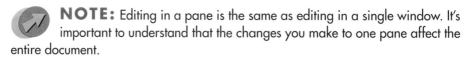 **NOTE:** Editing in a pane is the same as editing in a single window. It's important to understand that the changes you make to one pane affect the entire document.

5. In the bottom pane, scroll until the paragraph beginning "The summer months" is displayed. Click within the top pane and scroll until the two paragraphs beginning "By springtime" and "In November" are both displayed.

6. Go back to the bottom pane. Select the paragraph beginning "The summer months" and click the Cut button 🔪 . (Remember to include the blank line after the paragraph when selecting it.)

7. Move to the top pane, position the insertion point to the left of "In November," and click the Paste button 📋 . The paragraph is moved from one part of the document to another.

8. Drag the split bar to the top of the screen. This is another way to remove the split bar. The document is again displayed in one pane.

9. Apply the correct letter formatting to the document by adding the date, your reference initials, and an enclosure notation. Use the correct spacing between all letter elements (don't forget a 2-inch top margin), and insert nonbreaking spaces where appropriate.

 TIP: See Appendix B: "Standard Forms for Business Documents" for standard business-letter formatting.

10. Save the document as *[your initials]*8-12 in your Lesson 8 folder.

11. Print and close the document.

EXERCISE 8-13 **Open Multiple Documents**

In addition to working with window panes, you can work with more than one document file at the same time. This is useful if you keyed text in one document that you want to use in a second document.

1. Display the Open dialog box. Simultaneously open the noncontiguous files **Albuquer** and **Summer**. To do this, click Albuquer once, press Ctrl, and click Summer once. With both files selected, click Open.

> **NOTE:** Noncontiguous files are files that are not listed consecutively. You can open several noncontiguous files at the same time if you keep Ctrl pressed while selecting additional files.

2. Open the Window menu and notice that the two open files are listed at the bottom of this menu. The active file has a check next to it. Switch documents by clicking the file that is not active.

3. Press Ctrl + F6 to switch back.

4. Look at the taskbar at the bottom of your screen. Notice the two buttons that contain the names of your open documents. The highlighted Albuquer button shows that it is the active document. Click the Summer button to activate that document.

FIGURE 8-8
Window menu

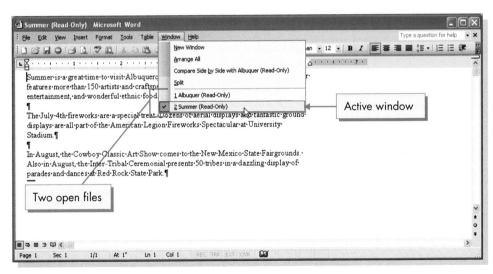

> **NOTE:** Be careful when you use the taskbar buttons to switch between documents. If you click the active document's highlighted taskbar button, you minimize that document. You can restore a minimized document by clicking its taskbar button.

5. From the <u>W</u>indow menu, choose <u>A</u>rrange All to view both documents at the same time. The two documents appear one below the other.

FIGURE 8-9
Two documents
displayed on
one screen

Active window →

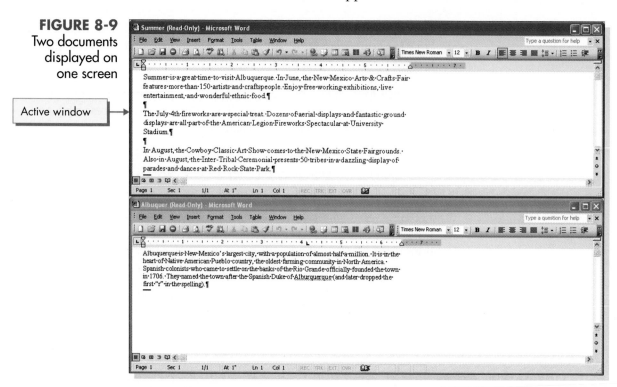

6. Press Ctrl+F6 to switch between documents. Press Ctrl+F6 again. Notice that the active window—the one containing the insertion point—has a darker blue (highlighted) Title bar.

7. Press Shift and choose <u>C</u>lose All from the <u>F</u>ile menu to close Summer and Albuquer.

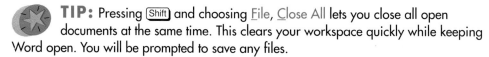

TIP: Pressing Shift and choosing <u>F</u>ile, <u>C</u>lose All lets you close all open documents at the same time. This clears your workspace quickly while keeping Word open. You will be prompted to save any files.

8. Click the Maximize button 🔲 to maximize the Word window.

9. Simultaneously open the four files, **Budget**, **Business**, **Caliente**, and **CarRent**, by accessing the <u>O</u>pen dialog box. If the files are contiguous, select the first file, Budget, and then press Shift and select the last file, CarRent. All four contiguous files should be highlighted. Click <u>O</u>pen. If they are noncontiguous, select the first file, Budget, and then press Ctrl and select the other three files. Click <u>O</u>pen.

10. Choose <u>A</u>rrange All from the <u>W</u>indow menu to display all four documents simultaneously.

11. Close Budget by clicking the Close button ⊠ on its Title bar. Close Caliente the same way, leaving only CarRent and Business on the screen.

EXERCISE 8-14 Rearrange and Resize Document Windows

You can rearrange the open documents in Word by using basic Windows techniques for minimizing, maximizing, restoring, and sizing windows.

1. Click the CarRent Title bar and drag this document's window to the top of the screen. Click the Maximize button 🗖 for CarRent. Click the Close button ⊠ for CarRent.

2. Minimize the Business window by clicking its Minimize button 🗕 . The document disappears from view. The Business button is on the taskbar, indicating that Word is still running.

3. Restore the Business document for viewing by clicking its taskbar button.

4. Drag a corner of the window's border diagonally out a few inches to make the window a different size.

5. Click the Maximize button 🗖 in the Business window to return the window to full screen.

6. Click the Close button ⊠ for Business. Remember, the Close button for a single open document appears just below the Close button for Word.

Moving and Copying Text among Windows

When you want to copy or move text from one document to another, you can work with either multiple (smaller) document windows or full-size document windows. Either way, you can use cut and paste or copy and paste. If you work with multiple windows, you can also use drag-and-drop. To use this technique, you must display both documents at the same time.

EXERCISE 8-15 Copy Text from One Document to Another by Using Copy and Paste

When moving or copying text from one document into another, the Paste command pastes text in the format of the document from which it was cut or copied. To control the formatting of pasted text, you can use the Paste Options button 🗐 or the Paste Special function. In this Exercise, you will use the Paste Special function to paste text without formatting.

1. Open the files **Albuquer**, **Climate**, and **Summer**. Click the Climate button on the taskbar to make it the active document.

2. In Climate, select the entire document and change the font to 14-point Arial. Select the first paragraph, beginning "Albuquerque is renowned," and click the Copy button 🖹.

3. Make Albuquer the active document.

4. Start a new paragraph at the end of the document. Click the Paste button 🖹 to insert the text copied from Climate. Notice the format of the new text does not match the format of the current document.

5. Click the Undo button 🔄 to remove the new text.

6. From the <u>E</u>dit menu, choose Paste <u>S</u>pecial and select Unformatted Text. Click OK. Now the format of the new text matches the format of the current document. (It may be necessary to expand the menu to display all options.)

7. Using the <u>W</u>indow menu, activate Climate again. Close this document without saving it.

> **TIP:** You can insert an entire file into the current document by using the Insert File command. Move the insertion point to the place in the document where you want to insert the file. Then from the <u>I</u>nsert menu, choose File and double-click the filename. The text from the entire file is inserted at the insertion point.

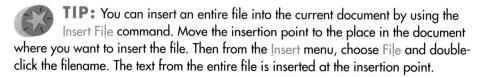

EXERCISE **8-16** **Move Text from One Document to Another by Using Drag-and-Drop**

1. Arrange the two open documents (Albuquer and Summer), so they are both displayed.

2. In the Albuquer window, move to the bottom of the document. Insert enough blank lines to prepare to start a new paragraph.

3. In the Summer window, select the first paragraph, which begins "Summer is a great time," and drag it to the bottom of the Albuquer window. (See Figure 8-10 on the next page.)

4. In the Summer window, select, and then cut the remaining text. Close this document without saving it.

5. Maximize the Albuquer document window and paste the text at the end of the document. Correct the spacing between paragraphs (if you have extra paragraph marks, for example).

6. At the top of the document, add the all-caps title **ABOUT ALBUQUERQUE**, formatted as 14-point bold centered. Add 24 points of paragraph spacing after the title. Change the top margin to 2 inches.

7. Save the document as *[your initials]***8-16** in your Lesson 8 folder; then print and close it.

FIGURE 8-10
Dragging a
paragraph between
document windows

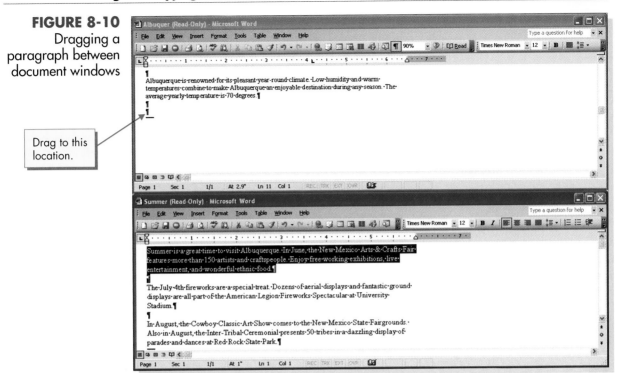

Drag to this
location.

USING ONLINE HELP

You've seen how the Office Clipboard can store cut or copied text from 24 separate sources. The Paste All button copies the contents of every Clipboard to the location of the insertion point. This enables you to collect 2 to 24 items from nonadjacent locations and insert them as a group in a new location. Another Word feature lets you remove multiple items from different sources and insert the new items as a group in a new location. This feature is called the *Spike*.

Use Help to find out more about the Spike:

1. Key **spike** in the Ask a Question text box and press [Enter].
2. Click the first topic, Use the Spike to move text and graphics from nonadjacent locations.
3. Click Spike to see a definition. Click anywhere in the definition to close it and review the rest of the help topic.
4. If desired, click the Print button in the Help window to print the topic.
5. Close the Help window.

> **NOTE:** There are some important differences between the Spike and the Office Clipboard. The Spike removes text or graphics from the original

location, and the Office Clipboard enables you to copy text or graphics from the original location (the Clipboard task pane). The Spike can hold more than 24 items. You can use the Spike only within Word, but the Office Clipboard is available in any Microsoft Office application.

LESSON Summary

➤ The most important tool for moving and copying text is the Clipboard, which is a temporary storage space.

➤ When you display the Clipboard task pane, you are activating the Office Clipboard, which can store up to 24 cut or copied items. With the Clipboard task pane open when you cut or copy text, the text appears as a new item in the task pane.

➤ You move text by cutting and pasting—cut the text from one location and paste it to another.

➤ Copy and paste is similar to cut and paste, but instead of removing the text from the document, you place a copy of it on the Clipboard.

➤ There are many methods for cutting, copying, and pasting text. Use buttons on the Standard toolbar, keyboard shortcuts, or the shortcut menu. Use the Clipboard task pane to paste stored text items.

➤ Use the Paste Options button 📋▾ to control the formatting of pasted text.

➤ You can use the drag-and-drop method to copy or move text from one location to another in a document or between documents.

➤ Split a document into panes to compare different parts of the document or to cut or copy text from one part of the document to another. Use the Window menu or the split box above the vertical scroll bar to split a document.

➤ Open multiple documents and arrange them to fit on one screen to move or copy text from one document to another.

LESSON 8	Command Summary		
FEATURE	**BUTTON**	**MENU**	**KEYBOARD**
Open Office Clipboard		Edit, Office Clipboard	Ctrl + C twice
Cut	✂	Edit, Cut	Ctrl + X

continues

LESSON 8   Command Summary   *continued*

FEATURE	BUTTON	MENU	KEYBOARD
Copy		Edit, Copy	Ctrl + C
Paste		Edit, Paste	Ctrl + V
Split a document		Window, Split	
Switch panes			F6
Arrange multiple windows		Window, Arrange All	
Next window		Window, [filename]	Ctrl + F6
Previous window		Window, [filename]	Ctrl + Shift + F6

Concepts Review

Each of the following statements is either true or false. Indicate your choice by circling T or F.

T F   **1.** Drag-and-drop stores text on the clipboard.

T F   **2.** The content of the Office Clipboard is replaced each time you copy or cut text.

T F   **3.** The keyboard shortcut for cut is Ctrl+C.

T F   **4.** You can drag text between two documents when they are maximized.

T F   **5.** The Cut, Copy, and Paste commands are all available from the shortcut menu.

T F   **6.** The only difference between cut and copy is that selected text remains in the document after copying.

T F   **7.** When you move a paragraph, you should select the blank line following it to preserve proper line spacing.

T F   **8.** When a document is split into panes, you can remove the split by double-clicking the split bar.

Write the correct answer in the space provided.

1. How can you copy text without using the Clipboard?

_____

2. Which buttons on the Standard toolbar do you use to move text?

_____

3. What is the keyboard shortcut for moving between two window panes?

_____

4. For which command is Ctrl+V the keyboard shortcut?

_____

5. Which menu command displays all open documents at the same time?

_____

6. What shortcut key combination can you press if you drag a sentence to the wrong location?

_____

7. What's different about the drag-and-drop pointer when you are copying, as opposed to moving, text?

_____

8. Where is the split box located?

_____

CRITICAL THINKING

Answer these questions on a separate page. There are no right or wrong answers. Support your answers with examples from your own experience, if possible.

1. Many people once wrote first-draft documents by hand or by using a typewriter. In either case, people would then literally cut and paste pieces of their document together and type a final draft. Some people say that word processing—specifically moving and copying text—has caused a basic change in the way people write. What do you think? Explain your answer.

2. You learned different methods for moving text by using cut and paste. You also learned the drag-and-drop method of moving text. Which method do you prefer? Why?

Skills Review

EXERCISE 8-17

Move text to a new location by using cut and paste and by dragging.

1. Open the file **Baggage**.
2. Display the Office Clipboard by following these steps:
   a. From the Edit menu, choose Office Clipboard.
   b. Click Clear All if there are any items in the task pane.

3. Use keyboard shortcuts to move the last paragraph by following these steps:

   a. Scroll to see the last two paragraphs.
   b. Select the last paragraph, which begins "Exclusions."
   c. Press Ctrl+X to cut the paragraph.
   d. Position the insertion point to the left of the word "Terms" in the preceding paragraph.
   e. Press Ctrl+V to paste the paragraph. Press Enter to add a line space between the two paragraphs.

4. Start a new paragraph at the end of the document and key in italics **Additional coverage can be purchased for**.

5. In the paragraph that begins "Exclusions," use the toolbar to cut and paste text by following these steps:

   a. Select the text "household furniture or furnishings; contact lenses, money, securities, tickets, or documents;" (including the semicolon).
   b. Click the Cut button .
   c. Position the insertion point at the end of the last paragraph.
   d. Click the Paste button .

6. Control the formatting of the pasted text by following these steps:

   a. Point to the Paste Options button .
   b. When you see the down arrow, click and choose the option Match Destination Formatting.

7. Correct the end-of-sentence punctuation in the last paragraph and change "or" to "and" in both cases where "or" appears.

8. Switch the information for paragraphs "A." and "B." by using the Office Clipboard:

   a. Select the information associated with paragraph "A." Start with the text "Baggage Loss or Damage," and include the period at the end of the sentence (don't include the paragraph mark).
   b. Click the Cut button to cut the text and place it on the Clipboard.
   c. Select the information associated with paragraph "B." Start with the text "Baggage Delay," and include the period at the end of the sentence (don't include the paragraph mark).
   d. Press Ctrl+X to cut the text and place it on the Clipboard.
   e. Position the insertion point to the right of the space after "A."
   f. Click the Clipboard item that contains the text "Baggage Delay" to paste the Clipboard contents at the location of the insertion point.
   g. Position the insertion point to the right of the space after "B."
   h. Click the Clipboard item that contains the text "Baggage Loss or Damage."

9. Select paragraphs "A." and "B." and the information associated with these two items. Do not include the blank line after paragraph "B." Click the Numbering button to change the paragraph formatting.

10. In the paragraph that begins "Terms," drag text by following these steps:

   a. Select the text "precious gems," (including the comma and the space character).

   b. Point to the selected text. Press and hold down the left mouse button to display the drag-and-drop pointer and the dotted insertion point.

   c. Drag the dotted insertion point to the left of the word "watches" in the same sentence and release the mouse button.

11. Clear and close the Office Clipboard by following these steps:

   a. Click Clear All to remove all Clipboard items.

   b. Click the task pane's Close button ⊠.

12. Set a 2-inch top margin. Center the title of the document, change it to uppercase, and add another blank line below it.

13. Save the document as *[your initials]*8-17 in your Lesson 8 folder.

14. Print and close the document.

EXERCISE 8-18

Copy text by using copy and paste and by dragging.

1. Open the file **Pueblos1**.

2. Display the Office Clipboard. Clear the Office Clipboard if it contains any items.

3. Start a new paragraph after the paragraph that begins "The San Juan Pueblo" by keying the word **Note:**.

4. Use the Office Clipboard to copy text to the new paragraph by following these steps:

   a. Select the text "The San Juan Pueblo."

   b. Click the Copy button to copy the text to a Clipboard.

   c. Position the insertion point to the right of "Note:".

   d. Paste the text by clicking the appropriate Clipboard.

5. After the copied text, complete the sentence by keying **also runs the Tiwa Indian Restaurant.** Italicize the sentence, including the word "Note."

6. Scroll to the paragraph that begins "The Santa Clara Pueblo." Start a new paragraph below it by keying **Note:**.

7. Use the drag-and-drop method to copy "The Santa Clara Pueblo" after "Note:" by following these steps:

   a. Select the text "The Santa Clara Pueblo."

      b. Point to the text. Press and hold down (Ctrl); then click and hold down the left mouse button.

      c. Drag the dotted insertion point to the right of the word "Note:" and release both the mouse button and (Ctrl).

8. Complete the sentence by keying **offers walking tours of the Puye Cliff Dwellings.** Italicize the sentence, including the word "Note".

9. At the top of the document, press (Enter) three times.

10. Use keyboard shortcuts to copy text to create a title by following these steps:

      a. Select the text "Native American pueblos in New Mexico" in the first sentence.

      b. Press (Ctrl)+(C) to copy the text to the Clipboard.

      c. Position the insertion point at the first paragraph mark.

      d. Press (Ctrl)+(V) to paste the text.

11. Clear and close the Office Clipboard.

12. Spell-check the document.

13. Format the title as centered, bold, all caps.

14. Set a 1.5-inch top margin and 1-inch left and right margins.

15. View the document in Print Preview and use the Shrink to Fit button 🔲 to fit the document on one page.

16. Save the document as *[your initials]*8-18 in your Lesson 8 folder.

17. Print and close the document.

EXERCISE 8-19

Split a document into panes.

1. Open the file **Special1**.

2. Split the document into two panes by double-clicking the split box above the vertical scroll bar.

3. In the top pane, use the scroll bar to display the paragraph that begins "Our local-area tours." Change "popular" to "sought after."

4. In the bottom pane, scroll to the list that begins with "Hawaii." Add "Australia" to the top of the list. Sort the list alphabetically.

 REVIEW: Use Sort from the Table menu to sort the list.

5. In the bottom pane, add 6 points of paragraph spacing before the top item in the list. Using the top pane, repeat this paragraph spacing for the text "The Great Sphinx."

6. Remove the split by double-clicking the split bar.

7. Scroll through the document to view the changes.

8. Add the bold, centered title **SPECIAL-INTEREST TOURS** to the document, adding the correct amount of space after the title.

9. Change the top margin to 2 inches.

10. Save the document as *[your initials]*8-19 in your Lesson 8 folder.

11. Print and close the document.

EXERCISE 8-20

Arrange windows to move and copy text.

1. Start a new document. Key the text shown in Figure 8-11 and format it as a standard business memo. Use single-spacing for the body of the memo and include today's date.

FIGURE 8-11

```
Memo to Susan Allen from Nina Chavez

Subject is Phoenix Convention Car Rentals
Regarding car rentals for the Phoenix convention next month, please
advise clients of the following rates and upgrade charges. Car rentals
are available for the entire stay and for additional days.

All rental agreements for this trip offer unlimited mileage.
```

2. Save the document as *[your initials]*8-20 in your Lesson 8 folder.

3. Simultaneously open the files **Memo3** and **CarRent**.

4. Arrange the documents by following these steps:

   a. Make sure Memo3 is the active document; then minimize it by clicking the Minimize button .

   b. Display the other two documents at the same time by selecting <u>A</u>rrange All from the <u>W</u>indow menu.

5. Drag a paragraph between documents by following these steps:

   a. In the *[your initials]*8-20 window, scroll to display the sentence beginning "All rental agreements for this trip."

   b. In the CarRent window, select the entire paragraph beginning "In the compact category" (include the blank line below the paragraph).

   c. Holding down Ctrl to copy, drag the paragraph to the *[your initials]*8-20 window until it precedes the paragraph beginning "All rental agreements."

6. Minimize CarRent and restore Memo3 by clicking its taskbar button.

7. Use <u>W</u>indow, <u>A</u>rrange All to display *[your initials]*8-20 and Memo3 at the same time, if necessary.

8. In Memo3, select the text beginning "Car rental upgrades" and the table following it. Copy and paste this text before the paragraph beginning "All rental agreements for this trip" in *[your initials]*8-20.

9. Close Memo3 without saving it. Maximize *[your initials]*8-20. Format the entire document in 12-point Times New Roman. Insert your reference initials at the bottom of the memo.

10. Save the document again. Then print and close it.

11. Close any other open documents without saving them.

 12. If the Word window is not maximized, click the Maximize button .

Lesson Applications

Move and copy text.

1. Open the file **Succeed**.
2. At the top of page 2, move the paragraph that begins "Businesses" to the first page, before the paragraph that begins "Churches and religious organizations."
3. At the end of the paragraph that begins "Sports clubs," key **Use your local directory for listings.** Copy the text you keyed to the end of each of the next three paragraphs.
4. Move the two paragraphs that begin "Recreation clubs" and "Singles clubs" before the paragraph that begins "Ethnic groups."
5. Copy the book title in the first line and use it to create a heading for the document. Format the heading as uppercase bold, no italics, centered, with 72 points of spacing before and 24 points of spacing after.
6. Change the font of the entire document to Arial.
7. Spell-check the document.
8. Save the document as *[your initials]*8-21 in your Lesson 8 folder.
9. Print and close the document.

Copy and move text in a memo.

1. Open the file **BikePrep**.
2. At the top of the document, insert a memo heading to the staff from you, using today's date. Move the first line in the document (which begins "Getting ready") to the subject line and make it italic. Set the top margin to 2 inches.
3. In the subject line, replace "bike race" with a copy of the words "Albuquerque Bike-a-Thon" found in the paragraph that begins "Here is." Match the destination formatting and apply the appropriate capitalization to the subject line.
4. Use copy and paste to replace "Bike-a-Thon" in the last sentence of the document with the words "Albuquerque Bike-a-Thon."
5. In the paragraph that begins "Once again," move the last two sentences to the end of the document, combining them with the paragraph that begins "As I'm sure you all know."

6. At the top of the document, combine the paragraphs that begin "Once again" and "Here is."

7. In the paragraph that begins "We need to," delete the words "We need to have a volunteer to" and capitalize the next word "go."

8. Format the three action paragraphs, beginning "Arrange," "Go through," and "Contact," as a numbered list.

9. Under the heading "<u>OTHER ISSUES</u>," make the following changes:

   ● Combine the first two paragraphs.

   ● In the second sentence, change the words "Tom will" to "He will."

   ● Move the sentence that begins "Tom's been working out" to a separate paragraph at the bottom of the document.

   ● At the beginning of that sentence, insert **PS:** followed by one space.

10. Format the two uppercase and underlined headings as uppercase bold (no underline) with 6 points of spacing after them.

 REVIEW: Remember to add your reference initials to the memo.

11. Save the document as *[your initials]*8-22 in your Lesson 8 folder.

12. Print and close the document.

EXERCISE 8-23

Copy text from multiple documents into an existing document.

1. Open the file **Vasquez**. Delete all the body text, leaving only the inside address, salutation, and closing. Format the letter as a standard business letter. Remember to add all the necessary letter elements and use proper spacing.

2. For the first paragraph of the new letter, key the text shown in Figure 8-12.

FIGURE 8-12

I'm looking forward to your visit here in October. You'll be staying at the Red River Hotel as our guest.

3. From the file **RedRiver**, copy the second paragraph, beginning "The hotel has 400 rooms," and paste it at the end of the paragraph you keyed in step 2. Delete the last sentence in the paragraph.

4. Start a new paragraph by keying **When you arrive, the balloon festival will be in full swing.**

5. From the file **Balloon**, copy the second paragraph and paste it at the end of the sentence you just keyed. In the fourth sentence of this paragraph, change the opening text from "A minimum of" to "Over."

6. Key the text shown in Figure 8-13 as a closing paragraph.

FIGURE 8-13

> Maria, I'm delighted that you're finally coming to Albuquerque! I know you're going to have a great time.

7. In the closing, add the title "President" below Susan Allen's name and add your reference initials.

8. Check for correct spacing and spell-check the document.

9. Save the document as *[your initials]*8-23 in your Lesson 8 folder. Then print and close it.

10. Close all other open documents without saving them. Maximize the Word window if needed.

EXERCISE 8-24   *Challenge Yourself*

Copy text from multiple documents and write paragraph headings.

1. Start a new document. Key the title **TRAVEL OFF THE BEATEN PATH** in 12-point Arial. Center the title and make it bold.

2. Insert two blank lines and key the text shown in Figure 8-14, including the corrections. Use 12-point Times New Roman (except where indicated) and single-spacing, and insert a blank line between paragraphs.

3. Save the document as a Web page named *[your initials]*8-24 in your Lesson 8 folder. Keep the document open and switch back to Normal view.

4. Copy all the text from files **Italy**, **Vienna**, and **England** to the end of *[your initials]*8-24.

TIP: Instead of opening each file and using the copy-and-paste method, you can use the Insert File command to insert all the text from each of the files. Move the insertion point to the bottom of the current document (be sure to leave a blank line above the insertion point). For each file, choose File from the Insert menu and double-click the filename.

FIGURE 8-14

Stroll Through the Streets of Budapest   (Bf Arial)

Stay in an elegant old Budapest inn, and wake up to a traditional
Hungarian breakfast. Take a leisurely walk through the city, visiting
the castle in city park, strolling the tree-lined boulevards, and
stopping at a sidewalk café for strudel and coffee. Tour the National
Museum, where the Crown of Hungary and other national treasures are
displayed. In the evening, after the cruise, you can try one of the
relaxing thermal baths offered in this charming city. Later, relax on
a sunset dinner cruise on the Danube River, viewing the magnificent
Gothic Parliament buildings, the 700-year-old Matthias Church, and
Castle Hill.

Explore the Turrets in Prague   (Bf Arial)

Prague is a city decked out with towers, turrets, steeples, and domes,
boasting a truly unique and fanciful skyline. While walking through "The City of a
Hundred Spires," you will feast your eyes on wonderful examples of
Medieval, Renaissance, Baroque, and Art Nouveau architecture. Guided
tours are available, taking you through the Hradcany Castle District,
Wenceslas Square, the Old Town, and the Jewish Quarter.

Experience local culture in Brittany   (Bf Arial)

Travel to the favorite vacation spot of the French. Brittany is filled
with its own style of old-world culture, beautiful beaches, secluded
coves, and quaint fishing villages. Stay in the medieval town of
Quimper, known for its rustic pottery. While touring this charming old
town, be sure to visit the farmers' market, the Museum of Breton Folk
Art, and the cathedral.

5. Make sure a blank line follows each new paragraph. Create a heading for
   each new paragraph. Use the same style (Arial bold) as the other headings
   in [your initials]8-24. A heading can be general (such as "The Treasures of
   Prague") or it can refer to a specific item in the paragraph (as in "Explore
   the Turrets in Prague").

6. Arrange the paragraphs and their headings in the following order:

    Budapest
    Prague
    Vienna
    England
    Brittany
    Florence
    Venice

7. Remove the blank line between each paragraph heading and the body text below it.

8. Add 72 points of spacing before the first line in the document (the title). Increase the size of the title to 14 points.

9. Spell-check the document. (Many of the proper names will not be in Word's dictionary. Check the names you keyed from Figure 8-14 against the spelling in the figure.)

10. Save the document again, print two pages per sheet, and close it. (Use the Print dialog box to print two pages per sheet.)

On Your Own

In these exercises, you work on your own, as you would in a real-life work environment. Use the skills you've learned to accomplish the task—and be creative.

EXERCISE 8-25
Write a short report proposing a change in your city, neighborhood, or school. Make the proposal at least three paragraphs long. At the end of the document, create a bulleted summary of the proposal, copying and pasting text from the proposal for the bulleted items. Save the document as *[your initials]*8-25 and print it.

EXERCISE 8-26
Copy text from the Internet about a person (present-day or historical) you admire. Use Paste Special to paste the text, without formatting, into a new document. Apply your own character and paragraph formatting. Save the document as *[your initials]*8-26 and print it.

EXERCISE 8-27
Write a summary about a TV show you have recently seen. Save it as *[your initials]*8-27a. Keep this document open and start a new document. Begin a letter to a friend, telling him/her about the TV show. Copy and paste or drag-and-drop text from the summary document into the letter. Save the document as *[your initials]*8-27b and print both documents.

LESSON 9

Find and Replace

OBJECTIVES

MICROSOFT OFFICE
SPECIALIST
ACTIVITIES
In this lesson:
  WW03S-1-3
  WW03S-3-1
See Appendix C.

After completing this lesson, you will be able to:

1. **Find text.**
2. **Find and replace text.**
3. **Find and replace special characters.**
4. **Find and replace formatting.**

 Estimated Time: 1¼ hours

When you create documents, especially long documents, you often need to review or change text. In Word, you can do this quickly by using the Find and Replace commands.

The *Find* command locates specified text and formatting in a document. The *Replace* command finds the text and formatting, and replaces it automatically with a specified alternative.

Finding Text

Instead of scrolling through a document, you can use the Find command to locate text or to move quickly to a specific document location.

Two ways to use Find are:

- From the Edit menu, choose Find.
- Press Ctrl+F.

You can use the Find command to locate whole words, words that sound alike, font and paragraph formatting, and special characters. You can search an entire document or only selected text and specify the direction of the search. In the following exercise, you use Find to locate all occurrences of the word "travel."

EXERCISE **9-1** **Find Text**

1. Open the file **Forman1**.
2. From the Edit menu, choose Find to open the Find and Replace dialog box. The Find tab is selected.
3. Delete any text in the Find what text box and key **travel**.
4. Click the More button, if it is displayed, to expand the dialog box.

FIGURE 9-1
Expanded Find and
Replace dialog box

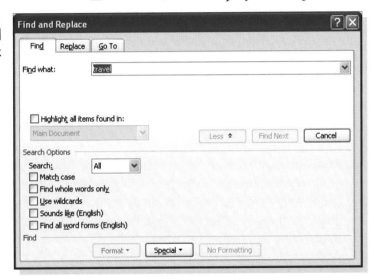

5. Click the No Formatting button (if it is active) to remove any formatting from previous searches. Then click the Less button. The dialog box should look like the one in Figure 9-2.

FIGURE 9-2
Using the
Find feature

6. Click Find Next. Notice that the first occurrence of "travel" found in the document is capitalized and italicized.

NOTE: To see more of the document text during a search, drag the dialog box by its Title bar to the bottom-right corner of the screen. It will move as you find different occurrences of "travel."

7. Continue clicking Find Next until you reach the end of the document. Notice that Word locates "travel" as a word and as text embedded in other words, such as "travelers."

8. Click OK in the dialog box that says Word finished searching the document.

9. Click Cancel to close the Find and Replace dialog box.

EXERCISE   **9-2**   **Find Text by Using the Match Case Option**

The Find command includes options for locating words or phrases that meet certain criteria. One of these options is Match case, which locates text that matches the case of text keyed in the Find what text box. The next exercise demonstrates how the Match case option narrows the search to locate "Travel" with an uppercase "T."

1. Move to the end of the document by pressing Ctrl+End. Position the insertion point to the right of "Chavez" in the closing.

2. From the Edit menu, choose Find. Click More to display an expanded dialog box that contains search options.

3. Key **Travel** in the Find what text box (or edit "travel" in the text box so it has an uppercase "T").

4. Click the Match case check box to select this option. Choose Up from the Search: drop-down list to reverse the search direction. Notice the Options that appear below the Find what text box.

FIGURE 9-3
Choosing
search options

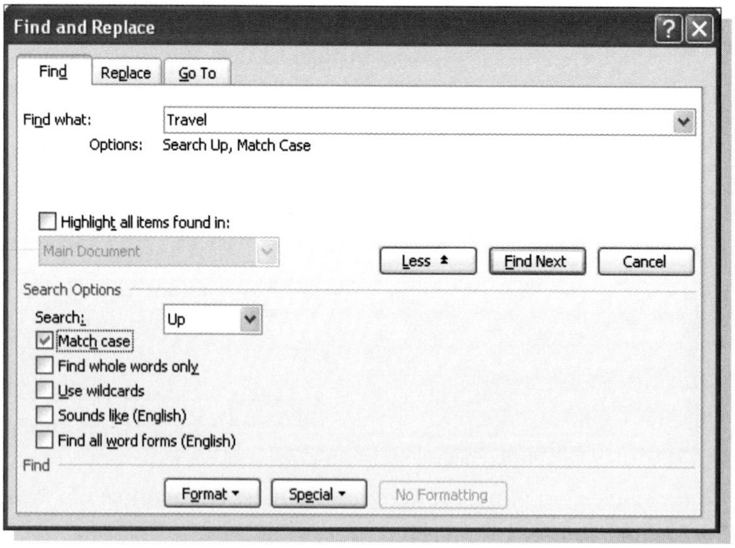

5. Click Less to collapse the dialog box. Click Find Next to begin the search. (If the dialog box is in your way, drag it to a preferred location.) Word ignores all occurrences of the word that do not match the search criteria.

6. Continue clicking Find Next until Word reaches the beginning of the document. Notice how the Match case option narrows the search.

7. Click No in the dialog box that asks if you want to continue searching.

8. Click Cancel to close the Find and Replace dialog box.

NOTE: The dialog box that appears when you end the search process is determined by the search direction and the position of the insertion point when you begin the search. When Word searches through the entire document, the dialog box tells you Word is finished searching and the insertion point returns to its original position. When you search from a point other than the top or bottom of the document and choose Up or Down as your search direction, Word asks if you want to continue the search. If you choose not to continue, the insertion point remains at the last occurrence found.

EXERCISE 9-3 Find Text by Using the Find Whole Words Only Option

The Find whole words only option is another way to narrow the search criteria. Word locates separate words, but not characters embedded in other words.

1. Move the insertion point to the beginning of the paragraph that starts "We have just celebrated." Press Ctrl+F to open the Find and Replace dialog box with the Find tab selected. Click More to expand the dialog box.

2. Click the down arrow next to the Find what text box and choose "travel" from the drop-down list. The last seven entries of the Find what text box are displayed in this list.

3. Click Find whole words only to select it, click Match case to clear it, and choose Down from the Search: drop-down list.

4. Click Less, and then click Find Next to begin the search. Word locates the word "travel," but not other word forms, such as "traveler."

5. Click Find Next several times.

6. Click Cancel to close the Find and Replace dialog box.

EXERCISE 9-4 Find Text by Using the Wildcard Option

You can use the Use wildcards option to search for text strings using special search operators. A *wildcard* is a symbol that stands for missing or unknown text. For example, the Any Character wildcard ? finds any character. Using it, a search for b?te would find both "bite" and "byte."

1. Position the insertion point at the beginning of the document. Open the Find and Replace dialog box with the Find tab displayed.

2. Display the expanded dialog box and click Use wildcards to select this option.

3. Select the text in the Find what text box and key **ca**.

4. Click the Special button and choose Any Character from the list. The code ^? is inserted.

FIGURE 9-4
Choosing a special
search operator

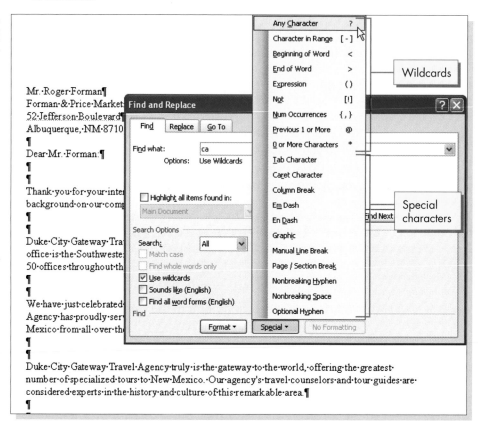

5. Choose All from the Search: drop-down list, if it is not already selected. Then click Less.

6. Click Find Next. The first occurrence appears in the word "location."

7. Continue clicking Find Next.

8. Click OK in the dialog box that says Word finished searching the document.

9. Click Cancel to close the Find and Replace dialog box.

EXERCISE 9-5 **Find Text by Using the Sounds Like Option**

To find a word that sounds similar to the search text but is spelled differently, or to find a word you don't know how to spell, use the Sounds like option. When you find the word, you can stop the search process temporarily to edit your document. Continue the search when you finish editing by clicking the scroll buttons at the bottom of the vertical scroll bar.

1. Move the insertion point to the beginning of the document and open the Find and Replace dialog box with the Find tab selected.

2. Key **insure** in the Find what text box. Expand the dialog box.

3. Click Sounds like to select this option, and choose Down from the Search: drop-down list. Click Less to reduce the dialog box.

4. Click Find Next. Word locates "insurance."

5. Click Find Next and notice that Word stops at "ensure," a word that sounds like "insure." Continue clicking Find Next.

6. Click OK when Word finishes checking the document.

7. Key **traveler** in the Find what text box.

8. Click Find Next until Word reaches "travellers" in the third paragraph (which begins "We have just").

9. Click Cancel to stop the search. The Find and Replace dialog box closes. Edit the document by deleting the second "l" in "travellers."

> **TIP:** You can also interrupt a search by clicking outside the Find and Replace dialog box, editing the document text, and then clicking the dialog box to reactivate it.

10. Continue the search by clicking the Next Find/Go To button ⏷ located at the bottom of the vertical scroll bar. As you continue the search, stop and edit every occurrence of "traveller."

FIGURE 9-5
Finding text without the Find and Replace dialog box

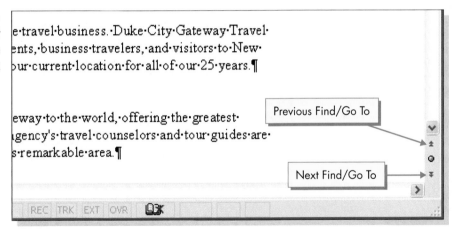

> **TIP:** After you initiate a find by using the Find and Replace dialog box, you can close the dialog box and use the Next Find/Go To button ⏷ and Previous Find/Go To button ⏶ to continue the search without having the dialog box in your way.

11. Click No in the dialog box that asks if you want to continue searching at the beginning. Check the document to make sure Word found all the incorrect occurrences of "traveller."

EXERCISE **9-6** **Find Formatted Text**

In addition to locating words and phrases, the Find command can search for text that is formatted. The formatting can include character formatting, such as bold and italic, and paragraph formatting, such as alignment and line spacing.

1. Position the insertion point at the beginning of the document. Choose Edit, Find.

2. Key **Duke City** in the Find what text box. Expand the dialog box and choose All from the Search: drop-down list. Click any checked search options to clear them.

3. Click the Format button and choose Font.

FIGURE 9-6
Format options

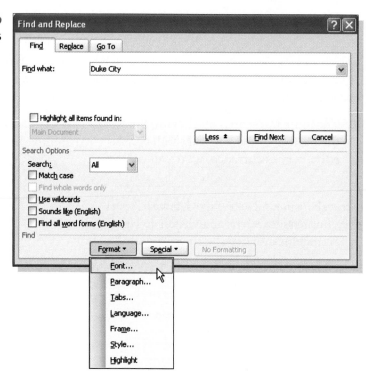

4. In the Find Font dialog box, choose Italic from the Font style list and click OK. Italic now appears below the Find what text box.

5. Click Less, and then click Find Next. Word locates "*Duke City.*"

6. Click Cancel to close the Find and Replace dialog box.

Finding and Replacing Text

The Replace command searches for specified text or formatting and replaces it with your specified alternative. You can replace all instances of text or formatting at once or you can find and confirm each replacement.

Two ways to replace text are:

● From the Edit menu, choose Replace.

● Press Ctrl+H.

EXERCISE   **9-7**   **Replace Text by Using Find Next**

---

1. Position the insertion point at the beginning of the document and choose Edit, Replace. The Replace tab is now selected in the dialog box.

2. Key **Travel Agency** in the Find what text box. Expand the dialog box and click the No Formatting button to remove formatting from previous searches. Make sure no options under Search: are selected.

3. Press Tab to move to the Replace with text box, and key **Travel**. Click the No Formatting button if it is active. Click Match case. Make sure it is the only search option checked.

> **NOTE:** Remember, pressing Tab in a dialog box moves the insertion point from one text box to another and highlights existing text. Pressing Enter executes the dialog box command.

FIGURE 9-7
Replacing text

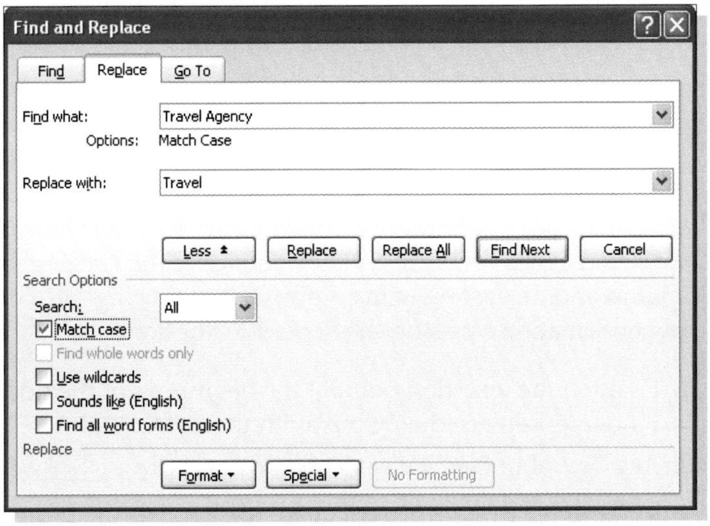

Find and Replace dialog box showing:
- Tabs: Find, Replace, Go To
- Find what: Travel Agency
- Options: Match Case
- Replace with: Travel
- Buttons: Less, Replace, Replace All, Find Next, Cancel
- Search Options:
  - Search: All
  - ☑ Match case
  - ☐ Find whole words only
  - ☐ Use wildcards
  - ☐ Sounds like (English)
  - ☐ Find all word forms (English)
- Replace: Format ▾, Special ▾, No Formatting

4. Adjust the position and size (click <u>L</u>ess) of the dialog box so you can see the document text. Click <u>F</u>ind Next. Click <u>R</u>eplace to replace the first occurrence of "Travel Agency" with "Travel."

5. Continue to click <u>R</u>eplace until Word reaches the end of the document.

6. Click OK when Word finishes searching the document.

7. Close the Find and Replace dialog box.

EXERCISE   9-8   **Replace Text by Using Replace All**

The Replace <u>A</u>ll option replaces all occurrences of text or formatting in a document without confirmation.

1. Move the insertion point to the beginning of the document and press [Ctrl]+[H] to open the Find and Replace dialog box with the Re<u>p</u>lace tab selected.

2. Key **Duke City** in the Fi<u>n</u>d what text box. Press [Tab] and key **Albuquerque** in the Replace wi<u>t</u>h text box.

3. Expand the dialog box, clear the Matc<u>h</u> case check box, and click Replace <u>A</u>ll. Word will indicate the number of replacements made.

4. Click OK and close the Find and Replace dialog box. Duke City Gateway Travel is now called "Albuquerque Gateway Travel" throughout the document.

5. Click the Undo button on the Standard toolbar to undo the Replace All command.

> **NOTE:** After replacing text or formatting, you can always undo the action. If you used Replace <u>A</u>ll, all changes are reversed at once. If you used <u>R</u>eplace, only the last change is reversed, but you can undo the last several changes individually by selecting them from the Undo drop-down list.

EXERCISE   9-9   **Delete Text with Replace**

You can also use the Replace command to delete text automatically. Key the text to be deleted in the Fi<u>n</u>d what text box and leave the Replace wi<u>t</u>h text box blank. You can find and delete text with confirmation by using the <u>F</u>ind Next option or without confirmation by using the Replace <u>A</u>ll option.

1. Position the insertion point at the beginning of the document and open the Find and Replace dialog box with the Re<u>p</u>lace tab selected.

2. Key **Travel** in the Fi<u>n</u>d what text box and press [Spacebar] once. The space character is not visible in the text box.

3. Press [Tab] to move to the Replace wi<u>t</u>h text box and press [Delete] to remove the previous entry.

4. Check the Match case option and click Replace All.

5. Click OK and close the dialog box. The word "Travel" followed by a space is deleted from the agency name throughout the document, but the word "Travel" followed by a punctuation mark is not deleted.

6. Scroll to the letter closing. Before you initiated Replace All, the closing title was "Senior Travel Counselor." Because Replace All deleted "Travel" from the title, key **Travel** between "Senior" and "Counselor" to restore the correct title. In this instance, you can see why using the Replace command with confirmation is often preferable to using Replace All.

7. Save the document as *[your initials]*9-9 in a new folder for Lesson 9. Leave the document open for the next exercise.

> **TIP:** The last option in the Find and Replace dialog box is Find all word forms. Use this option to find different forms of words and replace the various word forms with comparable forms. For example, if you key "travel" in the Find what text box and key "visit" in the Replace with text box, Word replaces "travel" with "visit" and "traveled" with "visited." Use Replace, rather than Replace All, when you choose this option to verify each replacement and ensure that correct word forms are used.

Finding and Replacing Special Characters

The Find and Replace features can search for characters other than ordinary text. Special characters include paragraph marks and tab characters. Special characters are represented by codes that you can key or choose from the Special drop-down list.

EXERCISE 9-10 Find and Replace Special Characters

1. Click the Show/Hide ¶ button ¶ on the Standard toolbar to display special characters in the document if they are not showing.

2. Position the insertion point at the top of the document. Open the Find and Replace dialog box with the Replace tab selected. Expand the dialog box, if it is not already.

3. Click the Special button and choose Paragraph Mark. A code (^p) is inserted in the Find what text box. Add two additional paragraph mark codes in the Find what text box to search for three consecutive paragraph marks in the document. (Use the Special drop-down list or key **^p^p**.)

4. Move to the Replace with text box and insert two paragraph mark codes.

5. Clear any Search Options check boxes and click Less.

6. Click Find Next. Word locates the extra paragraph mark after the salutation of the letter. (See Figure 9-8 on the next page.)

FIGURE 9-8
Replacing special
characters

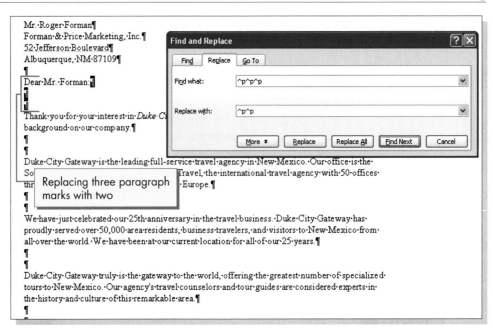

7. Click Replace. Notice the elimination of the extra paragraph mark. Continue to click Replace for each paragraph mark until you reach the paragraph marks just after "Sincerely."

8. Close the Find and Replace dialog box. The document paragraphs are now correctly spaced.

TABLE 9-1

Find and Replace Special Characters

FIND OR REPLACE	SPECIAL CHARACTER CODE TO KEY
Paragraph mark (¶)	^p (must be lowercase)
Tab character (→)	^t (must be lowercase)
Any character (Find only)	^?
Any digit (Find only)	^#
Any letter (Find only)	^$
Column break	^n
Clipboard contents (Replace only)	^c
Em dash	^+
En dash	^=
Field (Find only)	^d
Footnote mark (Find only)	^f

continues

TABLE 9-1 Find and Replace Special Characters *continued*

FIND OR REPLACE	SPECIAL CHARACTER CODE TO KEY
Graphic (Find only)	^g
Manual line break	^l
Manual page break	^m
Nonbreaking hyphen	^~
Nonbreaking space	^s
Section break (Find only)	^b
White space (Find only)	^w

TIP: If the text you want to find or use as a replacement already exists in a document, you can use the Clipboard to avoid re-keying it. First, copy the text to the Clipboard. Next, paste the contents of the Clipboard into the Find what or Replace with text box by pressing Ctrl + V.

Finding and Replacing Formatting

Word can search for and replace both character and paragraph formatting. You can specify character or paragraph formatting by clicking the Format button in the Find and Replace dialog box, using keyboard shortcuts, or using toolbar buttons.

EXERCISE 9-11 Find and Replace Character Formatting

1. Position the insertion point at the top of the document and open the Find and Replace dialog box with the Replace tab selected. Expand the dialog box.
2. Key **Duke City Gateway** in the Find what text box. Press Tab and delete the text in the Replace with text box.
3. Click the Format button and choose Font. Choose Bold for Font style and click OK.
4. Click Replace All.
5. Click OK when Word finishes searching the document and close the dialog box. "Duke City Gateway" appears bold throughout the document.
6. Reopen the Find and Replace dialog box with the Replace tab selected.
7. Highlight the text in the Find what text box, if it is not already. Click the Format button and choose Font. Choose Bold and click OK.
8. Press Tab to move the insertion point to the Replace with text box. Click No Formatting to clear existing formatting.

9. Click the Format button and choose Font. Choose the Not Bold style and click OK.

10. Press `Ctrl`+`I` (the keyboard shortcut for italic text). Now the format for the Replace with text box is Not Bold, Italic.

11. Click Replace All.

12. Click OK and close the Find and Replace dialog box. "Duke City Gateway" is now italic, and not bold, throughout the document.

EXERCISE   9-12   Find and Replace Paragraph Formatting

1. Position the insertion point at the end of the salutation (after "Dear Mr. Forman:"). Open the Find and Replace dialog box with the Replace tab selected.

2. In the Find what text box, insert two paragraph mark special characters (use the Special list or key **^p^p**). Clear existing formatting.

3. Move to the Replace with text box, enter two paragraph marks, and clear existing formatting.

4. Click the Format button and choose Paragraph. Click the Indents and Spacing tab if it is not active.

5. Choose First Line from the Special drop-down list. If 0.5" is not the measurement displayed in the By text box, select the text in the By box and key **0.5**. Click OK.

6. Click Find Next until Word highlights the paragraph marks after "Europe." Click Replace to format that paragraph.

7. Click Replace seven more times for the next seven paragraphs (through the paragraph ending "without a fee").

8. Close the Find and Replace dialog box. Scroll through the document to view the paragraph formatting changes. All these paragraphs should now have a 0.5-inch first-line indent.

9. Position the insertion point at the top of the document. Open the Find and Replace dialog box with the Replace tab selected.

10. Delete the text in the Find what text box and set the text box to look for a 0.5-inch first-line indent.

FIGURE 9-9
Defining paragraph
formatting

11. Delete the text in the Replace with text box, clear the formatting, and replace with 0.25-inch left and right indents and no first-line indent (choose (none) from the Special drop-down list in the Replace Paragraph dialog box).

FIGURE 9-10
Replacing
paragraph
formatting

12. Click Replace All and click OK. Close the dialog box.
13. Scroll through the document to observe the replacement of first-line indented paragraphs with 0.25-inch left- and right-indented paragraphs.
14. Enter the date at the top of the document, with the correct spacing before and after it. Replace "xx" with your reference initials.
15. Replace all straight apostrophes with smart apostrophes ('). (If you use the Find and Replace dialog box, simply key an apostrophe in the Find what text box without formatting and one in the Replace with text box without formatting. Word will automatically replace straight apostrophes with smart ones.)

16. Save the document as *[your initials]*9-12 in your Lesson 9 folder.

17. Print and close the document.

TABLE 9-2   **Find and Replace Formatting Guidelines**

GUIDELINE	PROCEDURE
Find specific text with specific formatting	Key the text in the Find what text box and specify its formatting (choose Font or Paragraph from the Format drop-down list, use a keyboard shortcut, or use toolbar buttons).
Find specific formatting	Delete text in the Find what text box and specify formatting.
Replace specific text but not its formatting	Key the text in the Find what text box. Click the No Formatting button to clear existing formatting. Key the replacement text in the Replace with text box and clear existing formatting.
Replace specific text and its formatting	Key the text in the Find what text box and specify its formatting. Delete any text in the Replace with text box and specify the replacement formatting.
Replace only formatting for specific text	Key the text in the Find what text box and specify its formatting. Delete any text in the Replace with text box and specify replacement formatting.
Replace only formatting	Delete any text in the Find what text box and specify formatting. Delete any text in the Replace with text box and specify the replacement formatting.

USING ONLINE HELP

Not only can you search for specific text within a document, but you can also search for documents that contain specific text.

Use Help to learn more about searching for documents:

1. In the Ask a Question text box key **find files** and press Enter.

2. Click Find a file in the list of topics.

3. Review the information, and then click Search for a file or Outlook item containing specified text.

4. Review the information; then close Help.

5. Try searching for documents that contain specific text. From the File menu, choose File Search.

6. Click Basic Search at the bottom of the task pane (unless the Basic File Search task pane is already displayed).

7. In the Search text box in the task pane, key **Golden Gate Bridge**.

FIGURE 9-12
Searching for
documents that
contain specific text

8. Open the Search in drop-down list. Word will search in locations that are preceded by a box with a check mark. Click an empty box to add a check mark or check Everywhere if you're not sure where a file is located.

NOTE: The files you're searching for are in the same location as the other files used in this course. Check with your instructor about where to search.

9. Open the Results should be drop-down list and make sure Word Files is checked. Clear all other check boxes, because you're only looking for Word documents.

10. Click Go. After some time, Word finds the files that contain the specified text. You can open these files by clicking a filename.

11. Close the task pane. Close any open files.

TIP: You can search for documents by filename, by date, and by more complex search criteria by using the Advanced Search task pane.

LESSON 9 Summary

➤ The Find command locates specified text and formatting in a document. The Replace command finds text and formatting, and replaces it automatically with specified alternatives.

➤ Use the Find command to locate whole words, words that sound alike, font and paragraph formatting, and special characters. Using the Find command, you can search an entire document or selected text. You can also specify the direction of the search.

➤ Use the Match case option to locate text that matches the case of document text. Example: When searching for "Travel," Word would not find "travel."

➤ When you want to locate whole words and not parts of a word, use the Find whole words only option. Example: When searching for the whole word "travel," Word would find only the whole word "travel," but not "traveler" or "traveling."

➤ Use the Use wildcards option to search for text strings by using special search operators. A wildcard is a symbol that stands for missing or unknown text. Example: A search for "b?yte" would find "bite" and "byte." See Table 9-1.

➤ Use the Sounds like option to find a word that sounds similar to the search text but is spelled differently or to find a word you don't know how to spell. When you find the word, you can stop the search process and edit your document.

➤ Use the Find command to search for text that is formatted. The formatting can include character formatting, such as bold and italic, and paragraph formatting, such as alignment and line spacing. Use the Replace command to replace any of this formatting. See Table 9-2.

➤ Use the Replace command to search for all instances of text or formatting at once, or to find and confirm each replacement.

➤ Use the Replace command to delete text automatically. Key the text to be deleted in the Find what text box and leave the Replace with text box blank.

LESSON 9   Command Summary

FEATURE	BUTTON	MENU	KEYBOARD
Find		Edit, Find	Ctrl + F
Replace		Edit, Replace	Ctrl + H

Concepts Review

TRUE/FALSE QUESTIONS

Each of the following statements is either true or false. Indicate your choice by circling T or F.

T F **1.** You can use keyboard shortcuts to specify formatting in the Find and Replace dialog box.

T F **2.** To find text or formatting, you must have the insertion point at the beginning of the document.

T F **3.** Line spacing and indents are two examples of paragraph formatting that you can specify in the Replace with text box.

T F **4.** The question mark represents a special character code used to search for any character.

T F **5.** You use the Match case option to specify only uppercase when finding or replacing text.

T F **6.** The keyboard command to find text is Ctrl+H.

T F **7.** The Undo command undoes all replacements made if you used the Replace All option.

T F **8.** You can use the Find command to search either selected text or an entire document.

SHORT ANSWER QUESTIONS

Write the correct answer in the space provided.

1. What is the special character code for a paragraph mark?

_____

2. Which button can you use to continue a Find operation when the Find and Replace dialog box is closed?

_____

3. With the insertion point in the Find what text box, how do you move to, and automatically highlight the contents of, the Replace with text box?

_____

4. Which Find option do you use to locate a specific word rather than all occurrences of the text?

_____

5. If the insertion point is in the Find what text box, what is the shortcut to insert text for which you previously searched?

_____

6. How do you clear previous formatting when it appears below the text boxes in the Find and Replace dialog box?

_____

7. Which button expands the Find and Replace dialog box to show more options, and which button reduces the dialog box to make it smaller?

_____

8. Which option, Replace or Replace All, allows for selective replacement of text?

_____

CRITICAL THINKING

Answer these questions on a separate page. There are no right or wrong answers. Support your answers with examples from your own experience, if possible.

1. Some people consider the Replace feature to be only a way to correct misspelled words. It can be much more than that, however. Can you describe some ways to use this feature to increase your efficiency?

2. The Replace All option can be very useful. It can also lead to occasional problems if you haven't thought through a specific Replace All operation. After you experiment with the feature, describe some precautions you would suggest for the use of Replace All.

Skills Review

EXERCISE 9-13

Find and replace text.

1. Open the file **Bike1**.
2. Use the Find command to locate the text "8 a.m." by following these steps:

    a. Position the insertion point at the beginning of the document and choose Edit, Find.

    b. Key **8 a.m.** in the Find what text box. Click More, if the dialog box is not already expanded.

    c. Click No Formatting to clear previous formats. Make sure no search options are selected. Click Find Next.

    d. Click Cancel to close the dialog box, and edit the found text to **8:30 a.m.**

3. Change the name "Rick's Cycle Shop" to "Rick's Bike Shop," using the Replace command, by following these steps:

    a. Move the insertion point to the beginning of the document. Choose Edit, Replace.

    b. Key **Cycle** in the Find what text box, press Tab, and key **Bike** in the Replace with text box.

    c. Click Match case and Find whole words only to select only these options. Click No Formatting if any formatting remains.

    d. Click Less to reduce the size of the dialog box and drag the dialog box to the bottom of the screen.

    e. Click Find Next, and then click Replace. Click Replace until Word reaches the end of the document. Click OK, and close the dialog box.

4. Change the date, using the Replace command, by following these steps:

    a. Position the insertion point at the beginning of the document and press Ctrl + H.

    b. Key **May 1** in the Find what text box.

    c. Press Tab and key **April 25** in the Replace with text box.

    d. Click More to expand the dialog box and click any checked search options to deselect them.

    e. Click Replace All. Click OK and close the dialog box.

5. Change the top margin to 2 inches.

6. Format the first line of text as 14-point bold.

7. Move to the end of the document and format "Rick's Bike Shop" as bold italic. Copy the formatting to the text "Duke City Gateway Travel Agency" located just above the address.

  REVIEW: Format "Rick's Bike Shop," select the formatted text, click the Format Painter button ✔, and then select "Duke City Gateway Travel Agency."

8. Spell-check the document.

9. Save the document as *[your initials]*9-13 in your Lesson 9 folder.

10. Print and close the document.

EXERCISE 9-14

Use Replace to replace special characters and delete text.

1. Start a new document.
2. Change the top, left, and right margins to 2 inches.
3. Key the text shown in Figure 9-12, using single-spacing. When keying the hyphens, do not insert space characters before or after the hyphen.

FIGURE 9-12

```
The People's Center
April 1 through September 30
7 days
9 a.m.-9 p.m.
Admission Free
October 1 through March 31
Monday-Friday
9 a.m.-5 p.m.
Native American fine arts and crafts
For more information, call (406) 555-0160
```

4. Center the entire document horizontally and change the font to Arial.
5. Change the first line to 16-point bold and the last line to bold italic.
6. Replace special characters by following these steps:
   a. Position the insertion point at the end of the document and choose Replace from the Edit menu. Click More to expand the dialog box, if it is not already expanded.
   b. Key a hyphen in the Find what text box.
   c. Press Tab to move to the Replace with text box.
   d. Click the Special button and choose En Dash.
   e. Choose Up from the Search: drop-down list and clear any search options that are selected.
   f. Click Less, and then click Find Next. Do not replace the hyphen in the telephone number.
   g. Click Find Next and click Replace to replace the hyphen in the time.
   h. Continue replacing hyphens until you reach the beginning of the document.
   i. Click OK and close the dialog box when the search is complete.

7. Use the Replace feature to delete text by following these steps:
   a. Position the insertion point at the top of the document and press [Ctrl]+[H].
   b. Key **Admission Free** in the Find what dialog box.
   c. Expand the dialog box and clear any formatting.
   d. Delete any text or formatting in the Replace with text box.
   e. Select All for the Search: direction.
   f. Click Replace All to delete the text. Click OK and close the dialog box.

8. Undo the replacement by clicking the Undo button on the Standard toolbar.

9. Change the spacing to 12 points after paragraphs for the entire document.

10. Spell-check the document.

11. Save the document as *[your initials]*9-14 in your Lesson 9 folder.

12. Print and close the document.

EXERCISE 9-15

Use Replace to delete special characters and replace text.

1. Open the file **Flights1**.

2. Use the Replace command to delete all tabs by following these steps:
   a. Position the insertion point at the beginning of the document and choose Edit, Replace.
   b. Key ^t (the code for a tab character) in the Find what text box and clear existing formatting.
   c. Delete any text in the Replace with text box and clear existing formatting and search options.
   d. Click Replace All, click OK, and click Close.

3. Undo the replacement.

4. Replace the text "am" with "a.m.," including a space character before "a.m.," by following these steps:
   a. Press [Ctrl]+[H].
   b. Key **am** in the Find what text box. Check Match case—this should be the only search option checked.
   c. In the Replace with text box, press [Spacebar] once and key **a.m.**
   d. Click Replace All. Click OK and click Close.

5. Replace the name "Southwest" with "SW Airlines."

6. Spell-check the document.

7. Change the heading of the document to uppercase bold. Change the top margin to 2 inches. Change the line spacing for all the tabbed text, including the tabbed heading, to double-spacing.

8. Save the document as *[your initials]*9-15 in your Lesson 9 folder.

9. Print and close the document.

EXERCISE 9-16

Replace character and paragraph formatting.

1. Open the file **WildWalk**.

2. Set a 2-inch top margin and 1.25-inch left and right margins. Format the first page of the document as a business letter. Insert the date as a field and key the inside address as shown in Figure 9-13. Add an appropriate salutation, your reference initials, and an enclosure notation (use the word "Attachment" instead of "Enclosure").

FIGURE 9-13

```
Mr. and Mrs. James Cratty

605 Cranberry Road

Albuquerque, NM 87111
```

3. Find and replace character formatting by following these steps:

a. Position the insertion point at the top of page 2 and open the Find and Replace dialog box with the Replace tab selected.

b. Delete existing text in the Find what text box and clear all search options and formatting. Press Ctrl+U to specify underline formatting.

c. Tab to the Replace with text box and delete all text. Click the Format button and choose Font.

d. In the Replace Font dialog box, choose Bold Italic and set the Underline style to (none). Click OK.

 NOTE: If you do not change the Underline style to (none), Word will keep the underline style when it replaces the formatting.

e. Change the search direction to Down and click Replace All. When Word reaches the end of the document, click No to end the task and close the Find and Replace dialog box.

4. Find and replace paragraph formats by following these steps:

a. Place the insertion point before the text "*Snowy Ice Caves*" on page 2, and open the Find and Replace dialog box with the Replace tab selected.

b. In the Find what text box, enter two paragraph mark codes by keying ^p^p or by using the Special button. Clear all search options and formatting.

c. In the Replace with text box, clear any text, formatting, and search options. Enter one paragraph mark code.

d. Click the Format button and choose Paragraph. Set Spacing After to 6 points (6 pt) and click OK.

e. Click Less, and then click Find Next. Replace the formatting on page 2 only. Close the dialog box.

5. Format the title on page 2 ("Wilderness Walks") as 14-point bold, uppercase, and centered. Format all the text below the title with a 1.75-inch left indent.

6. Spell-check the letter portion of the document.

7. Save the document as *[your initials]*9-16 in your Lesson 9 folder.

8. Print and close the document.

Lesson Applications

EXERCISE 9-17

Replace text and character formatting.

1. Open the file **Falls**.
2. Change the top margin to 2 inches and the left and right margins to 1 inch.
3. Format the document as a memo to the Office Staff from Frank Youngblood. Use today's date. The subject is Montana Waterfalls.
4. Replace the word "falls" with **Falls** throughout the document when it is part of the name of one of the falls. Use a selective search and choose the Find whole words only option.

 REVIEW: Remember to remove all formatting from previous search and replace actions.

5. Replace the underlined formatting of the waterfall names with no underline, bold, small caps formatting. Make sure no search options are checked before you begin replacing.

TIP: The Small Caps check box and all the other check boxes in the Find and Replace Font dialog boxes initially appear shaded. Click the Small Caps check box once to select it.

6. Add your reference initials to the document.
7. Save the document as *[your initials]*9-17 in your Lesson 9 folder.
8. Print and close the document.

EXERCISE 9-18

Find and replace text, special characters, and formatting.

1. Open the file **Memo2**. Enter today's date in the memo heading.
2. Use the Find command to locate "Record Keeping." Close the Find and Replace dialog box.
3. Revise the paragraph under the heading "Record Keeping" as shown in Figure 9-14 (on the next page).
4. Find the paragraph beginning "July 1." Key the text **By no later than** before "July 1."
5. Copy (or repeat) the text you just keyed to the beginning of the next four paragraphs.

FIGURE 9-14

Use the computer to keep

∧All correspondence between group participants and you ~~should be kept~~
                                                                        stored
up₌to₌date ~~in the computer~~. Each group is named and ~~kept~~ as its own

database, so you always have access to all the active participants.

There is a master list for each group that should be ∧constantly (updated)

to reflect any change in name, itinerary, accommodations, options
                                                        see the status of the group plans
bought, money paid, and so forth. This way, you can ~~tell~~ at a glance⊙
                                                          ∧
~~where you are in finalizing the details~~. Valerie keeps a hard copy of
                                                The complete files are stored on the network in
all these records in her master file.·the folder C:\Travel\Group.
                                        ∧

6. Replace hyphens with en dashes only where there are spaces before and after the hyphens. Do not remove the spaces.

7. Replace all occurrences of "group travel" with "group tours." Be sure to correct any case changes the replacement makes.

8. Replace all underlined formatting with no underline, bold italic formatting.

9. Replace "August" with **September** and "July" with **August** throughout the document.

10. Delete the text at the end of the document from "For more detailed information" through "IATA."

11. Alphabetize the list that begins "Airline tickets" and format it with checkmark bullets.

12. Format the first line of text in the document with 72 points of spacing before the paragraph.

13. Add your reference initials and spell-check the document.

14. Save the document as *[your initials]*9-18 in your Lesson 9 folder.

15. Print and close the document.

EXERCISE 9-19

Find and replace text and formatting.

1. Open the file **Special1**.

2. Alphabetize the lists that begin "The Great Sphinx" and "Hawaii" and apply a 0.5-inch left indent to each. (Hint: To sort, select the list and choose Sort from the Table menu.)

3. Delete "Our" in the second paragraph and replace it by keying **Duke City Gateway's**.

4. Copy the keyed text to the Clipboard.

5. Use the Replace command to replace uppercase and lowercase "our" with the Clipboard contents throughout the document, *except* in "among our most popular" and in the last paragraph. Confirm each replacement.

6. Replace the whole word "tours" with "expeditions," except when the word occurs in "special-interest tours."

7. In the last paragraph only, replace "of" with "to."

8. In the two alphabetized lists, replace the paragraph mark with a manual line break for each line except the last line.

9. Format each list with 6 points of spacing before paragraphs.

10. Add a title to the document by keying **INTERESTING TOURS FOR TRAVELERS WITH INTERESTS** in uppercase bold.

11. Include two blank lines below the title and set a 2-inch top margin.

12. Add a 1½ point shadowed box around the title, with a left and right indent of 0.75 inch, and center the text in the box.

13. Spell-check the document, using Word's suggested spelling for "Colosseum."

14. Save the document as *[your initials]*9-19 in your Lesson 9 folder.

15. Print and close the document.

EXERCISE 9-20 *Challenge Yourself*

Find and replace text, special characters, and formatting.

1. Open the file **Gateway**.

2. Replace each single paragraph mark with two paragraph marks. Do not replace the paragraph marks before "One-Stop Shopping."

3. Find the word "Rome" and replace it with "Florence."

4. Use the Find command to locate "In Print." In this section of the report, replace the underline formatting of the publication names with italic, no underline formatting.

5. Replace with confirmation the following bold headings with 14-point bold small caps formatting:

   Duke City Gateway Travel—Your Gateway to the Southwest
   Duke City Gateway Travel—Your Full-Service Travel Agency
   Corporate Travel
   Group Travel
   Automation
   Information
   One-Stop Shopping

6. Copy the text "Duke City Gateway Travel" from the first line and paste it as a title at the top of the document. Keep the existing character formatting.

7. Center the title and add 72 points of spacing before and 24 points of spacing after it.

8. Spell-check the document.

9. Save the document as *[your initials]*9-20 in your Lesson 9 folder.

10. Print and close the document.

On Your Own

In these exercises, you work on your own, as you would in a real-life work environment. Use the skills you've learned to accomplish the task—and be creative.

EXERCISE 9-21
Key a song lyric you know, preferably one with a repetitive chorus. Copy the lyric and paste it below the original. In the copy of the lyric, find an important word that is used repeatedly in the lyric and replace it with its opposite. Save the document as *[your initials]*9-21 and print it.

EXERCISE 9-22
Write a summary about a book you recently read. Replace paragraph marks that begin new paragraphs with 6 points of spacing after paragraphs. Replace any occurrence of two spaces with one space. Save the document as *[your initials]*9-22. Print the document.

EXERCISE 9-23
Log on to the Internet and find a Web site about one of your favorite hobbies or interests. Copy and paste information from the Web site to a new document. Find and replace any formatting you do not want in the document. Give the document a title. Save the document as *[your initials]*9-23 and print it.

10

Sharing Your Work and Hyperlinks

After completing this lesson, you will be able to:

1. **Create comments.**
2. **Use the Track Changes feature.**
3. **Compare and merge documents.**
4. **Create versions of the same document.**
5. **Review a document.**
6. **Distribute documents via e-mail.**
7. **Insert hyperlinks.**

In this lesson:
- WW03S-1-3
- WW03S-2-3
- WW03S-4-1
- WW03S-4-2
- WW03S-4-3
- WW03S-4-4
- WW03S-5-7
- WW03E-4-1
- WW03E-4-3
- WW03E-4-4

See Appendix C.

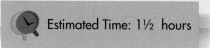

Estimated Time: 1½ hours

In most offices, more than one person works on a document. A document might have a primary author and several editors. Word provides tools that make it easy to collaborate with others—on your computer instead of on paper—to produce a finished document. After you have a finished document, you might take it one step further—sharing it with others on the World Wide Web. You will also learn to work with hypertext to link documents.

Creating Comments

Comments are notes or annotations you add to a document. Each person who adds a comment is called a *reviewer*. Comments are color-coded by each reviewer.

Word refers to comments and any other type of document revision marks as *markup.* You can display, hide, or print markup.

In Normal view, you insert, edit, and view comments in the *Reviewing pane*—a narrow horizontal pane that opens at the bottom of the screen. In Print Layout view, you can enter, edit, and view comments in *markup balloons* that appear in the document margin with a line leading to where the comment was inserted.

EXERCISE 10-1 Add Comments to a Document

The easiest way to add a comment to a document is to use the Reviewing toolbar. You can also use the keyboard shortcut [Alt]+[Ctrl]+[M] or choose Comment from the Insert menu.

1. Open the file **CruiseFAQs2**. Make sure the document is in Normal view.

2. Display the Reviewing toolbar by right-clicking any toolbar button and then choosing Reviewing.

     NOTE: If the toolbar appears in the document text area, drag it above the ruler to dock it.

3. Choose Options from the Tools menu and click the User Information tab.

4. Enter your name and initials in the appropriate data fields. Click OK.

    TIP: If you review a document by using another person's computer, your comments will be attributed to the person whose name appears in the User Information tab. Make sure to change the name in the User Information tab to your name when you insert comments, and then back to the original user's name when you are finished.

5. Position the insertion point at the beginning of the document.

6. Click the Insert Comment button 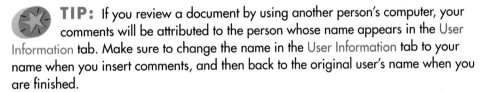 on the Reviewing toolbar. The Reviewing pane opens, showing your name, followed by the date and time you are inserting the comment. The entry in the Reviewing pane is color-coded to the reviewer. *Comment reference marks* also appear in the document where you inserted the comment. The reference marks are color-coded to the reviewer.

7. In the Reviewing pane, at the insertion point, key:

    This is an abbreviated version of the longer CruiseFAQs document. Make sure Frank Youngblood reviews both documents.

8. Scroll within the document to the paragraph below *"What kinds of people take a cruise?"* Select "$100" and click the Insert Comment button to insert another comment.

    NOTE: When you comment on a selected word, sentence, or paragraph, the text is enclosed by two comment reference marks. You can also just position the insertion point at a particular location and insert a comment.

FIGURE 10-1
Inserting a comment

Reviewing toolbar

Comment reference marks

Reviewing pane

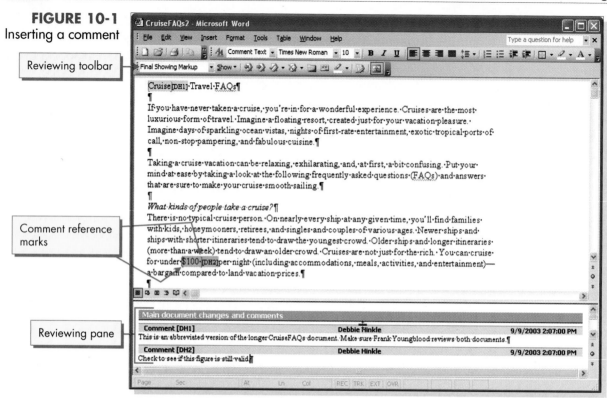

9. In the Reviewing pane, key **Check to see if this figure is still valid.** Notice the comment reference marks around "$100" in the document.

10. Switch to Print Layout view, and use the View menu to change the zoom to 75%. Notice that your comments appear as balloons, with lines leading to where the comments were inserted. (See Figure 10-2 on the next page.)

11. Move the insertion point over a balloon. A ScreenTip displays the name of the reviewer, and the time and date when the reviewer entered the comment.

12. Click the drop-down arrow next to the Show button on the Reviewing toolbar. Choose Balloons and select Never. Choose Never to hide balloons in the document. When the balloon option is Never, use the Reviewing pane to view comments.

13. Click the drop-down arrow next to the Show button on the Reviewing toolbar and choose Options. The Track Changes dialog box opens and includes additional options for formatting balloons, which include changing the width of the balloon, the margin setting, and the display of lines connecting to text. Click the drop-down arrow next to the option Use Balloons (Print and Web Layout) and choose Always. Click OK.

14. Click the Reviewing Pane button on the Reviewing toolbar to close the Reviewing pane.

TIP: The Reviewing Pane button opens and closes the Reviewing pane.

FIGURE 10-2
Comments
displayed as
balloons in Print
Layout view

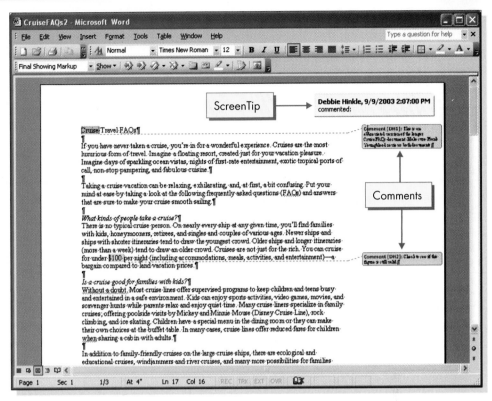

15. Increase the zoom back to 100%. On the second page, locate the paragraph below the text "*How long are cruises?*" Select the entire paragraph (which starts "Cruise length varies").

16. Press Alt+Ctrl+M to insert a new comment. Word inserts a balloon in the right margin area.

17. Use the horizontal scrollbar to see the entire balloon, and then key the comment **Check the number of days for each cruise mentioned here.** The document now has three comments.

TABLE 10-1 Reviewing Toolbar Buttons

BUTTON		DESCRIPTION
Show ▼	Show	Displays a menu to control which reviewer corrections or comments are displayed.
	Previous	Goes to the previous comment.
	Next	Goes to the next comment.

continues

TABLE 10-1   Reviewing Toolbar Buttons   *continued*

BUTTON		DESCRIPTION
	Accept Change	Accepts selected change and removes highlighting.
	Reject Change/ Delete Comment	Rejects selected change and removes highlighting; deletes one or all comments.
	Insert Comment	Inserts a comment at the insertion point.
	Insert Voice	Inserts a voice comment at the insertion point.
	Highlight	Marks text with a color highlighter.
	Track Changes	Marks changes in the document and keeps track of all changes by reviewer name.
	Reviewing Pane	Displays or hides the Reviewing pane.

NOTE: The Display for Review drop-down list appears at the left of the Reviewing toolbar. Use this list box to view the document before and after reviewer changes.

EXERCISE 10-2 Edit and Delete Comments

You can edit and delete comments by working directly with the balloon text in Print Layout view or by using the Reviewing pane in Normal view.

1. Click the comment in the first balloon. The comment appears highlighted (the balloon's background color darkens and the dotted connector line becomes solid).

2. Select the name "Frank Youngblood" and change it to **Susan Allen**.

3. Click the Next button on the Reviewing toolbar. The second comment mark is highlighted.

TIP: The Next and Previous buttons are helpful for jumping from one comment to another, particularly in a long document. The insertion point moves to the next or previous comment.

4. Click the Reject Change/Delete Comment button on the Reviewing toolbar. The comment is removed from the document.

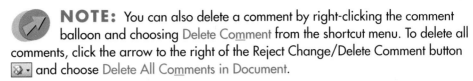

NOTE: You can also delete a comment by right-clicking the comment balloon and choosing Delete Comment from the shortcut menu. To delete all comments, click the arrow to the right of the Reject Change/Delete Comment button and choose Delete All Comments in Document.

5. Switch to Normal view and click the Reviewing Pane button 🔳 to display the Reviewing pane. The two remaining comments are listed.

6. In the Reviewing pane, change the word "here" in the second comment to **in this paragraph**.

FIGURE 10-3
Edited text of two remaining comments

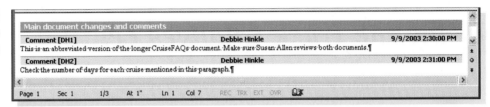

7. Close the Reviewing pane.

8. Save the document as *[your initials]*10-2 in a new folder for Lesson 10.

 TIP: You can set the Select Browse Object button 💿 on the vertical scroll bar to browse by comments. Then the double-arrow scroll buttons become the Previous Comment and Next Comment buttons.

EXERCISE 10-3 **Print Comments**

You can print a document with its balloon comments, as they appear in Print Layout view, or you can print just a list of a document's comments. You can also print a document without comments.

1. Switch to Print Layout view, and then to Print Preview. Notice that comments appear in Print Preview. (Scroll to view the two comments, or use the Multiple Pages button 🔳 to display all pages in the document.)

2. Click the Print button 🖨. The document prints with comments.

 NOTE: Word sets the zoom level and page orientation to best display the comment balloons in your printed document.

3. Open the Print dialog box. Notice that Print what is automatically set to Document showing markup. This setting will appear when you print from any view that displays balloon comments. Click Cancel.

4. Close the Print Preview window and switch to Normal view. Notice that balloon comments are not displayed.

5. Open the Print dialog box. Notice that Print what is set to Document. With this setting, the document would print without comments.

TIP: From the Print what drop-down list in the Print dialog box, choose Document to print a document without comments, choose Document showing markup to print a document with its comments in balloons, or choose List of markup to print the document's comments only.

6. Open the Print what drop-down list box and choose List of markup. Click OK. Word prints your comments as a list, similar to how comments appear in the Reviewing pane.

7. Close the document.

Using the Track Changes Feature

When the job of revising a document is shared by more than one person, you can use the Track Changes feature to mark each person's changes. When this feature is turned on, Word tracks changes as you edit the document. These *tracked changes,* also called *markup,* are revision marks that show where deletions, insertions, and formatting changes were made. When you later review the document, you can either accept the changes or return to the original wording.

In Normal view, Word displays tracked changes in the following way:

- Underline
  New text is underlined.

- Strikethrough
  Deleted text has a horizontal line through it.

- Revision bar
  A vertical bar appears in the margin next to revised text.

In Print Layout view, tracked changes appear in the text and also in markup balloons in the document margin. These balloons make it easy to see and respond to document changes, additions, and comments.

EXERCISE   10-4   Use the Track Changes Feature to Enter Revisions

To track changes in a document, turn on Track Changes mode by double-clicking TRK on the Status bar or by choosing Tools, Track Changes.

1. Open the file **AlbuqClimate**. Tracked changes already appear in the document, indicating that another reviewer has edited the document. You're going to make additional revisions to the document.

 NOTE: If tracked changes do not appear, choose View, Markup.

 2. Make sure the Show/Hide ¶ button ¶ is turned on.

3. Double-click TRK on the Status bar. The Reviewing toolbar opens and the letters TRK appear in black on the Status bar, indicating that Track Changes mode is turned on.

4. In the first paragraph, key the word **vacation** between the words "enjoyable" and "destination." The new word appears underlined and in another color.

 NOTE: Word can display as many as eight different colors for tracked changes. The first eight reviewers of a document are automatically assigned different colors.

5. In the first paragraph, select "Albuquerque" and make it bold. No new revision marks appear, but the revision bar displays to the left of the first and second lines of the paragraph.

6. Change to Print Layout view. A colored line appears under the first line of text and the word "formatted" appears in a balloon in the margin. Deselect "Albuquerque" and the line changes to a dotted line.

FIGURE 10-4
Tracked changes in
Print Layout view

Reviewing toolbar

Revision bars
indicate paragraphs
with changes.

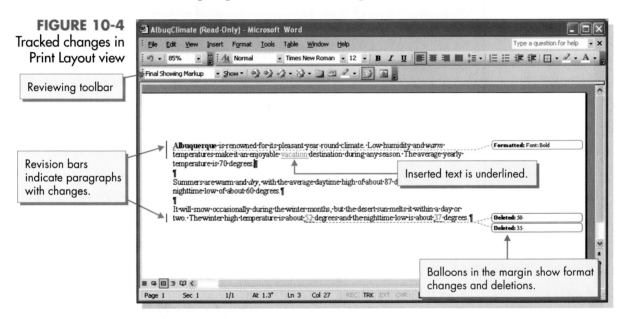

 NOTE: To see tracked changes for formatting revisions, you must be in Print Layout view.

7. In the second paragraph, select "warm" in the first sentence and make it italic.

8. Change back to Normal view. Select the last sentence in the first paragraph and press (Delete) to remove it. The sentence appears as strikethrough text (with a line going through it) and has the revision color applied.

9. Select the second paragraph and the paragraph mark below it. Cut the selection.

REVIEW: When you're making changes to a document that will be reviewed by others, a good idea is to insert comments to explain why you made specific changes. To add a comment, use the Reviewing toolbar. You can also use the keyboard shortcut Alt+Ctrl+M or choose Insert, Comment.

10. Press Enter twice at the end of the document and paste the cut text. Vertical bars appear to the left of the paragraphs to show they were changed.

11. Place the insertion point at the beginning of the first paragraph and press Enter.

12. Position the insertion point on the new blank line, and key **Albuquerque Climate**. The new title appears underlined.

13. Select the title and apply bold, all caps, and 18-point formatting.

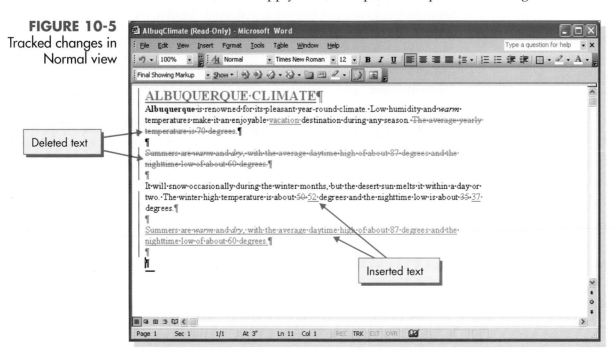

14. Point to any revision you have made in the document. A ScreenTip identifies your name, the date, and the type of revision.

15. Point to the revision marks "5052." A ScreenTip identifies Nina Chavez as the reviewer, and shows the date and type of revisions she made.

NOTE: Not only can you track changes made to the main body of a Word document, but you can also track changes made to headers and footers, and to footnotes and endnotes.

16. Click the Track Changes button on the Reviewing toolbar to turn off Track Changes mode.

17. Save the document as *[your initials]*10-4 in a new folder for Lesson 10.

18. To print the document with tracked changes, open the Print dialog box. Make sure Document showing markup appears in the Print what drop-down list box. Click OK. Tracked changes appear on the printed page just as they did in Print Layout view. Leave the document open for the next exercise.

EXERCISE **10-5** **Accept and Reject Revisions**

You can review a document after it is edited and decide if you want to accept or reject the revisions. You can review changes by the type of change made, or review only changes made by a particular reviewer. If you accept the revisions, Word deletes the strikethrough text, removes the underlining from the new text, and removes the revision bars. If you don't want to make the revisions, you can reject them and restore the original document. You can accept or reject changes individually or all at once.

To accept or reject revisions, you can do one of the following:

● Use the Reviewing toolbar.

● Right-click the revision mark and use the shortcut menu.

You can also merge more than one revised file from other reviewers into the same document. Then you can protect the document from revisions and allow only the author to turn off the tracking or accept or reject revisions.

1. Select the title at the beginning of the document. Click the Accept Change button on the Reviewing toolbar to accept the change.

2. Click the Show button and deselect the Insertions and Deletions option, leaving the Formatting option checked. Switch to Print Layout view. Now only the changes to formatting appear in the document.

> **NOTE:** You can leave the Comments option turned on because there are no comments in the document.

3. With the title still selected, click the Next button to move to the next formatting change in the document.

4. Click the Reject Change/Delete Comment button on the Reviewing toolbar to reject the formatting change.

5. Return to Normal view. Click the Show button and choose Insertions and Deletions. Now the other changes you made to the document appear.

> **NOTE:** To review only those changes made by a particular reviewer, click the Show button and choose Reviewers. Deselect each reviewer name except the one whose changes you want to review. To display all reviewer changes again, click the Show button , choose Reviewers, and then choose All Reviewers.

6. Click the inserted word "vacation" in the first paragraph. Click the Reject Change/Delete Comment button . The word is deleted from the document.

7. Right-click the sentence with the strikethrough text at the end of the same paragraph. Choose Accept Deletion from the shortcut menu. The sentence is deleted.

8. Click the Next button ⏩ to select the next revision. Click the Accept Change button 🗸▾ to accept the change.

9. Click the Next button ⏩ to move to the "5θ" revision and click the Accept Change button 🗸▾ twice to accept it and the next insertion, "52."

10. Repeat this process to accept the remaining revisions. When you reach the end of the document, continue searching from the beginning of the document to accept the paragraph mark for the title insertion.

11. Close the document without saving it, and then reopen *[your initials]*10-4.

12. Click the drop-down arrow beside the Display for Review button [Final Showing Markup ▾] on the Reviewing toolbar. Choose Final to see the document with all changes accepted. Choose Original to see the original document without changes. Choose Final Showing Markup to display deleted text and inserted or formatted text in the paragraph.

13. Choose Tools, Track Changes and click the down arrow next to the Accept Change button 🗸▾. Choose Accept All Changes in Document.

FIGURE 10-6
Accepting all changes in a document

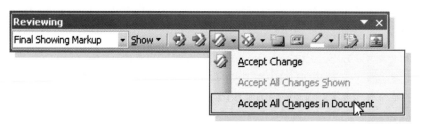

NOTE: To reject all changes in a document, click the arrow next to the Reject Change/Delete Comment button 🗷▾ and choose Reject All Changes in Document.

14. Save the document as *[your initials]*10-5 in your Lesson 10 folder.

15. Print and close the document.

Comparing and Merging Documents

Suppose you have two similar documents—an edited copy of a document and the original document. You can use the Compare and Merge feature to compare the edited document with the original. Word will show the differences as tracked changes.

When comparing two documents, use the *legal blackline* feature, which compares the documents and creates a third document that shows the changes. Law firms use this feature to maintain records of all document revisions and versions.

EXERCISE **10-6** Compare and Merge Documents

1. Open the file **Turquoise**. This is a revised version of a document called **Trail**. You're going to compare both files to see how they are different.

2. Turn on Track Changes if necessary. Before comparing documents, make your own revisions to this document. Key **in Cibola National Forest** after the text "the tour begins" in the second paragraph. In the same paragraph, delete the second sentence (which begins "You can reach").

3. Choose Compare and Merge Documents from the Tools menu.

4. In the Compare and Merge dialog box, click the Legal blackline check box to select it.

FIGURE 10-7
Compare and
Merge dialog box

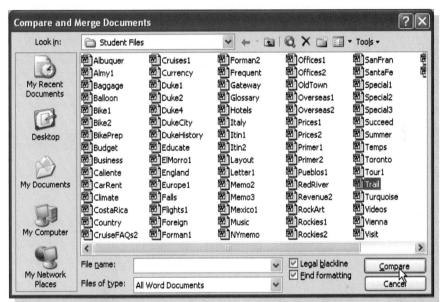

NOTE: The Legal blackline feature compares the two documents and opens a new document to display the changes. When comparing more than two documents, do not check this box. Instead, click Merge or open the Merge drop-down list and choose one of the Merge options.

5. Select the file **Trail**, and then click Compare. Word opens a new document that compares the edited file, Turquoise, with the original file, Trail. All these changes appear as revision marks. Deleted text appears as strikethrough text; new text is underlined. Revisions are color coded by reviewer. A revision bar appears to the left of each line with a revision. (See Figure 10-8 on the next page.)

6. Switch to Print Layout view. The deleted text appears in balloons in the margin.

FIGURE 10-8
Merged documents
showing revision
marks

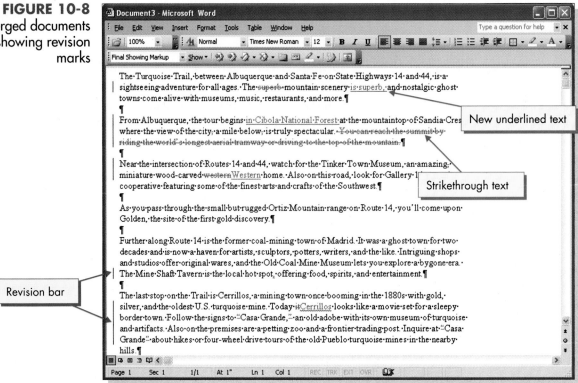

> ★ **TIP:** In Normal view, revisions also appear in the Reviewing pane.

7. Save the document as *[your initials]*10-6 in your Lesson 10 folder.

8. Print and close the document. Word prints the document with the revisions, changing the zoom level to best display the revision balloons.

> **NOTE:** Printing a document with revision marks is a good way to keep a record of changes made to a document.

EXERCISE 10-7 Protect a Document from Revisions

You can protect a document from changes in several ways. One is to require a password to open or modify the document. Another way is to designate the document as a *read-only* file. *Read only* allows a user to open or copy a file, but not to change or save the file. Other ways are to allow users to insert comments or you can ensure that any changes appear only as tracked changes.

1. Open the file **Turquoise**.

2. Open the Reviewing toolbar, if it is not already open.

3. Save the document as *[your initials]***10-7** in your Lesson 10 folder.

4. Switch to Print Layout view and turn on Track Changes.

FIGURE 10-9
Protect Document
task pane

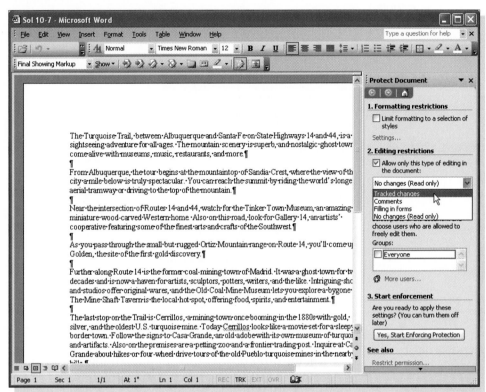

5. Choose Protect Document from the Tools menu. The Protect Document task pane displays.

6. Click the check box under Editing Restrictions and choose Tracked Changes from the drop-down list.

7. Click the Yes, Start Enforcing Protection command button. Do not enter a password, but click OK to apply the editing restrictions.

8. Locate the word "Western" in the third paragraph and delete it. The deletion appears as a tracked change.

9. Right-click the markup balloon. Notice that the Accept Deletion and Reject Deletion options are not available. As long as the protection is turned on, revisions cannot be accepted or rejected.

 NOTE: The Accept Change 🖉▾ and Reject Change/Delete Comment 🖉▾ buttons are not available on the Reviewing toolbar. (See Figure 10-10 on the next page.)

10. Click the Stop Protection command button in the Protect Document task pane.

FIGURE 10-10
Revision attempt
with protection
enabled

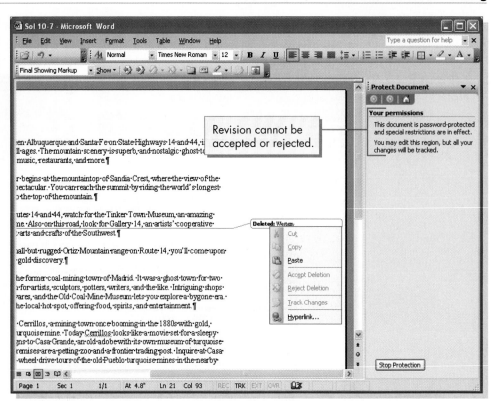

11. Click the drop-down arrow in the Editing restrictions section of the Protect Document task pane and choose Comments.

12. Select the first and second paragraphs of the document, and select the Everyone check box in the Groups box.

> **NOTE:** Words, sentences, or paragraphs in a document can be designated as unrestricted for editing. You can also specify which individuals have permission to modify the unrestricted areas of the document. Click the More users link to enter the names of the users who have permission to edit the document.

13. Click the Yes, Start Enforcing Protection command button. Do not enter a password and click OK. The unrestricted areas of the document are highlighted.

14. Delete the text "**is superb,**" in the first paragraph. Select and delete the last sentence in the second paragraph.

15. Select the fourth paragraph of the document. Press the Delete key. The Status bar displays a message indicating that you cannot modify this paragraph.

16. Close the document without saving.

17. Reopen *[your initials]*10-7.

18. Choose Options from the Tools menu and select the Security tab. Select the check box to apply the Read only recommended option. Click OK.

19. Save, close, and reopen the document. Word displays a message box indicating that the document will open as read-only and changes cannot be made. Click Cancel. If you click Yes and choose to edit the document, you would have to use the Save As command and rename the document.

20. Close the Reviewing toolbar if it is still open.

 NOTE: The Security tab in the Options dialog box is used to assign a password to a document.

Creating Versions of a Document

Suppose you want to revise a document in different ways so it can be used for different purposes. You can save multiple versions of the same document under a single filename by using the File, Versions command. You can add detailed comments about each version to keep a record of the changes made to the document. In longer documents, you can use bookmarks to help you quickly locate the places where two documents have been changed.

EXERCISE **10-8** **Create Two Versions of a Document**

1. Open the file **CruInfo** and save it as *[your initials]*10-8 in your Lesson 10 folder.

2. Select the entire document and change the font to 11-point Tahoma.

3. Save this version of the document by choosing Versions from the File menu. Click the Save Now button and key the following comment:

Includes sample itineraries.

(See Figure 10-11 on the next page.)

4. Click OK. Notice the Versions icon at the right of the Status bar. The icon appears after you've created one or more versions of a file. (See Figure 10-12 on the next page.)

5. Create another version of the document by deleting the three sample itineraries at the end of the document (under the headings "Short Cruises," "Weeklong Cruises," and "14-Day Cruises").

6. At the end of the document, press Enter twice, and then key the following three lines of text. Format the three lines as 12-point Arial, bold italic, centered:

Call for a complete guide to cruise travel and sample itineraries.
Duke City Gateway Travel
(505) 555-1234

7. Choose File, Versions. Then click Save Now and enter the comment:

Deleted sample itineraries, added note to call agency.

FIGURE 10-11
Saving versions
of a document

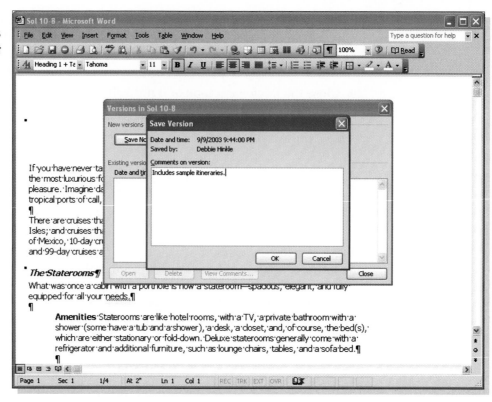

FIGURE 10-12
Versions icon on
the Status bar

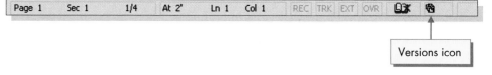

8. Click OK. Close the document, and then reopen it. The latest version always opens first.

 9. To view the other version, choose File, Versions or double-click the Versions icon 🗃 on the Status bar.

10. In the Versions dialog box, select the version that contains the sample itineraries.

 NOTE: To delete a version, select it in the Versions dialog box and click Delete.

11. Click Open. Word opens the version in a separate pane so you can view both versions.

12. Print the latest version, the one without the itineraries. Close the version with the itineraries.

NOTE: If you open a document version, and then save it, Word saves it as a new file with a filename that includes the version date. You can then open the new file like any other document, instead of using the Versions dialog box.

EXERCISE 10-9   Use a Bookmark to Locate Document Changes

In longer documents where you have several versions, bookmarks can be useful for helping you quickly locate where changes have been made.

1. Maximize the window of the current version of the file *[your initials]*10-8. This is the version without the sample itineraries.

2. Locate the heading on the third page of the document, "Types of Cruises," and select the subheading "Short Cruises."

3. Choose **Bookmark** from the **Insert** menu, and key **Begin_Itineraries** in the **Bookmark name** text box. Click **Add** to save the bookmark.

4. Position the insertion point at the beginning of the document and test the bookmark by using the Go To command (press Ctrl+G, choose the bookmark, click **Go To**).

5. Close the Find and Replace dialog box.

6. Create a new version, with the following comment:

   Deleted sample itineraries, added note to call agency, bookmarked location of deletions.

7. Close the file.

Review a Document

When a document has been revised and reviewed several times, you can choose Reading Layout view to read the document and to verify its accuracy and content. Reading Layout view hides all toolbars, except the Reading Layout and Reviewing toolbars, and displays the Document Map.

There are three ways to switch to Reading Layout view:

- Click the Read button 🕮 Read on the Standard toolbar.
- Choose **Reading Layout** from the **View** menu.
- Press Alt+R.

EXERCISE 10-10   Review a Document

1. Open the file Primer2 and save the document as *[your initials]*10-10.

2. Click the Read button 🕮 Read on the Standard toolbar. Select the Document Map button on the Reading Layout toolbar. The Document Map pane appears on the left of the screen.

FIGURE 10-13
Reading Layout view

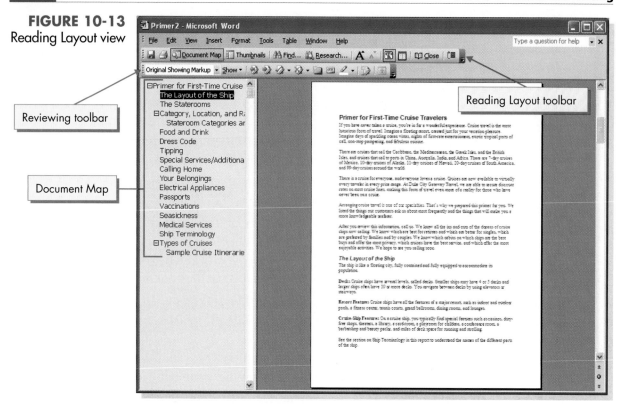

Reviewing toolbar

Document Map

Reading Layout toolbar

NOTE: The Document Map displays headings in the document. Click a heading to move to that part of the document. (See Table 10-2 on the next page.)

3. Click the Thumbnails button 🔲 Thumbnails on the Reading Layout toolbar. The Document Map pane is replaced with miniature representations of each page.

NOTE: Reading Layout view uses the term screen rather than pages because the screens do not display an entire page of text. The document does not display as it would in Print Layout view.

4. Scroll through the thumbnails and click on screen 8. The thumbnail is highlighted and displays in an enlarged view.

5. Click the Thumbnails button 🔲 Thumbnails to remove the pane.

6. Click the Actual Page 🔲 button to change the magnification level. The document display resembles Print Layout view. Click the Actual Page button again to display the entire page.

7. Scroll to Screen 1. Select all the text on screen 1 and press Delete. Documents can be edited in Reading Layout view, and the track changes feature can be turned on using the Reviewing toolbar.

TABLE 10-2   Reading Layout Toolbar Buttons

BUTTON		FUNCTION
💾	Save	Saves changes to the document.
🖨	Print	Prints the document.
Document Map	Document Map	Displays document headings that can be used to navigate in a document.
Thumbnails	Thumbnails	Displays miniature representations of each screen in the document.
Find...	Find	Opens the Find and Replace dialog box.
Research...	Research	Displays the Research task pane.
🔍+	Increase Text Size	Increases the text size for easier reading. The font size of the document is not affected.
🔍-	Decrease Text Size	Decreases the text size without changing the font size of the document.
🗏	Actual Page	Displays the page as it would appear in Print Layout view.
📖	Allow Multiple Pages	Toggles between viewing two pages or one page.
Close	Close	Returns to Normal view.
🗒	Start of Document	Moves to the first screen of the document.

8. Click the Allow Multiple Pages button 🗏. The document displays as a two-page spread.

9. Click the Close button Close on the Reading Layout toolbar to return to Normal view.

10. Choose Properties from the File menu and select the Statistics tab to view document information including word count and number of pages. Click OK.

NOTE: An alternative to showing document statistics is to choose Options from the Tools menu. Select the Spelling & Grammar tab and verify that the Check grammar with spelling feature and the Show readability statistics feature are both selected. When the spell and grammar check is complete, the Readability Statistics dialog box displays and lists readability and word count information. (See Figure 10-14 on the next page.)

FIGURE 10-14
Spelling & Grammar
options

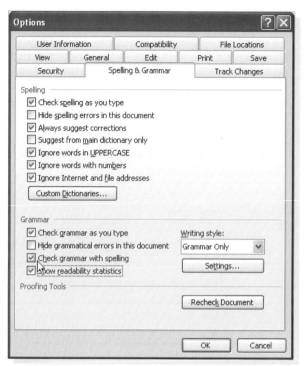

11. Close the document without saving changes.

Distribute Documents via E-mail

The Internet enables communication in the workplace to be more efficient and faster than traditional methods of communication. Sending a document via e-mail does not require the recipient to be in the office, and e-mail is not limited by time—it can be sent and read at any time. Documents can be sent as the body of an e-mail message, sent as an attachment, or sent for review. When a document is sent for review, the creator of the document accepts or rejects the proposed revisions.

EXERCISE 10-11 **Distribute Documents via E-mail**

1. Start a new document.
2. Change the top margin to 2 inches and set a left tab at one inch.
3. Create a memo to the Staff from Frank Youngblood. Key **September 20** as the date and key **PDA training** in the subject line.
4. Insert two blank lines after the subject line and key the text from Figure 10-15 as the body of the memo.

FIGURE 10-15

```
The training session for our new PDAs will be held on
Wednesday, October 2, in the conference room at 3 p.m.


Please let me know if you are unable to attend.
```

5. Save the memo as *[your initials]***10-11** in your Lesson 10 folder. Do not close the document.

6. Choose Send To from the File menu and choose Mail Recipient. The e-mail header appears with the filename listed in the Subject box, and the E-mail toolbar displays.

 NOTE: You can also click the E-mail button 🖃 on the Standard toolbar to open the e-mail header.

TABLE 10-3   E-mail Toolbar Buttons

BUTTON		FUNCTION
🖃 Send a Copy	Send a Copy	Sends a copy of the document to the e-mail address specified.
📎 ▾	Insert File	Opens the Insert File dialog box from which you can choose a file to attach.
📖	Address Book	Displays the Address Book entries added to Outlook.
👤✓	Check Names	Checks the recipient name against the e-mail address listed in the Address Book.
⊖	Permission	Apply restrictions to documents.
❗	Importance: High	Sets the importance level of the e-mail as high.
⬇	Importance: Low	Sets the importance level of the e-mail as low.
⚑	Message Flag	Displays the Flag for Follow Up dialog box. Sender can choose follow up actions and due dates.
✉	Create Rule	Opens the Create Rule dialog box.
📋 Options... ▾	Options	Displays the Message Options dialog box, which includes message and delivery options.

7. Enter an e-mail address in the To box.

8. Click Send a Copy ⊟Send a Copy. Word sends the e-mail to the specified address and closes the e-mail header.

9. Print and close the document.

10. Open the file *[your initials]***10-11**.

11. Choose Send To from the File menu and choose Mail Recipient (as Attachment). The e-mail header appears with the file listed in the Attach box.

12. Key **sallen@gwaytravel.com** in the To box.

13. Click in the message area, and key the text from Figure 10-16.

FIGURE 10-16

Attached is the memo to the staff announcing the PDA training session. Let me know if you would like to include additional comments before the memo is distributed.

14. Save the document as *[your initials]***10-11b** in your Lesson 10 folder. Print and close the document.

15. Open the document *[your initials]***10-11** and choose Save As from the File menu.

16. Change the File name to *[your initials]***10-11c**. In the Save as type text box, click the drop-down arrow and choose Rich Text Format. Rich Text Format is used to ensure document compatibility with other applications or operating systems. Click Save and close the document.

 NOTE: You can also choose Plain Text to save a file. Plain Text does not support bold, italic, and other formatting.

Creating Hyperlinks

A hypertext link, called a *hyperlink,* is text you click to move to another location. The location to which you move can be within the same document, in another document, somewhere on the World Wide Web, or your company's intranet. Readers use hyperlinks to jump to related information. You can create a series of hyperlinks, thereby creating your own "web" of locations.

In a document, hyperlinks can be a word or phrase. A hyperlink is usually blue and underlined.

EXERCISE **10-12** Create a Hyperlink Within the Same Document

In a long document, you can use a hyperlink to jump quickly from one page to another. You can do this in one of three ways:

- Create a bookmark at a particular location in the document, and then insert a hyperlink to the bookmark.
- Apply a heading style to text in the document, and then insert a hyperlink to the styled text.
- Copy text in the document, and then use the Paste as Hyperlink command.

1. Open the file **DesertSW**.

2. Delete the text on page 1, and delete the page break. Make the first heading ("Desert Southwest Travel Guide") bold. Deselect the heading.

3. Choose AutoFormat from the Format menu and click OK.

4. Copy the format of the first heading "Desert Southwest Travel Guide" to the heading "Events" on page 7.

5. Save the document as *[your initials]***10-12** in your Lesson 10 folder.

 NOTE: You should always save a document before creating a hyperlink.

6. Using a bookmark, you'll create a hyperlink from the text "Grand Canyon" on page 1 to the description of the canyon on page 4. Locate the second paragraph below the bold heading "**Parks and Monuments**" on page 4. Position the insertion point to the left of the text "The most popular national park" and insert a bookmark named **GrandCanyon**. (Choose Insert, Bookmark. Remember, bookmark names cannot have spaces.)

7. Below the first heading on page 1, at the end of the paragraph that begins "One of the most unique," select the text "Grand Canyon."

 8. Click the Insert Hyperlink button on the Standard toolbar to open the Insert Hyperlink dialog box.

 TIP: You can also use the menu command Insert, Hyperlink or the keyboard shortcut Ctrl+K to open this dialog box.

9. Under Link to, click the Place in This Document button. Notice that the document headings appear in the dialog box. Below the document headings is the bookmark you created.

10. Scroll to the bottom of the document headings. Under Bookmarks, click GrandCanyon to choose this bookmark as the "jump to" location. (See Figure 10-17 on the next page.)

11. Click OK. The words "Grand Canyon" are blue and underlined, indicating you have created a hyperlink.

FIGURE 10-17
Creating a hyperlink
to a bookmark

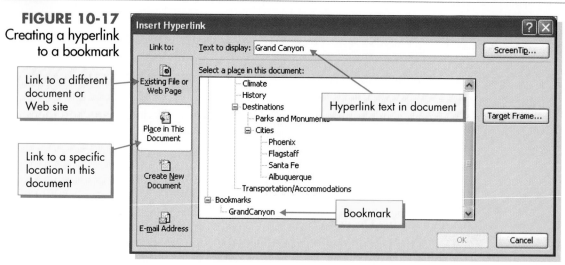

Link to a different
document or
Web site

Link to a specific
location in this
document

12. Point to the blue hyperlink text. A ScreenTip displays the bookmark name with the instructions CTRL + click to follow link. Press Ctrl and the pointer changes to a pointing hand.

13. Test the hyperlink by clicking it. The insertion point jumps to the description of the Grand Canyon on page 4, and the Web toolbar opens.

NOTE: If the Web toolbar does not open automatically, choose View, Toolbars, Web.

14. On the Web toolbar, click the Back button 🔘 to return to the hyperlink text. Notice the change in color, indicating you have used the hyperlink.

NOTE: In a document with many hyperlinks, this change in color helps you remember which hyperlink locations you already visited.

15. On page 1, below the heading "**The Region**," select the word "Phoenix" toward the end of the third paragraph.

16. Open the Insert Hyperlink dialog box. Click the heading Phoenix, and then click OK.

17. Press Ctrl and click the hyperlink to test it. The insertion point jumps to the heading "**Phoenix**" on page 5. (See Figure 10-18 on the next page.)

18. Locate the heading "**Santa Fe**" on page 6. Select and copy the heading (but not the paragraph mark). In the following steps, you create a hyperlink in this heading by using the Paste as Hyperlink command.

19. Return to the beginning of the document. Find and select the first instance of the text "Santa Fe" (in the paragraph under the heading "**History**" that begins "In 1803").

TIP: Instead of scrolling back and forth to create hyperlinks within a document, you can split the document and display the appropriate text within each pane. Choose Window, Split (or double-click the split bar on the vertical scroll bar).

FIGURE 10-18
Creating a hyperlink
to a heading

Web toolbar

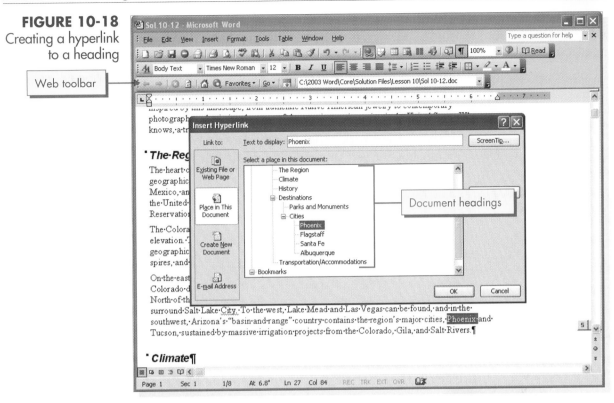

20. From the <u>E</u>dit menu, choose Paste as <u>H</u>yperlink. Word replaces the selected text with hyperlink text.

21. Press Ctrl and click the hyperlink to jump to the Santa Fe heading on page 6.

> **NOTE:** You can also drag-and-drop to insert hyperlinks. Select the text to which you want to jump, drag it to another location by using the right mouse button, and then choose Create <u>H</u>yperlink Here to insert the text as a hyperlink.

22. Using one of the above-practiced methods, find the first instance of the word "Albuquerque" and create a hyperlink from it to the heading "**Albuquerque**."

23. Add page numbers to the document, using the default position.

24. Save the document. Print the pages with hyperlinks (pages 1–3), and then close the document.

> **NOTE:** After creating a hyperlink, you can change the blue hyperlink text as well as the link destination. To change the text, right-click the hyperlink, choose <u>S</u>elect Hyperlink from the shortcut menu, and then key different text. To change the hyperlink destination, right-click the hyperlink, choose Edit <u>H</u>yperlink, and then choose a different heading or bookmark in the document to which you want to link. To remove a hyperlink, right-click the hypertext and choose <u>R</u>emove Hyperlink.

EXERCISE 10-13 Create a Hyperlink to Another Document

You can create hyperlinks to another Word document or to Office documents such as an Excel worksheet or a PowerPoint presentation.

1. Start a new document. Create a memo to Alice Fung from Valerie Grier. The subject is **New York "Night on the Town" Weekend**. For the body of the memo, key the text shown in Figure 10-19.

FIGURE 10-19

> Travel materials for the New York weekend package have been sent to all participants.
>
> Participants should receive their materials by Friday.

2. Add your reference initials to the memo and save it as *[your initials]*10-13 in your Lesson 10 folder.
3. Select the words "all participants" in the first paragraph.
4. Click the Insert Hyperlink button ![icon] or press Ctrl+K.
5. In the Insert Hyperlink dialog box, under Link to, click Existing File or Web Page.
6. Click Current Folder if it is not active.
7. In the Look in text box, locate and click the file **NYwkend** on the student disk. The filename appears in the Address text box.

FIGURE 10-20
Linking to another document

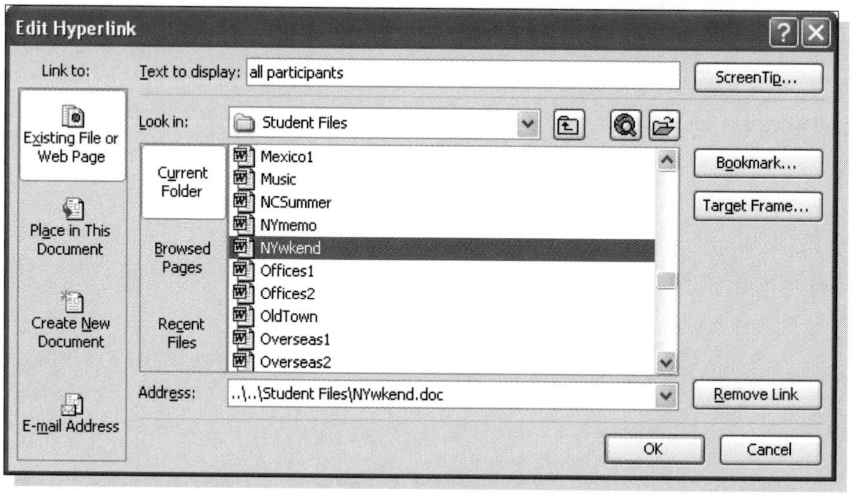

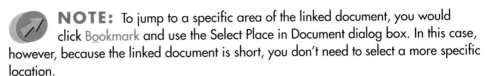

NOTE: To jump to a specific area of the linked document, you would click Bookmark and use the Select Place in Document dialog box. In this case, however, because the linked document is short, you don't need to select a more specific location.

8. Click OK in the Insert Hyperlink dialog box. The text "all participants" is blue and underlined, indicating you have created a hyperlink.

9. Press (Ctrl) and click the hyperlink to test it. Click Yes if a message box appears. Word opens the linked document.

NOTE: After inserting a hyperlink to another document, you can select the hyperlink text and modify it. You can also change the document or location to which you want to link. Right-click the hyperlink, choose Edit Hyperlink, and choose a different location.

EXERCISE 10-14 Use the Web Toolbar

The Web toolbar is designed to navigate between locations on the World Wide Web or on your company's intranet. It also provides some helpful buttons for managing Web pages.

1. Click the Back button 🌐 on the Web toolbar to return to the memo.

2. Click the Show Only Web Toolbar button 🔳. The Standard and Formatting toolbars disappear and the Web toolbar is the only toolbar open.

3. Click the Refresh button 🔳. When the dialog box opens asking if you want to discard changes, click Yes. The version of the document without the hyperlink reloads.

FIGURE 10-21
Word dialog box for discarding changes

4. Click the Show Only Web Toolbar button 🔳 so the Standard and Formatting toolbars reappear.

5. Select "New York weekend" in the first paragraph and click the Insert Hyperlink button 🔳.

6. In the Look in text box, locate and click the file **NYwkend**. Click OK.

7. Test the hyperlink, and then close the linked file.

8. Save the memo, print, and close it.

TABLE 10-4 Web Toolbar Buttons

BUTTON		FUNCTION
⊕	Back	Displays the document location or Web page from your previous hyperlink jump.
⊕	Forward	Displays the document location or Web page from your next hyperlink jump.
⊗	Stop	Stops the connection to a document or Web page that is in progress.
🗎	Refresh	Reloads the current document or Web page.
⌂	Start Page	Displays the Microsoft Start Page or the Web page you specified as your Start Page.
🔍	Search the Web	Displays the Search Page, where you can enter a keyword for a search.
Favorites ▾	Favorites	Displays a menu of favorite or frequently used Web pages or documents you specify.
Go ▾	Go	Displays a menu of commands for navigating and setting the Start and Search pages.
▣	Show Only Web Toolbar	Toggles between showing and hiding all toolbars except the Web toolbar.
[▾]	Address	Shows the current location and enables you to enter or select a file or Web site to access.

EXERCISE 10-15 Create a Hyperlink to a Web Site

In addition to creating hyperlinks between documents, you can create a link to a location on the World Wide Web. For example, suppose you're sending an e-mail memo to Duke City Gateway travel counselors about how to convert foreign currency to dollars. You can include a hyperlink to a Web site that lists up-to-date exchange rates for foreign currency.

To link to a Web site, you need to know the site's "address" or URL (Uniform Resource Locator). A *URL* is a combination of characters that is recognized by the Internet. For example, the Web address for Microsoft is **http://www.microsoft.com**. The prefix **http://** (which stands for Hypertext Transfer Protocol) is used for Web addresses, but is understood and is no longer necessary when keying a Web address.

NOTE: You cannot link to the Web without an Internet connection. Your computer must be equipped with appropriate hardware and software for navigating the Internet.

1. Start a new document. Create the memo shown in Figure 10-22. When you key the Web address **www.weather.com**, Word will automatically apply the hyperlink character formatting.

FIGURE 10-22
Creating automatic hypertext as you type

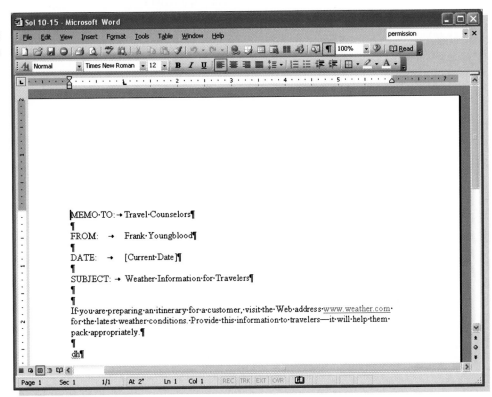

2. Add your reference initials to the memo and save it as *[your initials]***10-15** in your Lesson 10 folder.

3. Press Ctrl and click the new hyperlink to test it. Word jumps to The Weather Channel home page on the Web if your Internet connection is activated. If it is not, you might be asked to sign on.

 NOTE: If you receive an error message, ask your instructor how to proceed.

4. Scroll to the bottom of the Web page. Notice that the Web page contains both text and graphics. Scroll back to the top of the Web page.

5. Locate the text box where you can enter your city or ZIP Code for a local weather report. Enter your ZIP Code and click the "GO" button.

6. View the information for the current weather conditions in your area.

7. Close the Internet connection.

8. Use the taskbar to return to the document to print and close it.

TIP: Instead of using a Web address as a hyperlink in a document, you can use any word or phrase in the document. Simply select the document text, open the Insert Hyperlink dialog box, and enter the URL in the Address text box (just as you entered the path to link a document).

USING ONLINE HELP

Comments and tracked changes are both important collaboration features in Word. If you use a tablet computer, you can insert ink comments.

Use Help for more information about ink comments:

1. Key **ink comment** in the Ask a Question box.
2. Click the topic About using ink in Word.
3. Review the information on handwritten comments, writing directly in a document, and annotating a document with handwriting.
4. Return to the Search Results task pane and click the topic Insert an ink comment. Read the information on inserting comments.
5. Close Help when you finish.

LESSON 10 Summary

➤ Use comments to insert notes in a document. Comments are color-coded by reviewer. Make sure your name and initials appear in the Options dialog box on the User Information tab before inserting your comments.

➤ In Normal view, comments appear in the Reviewing pane (open the pane by first displaying the Reviewing toolbar). In Print Layout view and Print Preview, comments appear as balloons in the margins.

➤ Edit comment text in either the comment balloon or the Reviewing pane. Use the Next button 🔳 and the Previous button 🔳 on the Reviewing toolbar to move from comment to comment. See Table 10-1.

➤ To delete a comment, click within the comment text and click the Reject Change/Delete Comment button 🔳, or right-click the comment text and choose Delete Comment from the shortcut menu.

➤ To print comments with a document, use the Print command from Print Layout view or Print Preview. Or, open the Print dialog box and choose Document showing markup from the Print what drop-down list. Word scales the document text of the printed document to make room for the comment balloons.

➤ To print comments separate from a document, open the Print dialog box and choose List of markup from the Print what drop-down list.

➤ The Track Changes feature makes it easy to collaborate with others when revising documents.

➤ Tracked changes appear as revision marks that show where deletions, insertions, and formatting changes were made. In Normal view, inserted text appears underlined, deleted text has a horizontal line through it, and a revision bar appears to the left of every revised paragraph. In Print Layout view, tracked changes appear in the text and also in balloons in the right margin. Formatting changes only appear in Print Layout view.

➤ You can review a document after it is edited and decide if you want to accept or reject the revisions. Use the Reviewing toolbar buttons to accept or reject revisions, or right-click a revision mark and use the shortcut menu.

➤ You can review changes by the type of change made. You can also review only the changes made by a particular reviewer.

➤ In Normal view, revision marks appear as underlines for inserted text, strikethroughs for deleted text, and revision bars to the left of each revised line. In Print Layout view or Print Preview, revision marks appear in the text and in balloons in the margin.

➤ To accept all revisions at once, click the arrow to the right of the Accept Change button 🖉▾ and choose Accept All Changes in Document, or open the Display for Review drop-down list on the Reviewing toolbar and choose Final.

➤ When several people have reviewed a copy of a document, you can use the Compare and Merge Documents feature to consolidate revisions into one file.

➤ You can compare two documents, such as an edited copy of a document and the original document. Open the edited file, open the Compare and Merge dialog box, check Legal blackline, and then click Compare. Word opens a new document that shows the differences as color-coded revision marks.

➤ When you compare more than two documents, open the edited file, open the Compare and Merge dialog box, uncheck Legal blackline, and then click Merge. Word combines the documents and displays the differences as color-coded revision marks.

➤ You can save multiple versions of the same document under a single filename. Each version can contain detailed comments that describe the changes made to that version of the document.

➤ In longer documents, you can bookmark a location in a document version to show where a change was made.

➤ Send a document as an e-mail message or as an attachment to an e-mail message.

➤ To review the content of a document, switch to Reading Layout view. The document appears as a series of screens for easier readability. You can change the view of the document and navigate easily through the screens.

➤ A hypertext link, called a hyperlink, is text you click to move to another location. The location to which you move can be within the same document, in another document, somewhere on the World Wide Web, or your company's intranet.

➤ In a long document, you can use a hyperlink to jump quickly from one page to another. You can do this by creating a bookmark in a document and inserting a hypertext link to the bookmark, by inserting a hypertext link to a heading, or by using the Paste as Hyperlink command.

➤ You can create hyperlinks to another Word document or to Office documents, such as an Excel worksheet or a PowerPoint presentation.

➤ The Web toolbar is designed to navigate between locations on the World Wide Web or on your company's intranet. It also provides a quick way to navigate between hyperlinks. See Table 10-4.

➤ In addition to creating hyperlinks between documents, you can create a link to a location on the World Wide Web. To link to a Web site, you need to know the site's "address" or URL (Uniform Resource Locator). A URL is a combination of characters that is recognized by the Internet.

LESSON 10 Command Summary

FEATURE	BUTTON	MENU	KEYBOARD
Comments		Insert, Comment	Alt + Ctrl + M
Comments, print		File, Print, Print what	
Track Changes		Tools, Track Changes	Ctrl + Shift + E
Accept Change			
Reject Change			
Compare and Merge Documents		Tools, Compare and Merge Documents	
Create Versions		File, Versions	
Reading Layout view	Read	View, Reading Layout	Alt + R
Send document as e-mail		File, Send to, Mail Recipient	
Insert a hyperlink		Insert, Hyperlink	Ctrl + K

Concepts Review

Each of the following statements is either true or false. Indicate your choice by circling T or F.

T   F   *1.* In Normal view, comments appear in balloons.

T   F   *2.* You can access the Reviewing pane from the Formatting toolbar.

T   F   *3.* You can print a document with or without revision marks.

T   F   *4.* Comments and revision marks both appear in Print Preview.

T   F   *5.* A hypertext link is usually blue and underlined.

T   F   *6.* You can send a Word document as the body of an e-mail message.

T   F   *7.* When you protect a document from revisions, no one can add a revision to the document without entering a password.

T   F   *8.* You can turn off tracked changes by double-clicking TRK on the Status bar.

SHORT ANSWER QUESTIONS

Write the correct answer in the space provided.

1. When you're checking revisions, what does the button 🔲▾ do?

_____

2. When you compare and merge two documents by using the legal blackline feature, how does Word display the differences between the two documents?

_____

3. When you point to a comment balloon in Print Layout view, what does the ScreenTip display?

_____

4. Where are comments displayed if you want to view them in Normal view?

_____

5. How do you use the shortcut menu to delete a comment in Print Layout view?

_____

6. Which view displays document pages as a series of screens?

_____

7. Which toolbar do you use for comments and tracked changes?

_____

8. What does the mouse pointer change to when you point to a hyperlink and press Ctrl?

_____

CRITICAL THINKING

Answer these questions on a separate page. There are no right or wrong answers. Support your answers with examples from your own experience, if possible.

1. Think of a scenario when three of the new features introduced in this lesson could be helpful. Discuss Comments, Track Changes, and Compare Documents. Include advantages and disadvantages for using each feature.

2. What are the advantages of creating multiple versions of a document within the same document? Can you think of a specific situation in which this could be useful?

Skills Review

EXERCISE 10-16

Track changes in a document and insert, delete, edit, and print comments.

1. Open the file **Summer**.

2. Prepare the document for the insertion of comments by following these steps:

   a. Choose Options from the Tools menu and click the User Information tab.

   b. Make sure your name and initials are in the appropriate data fields.

   c. Click the Track Changes tab.

   d. Locate the Comments color: option in the Markup section, and click the down arrow to display the list of colors. Select dark blue.

   e. Verify the balloon options. The Preferred width should be 2.5 inches and balloons should display on the right margin.

   f. Click OK to close the dialog box.

3. Switch to Print Layout view and display the Reviewing toolbar.

4. Add comments by following these steps:

   a. Select Albuquerque in the first sentence.

   b. Click the Insert Comment button on the Reviewing toolbar.

c. Key **Check average temperature for June.** in the Comments balloon.

d. Select "July 4th" in the document, and click the Insert Comment button . Key the comment **Call to inquire about admission fees.** in the Comments balloon.

e. Create a comment for August in the third paragraph, using the text **Verify the dates for these events.**

5. Delete the Albuquerque comment by clicking within the balloon, and then clicking the Reject Change/Delete Comment button .

6. Edit the comment text for July 4th by clicking within the balloon and typing **the American Legion** after "Call."

7. Click the Track Changes button to turn on Track Changes mode.

8. In the first paragraph, select and delete the text "**, live entertainment,**" (be sure your selection begins with the first comma and ends with the second comma).

9. At the end of the second paragraph, key the sentence **This will be the 25th year that this event has been held**.

10. In the third paragraph, change the number 50 to **60**.

11. Turn off Track Changes mode by double-clicking TRK on the Status bar.

12. Save the document as *[your initials]***10-16** in your Lesson 10 folder.

13. Print the comments and track changes as balloons with the document text and as a separate list by following these steps:

a. Open the Print dialog box.

b. Make sure that the Print what drop-down list box is set to Document showing markup. Click OK.

c. Open the Print dialog box again.

d. Change the Print what drop-down list box to List of markup. Click OK.

14. Close and save the document.

EXERCISE 10-17

Merge three revisions of the same document and protect a document.

1. Open the file **Summerev1**. This is another revision of the original file Summer.

2. Compare the Summerev1 file with the original by following these steps:

a. Choose Tools, Compare and Merge Documents.

b. In the Compare and Merge Documents dialog box, locate and select the file **Summer**.

c. Make sure Legal blackline is unchecked. Click the arrow next to the Merge button and choose Merge into current document.

d. Use the down arrow next to the Accept Change button to accept all changes in the document.

e. Repeat steps a through c, merging the file **Summerev2**. Reject all the changes.

3. Protect the document from further revisions by following these steps:

   a. Choose <u>T</u>ools, <u>P</u>rotect Document.

   b. In the Protect Document task pane, click the check box below Editing Restrictions to select it. Choose <u>T</u>racked changes from the drop-down list and click Yes, Start Enforcing Protection.

   c. Click OK in the Start enforcing protection dialog box.

   d. In the first paragraph, delete the word "ethnic." Check the document protection by attempting to accept the revision.

4. Save the document as *[your initials]***10-17** in your Lesson 10 folder.

5. Print the document with revision marks; then close the document.

EXERCISE 10-18

Create two versions of a document.

1. Open the file **WildWalk**.

2. Save it as *[your initials]***10-18** in your Lesson 10 folder.

3. Save this version of the document by following these steps:

   a. Choose <u>F</u>ile, Ve<u>r</u>sions.

   b. Click the <u>S</u>ave Now button and key the comment **Letter format for first page.**

   c. Click OK.

4. Select the paragraph that begins "Thank you" (do not include the blank paragraph mark below it).

5. Cut the paragraph and paste it on the second page, below the text "Wilderness Walks."

6. Change the second sentence of this paragraph so it ends **is as follows** instead of "is attached."

7. Delete all the text from the beginning of the document through "Wilderness Walks" so the document fits on one page. One blank line should follow the paragraph.

8. Save this version of the document, with the comment **One-page handout.**

9. Display the previous version by opening the Versions dialog box, selecting the letter version, and clicking <u>O</u>pen.

10. Close the letter version.

11. Maximize the one-page handout version and add the heading **MONTANA WILDERNESS WALKS**. Center the heading and apply 16-point bold formatting. Change the spacing after to 24-points.

12. Format the underlined text as 14 point bold italic with no underline.

13. Print the one-page version of the document.

14. Save and close the document.

EXERCISE 10-19

Create hyperlinks within a document, to another document, and to a Web site.

1. Open the file **Duke2**.

2. Choose AutoFormat from the Format menu, and click OK.

3. Display the Web toolbar (View, Toolbars, Web).

4. Save the document as *[your initials]*10-19a in your Lesson 10 folder.

5. Create a link between the word "Southwest" on page 2 and a heading on page 7 by following these steps:

   a. At the bottom of page 2 or top of page 3, locate the paragraph that begins "Duke City Gateway truly is your gateway to the Southwest." Select the word "Southwest" and click the Insert Hyperlink button on the Standard toolbar.

   b. In the Insert Hyperlink dialog box, click the Place in This Document button if it is not active. Locate and select the heading Travel in the Southwest, and click OK.

   c. Press Ctrl and click the hyperlink to test it. Click the Back button on the Web toolbar to return to the hyperlink text.

6. Open the file **Offices**. Change the title to **Gateway Flagship Locations**.

7. Save the document as *[your initials]*10-19b in your Lesson 10 folder, and then close it.

8. Create a hyperlink to *[your initials]*10-19b by following these steps:

   a. On page 2, locate the paragraph under the heading "**Gateway Travel— Your Gateway to the World**." In the first sentence, change "50 offices" to **locations**. Delete the next paragraph, which begins "Of our 50 offices" and the office locations below it.

   b. Select the word "locations" you keyed in the previous step. Press Ctrl+K to open the Insert Hyperlink dialog box.

   c. Click the Existing File or Web Page button.

   d. Click Current Folder if it is not active.

   e. In the Look in text box, locate your Lesson 10 folder. Click the file *[your initials]*10-19b and click OK.

   f. Test the hyperlink. Click the Back button to go back to the first document.

9. Create a hyperlink to a Web site by following these steps:

   a. Locate the heading "**The Climate**." In the paragraph below the heading, select the text "weather conditions."

   b. Open the Insert Hyperlink dialog box. In the Address text box, key the URL **www.weather.com**. Click OK.

   c. Make sure your Internet connection is active, and click the hyperlink to test it.

> **d.** When the Web page loads, locate the text box where you can enter a city or ZIP Code for a local weather report. Key **Albuquerque** and click the "GO" button.
>
> **e.** Review the information, and then disconnect from the Internet.

10. To all pages in the document, add a footer that contains the filename as an AutoText entry, left-aligned, and "Page" followed by the page number, right-aligned.

11. Review the document in Reading Layout view by following these steps:

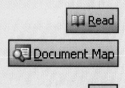

> **a.** Click the Read button on the Standard toolbar, and click the Document Map button if the Document Map pane is not displayed.
>
> **b.** Click The Climate heading to navigate to the screen containing this text. Notice the hyperlink in the paragraph.
>
> **c.** Click the Document Map button to close the Document Map pane, and click the Allow Multiple Pages button. Scroll through the document.
>
> **d.** Click the Thumbnails button and click the thumbnail for screen 4.
>
> **e.** Close Reading Layout view by clicking the Close button.

12. Print the pages that contain hyperlinks.

13. Save the document and close it. Close the linked document.

Lesson Applications

EXERCISE 10-20

Create comments, track changes, accept revisions, and create versions of a document.

    1. Start a new document and key the text shown in Figure 10-23.

FIGURE 10-23

```
Texas Travel

Duke City Gateway Travel is pleased to announce that its popular
Texas bus tour has been expanded this year to include more departure
dates and longer stays. A comfortable, air-conditioned bus becomes
your daytime "home on the road" as we travel through Texas. Hotel
accommodations throughout the trip are in charming, mid-sized hotels.
As we make our six-day journey through the Lone Star State, a trained
and knowledgeable tour guide will describe the sights along the way.
You will learn about the history and traditions of Texas and have a
chance to explore many different parts of the state.
```

    2. Center "Texas Travel" and make it uppercase bold. Format the heading with 72-point spacing before and 24-point spacing after. Delete the blank paragraph mark before the first paragraph if necessary.

    3. In the first sentence, select "departure dates" and insert the comment:

        Get complete list of departure dates, but do not include in promotional material.

    4. At the end of the paragraph, insert the comment:

        Ask Steve to write a summary about each stop on the tour.

    5. Turn on track changes.

    6. Open the document **Special1**.

    7. Copy the first two paragraphs and paste them after the last paragraph in the Texas Travel document.

    8. Above the pasted text, key the uppercase heading **OTHER TRIPS** in bold. Center the heading and apply 48-point spacing before and 24-point spacing after.

    9. Return to Special1. Find the paragraph that begins "Food lovers." Copy the paragraph and paste it at the end of the Texas Travel document.

10. Create a new version, with the comment **Original copy with proposed additions.** Save the document as *[your initials]*10-20 in your Lesson 10 folder when the Save As dialog box opens.

11. Turn off Track Changes mode. Delete all comments and accept all changes.

12. Justify all the paragraphs under the two headings.

13. Format both headings ("Texas TRAVEL" and "OTHER TRIPS") in a larger font size. Make any other revisions appropriate to the layout.

14. Save the revised document as a version, with the comment **Proposed additions with formatting.**

15. Print both versions, making sure revision marks and comments appear in the first version. Close the file.

16. Close Special1 without saving it.

EXERCISE 10-21

Create a hyperlink from one document to another.

1. Open the file **RedRiver**.

2. AutoFormat the document. Add the title **Red River Hotel and Conference Center**. Center the heading and apply 16-point bold formatting. Change the paragraph spacing for the title to 72 points before and 24 points after.

3. Spell-check the document and save it as *[your initials]*10-21a in your Lesson 10 folder.

4. Copy the first paragraph under the title. Save and close the document.

5. Start a new document and create a memo to Travel Counselors from Frank Youngblood regarding the Red River Hotel. Paste the paragraph below the memo heading.

6. Delete the last sentence of the paragraph you pasted. Add the following closing paragraph:

   More information is available about the hotel, if needed. The hotel is developing a Web site, which we will be able to visit soon.

7. Add your reference initials to the memo and save it as *[your initials]*10-21b in your Lesson 10 folder.

8. Create a hyperlink between the text "More information" and the document *[your initials]*10-21a.

9. Test the hyperlink. Print the linked document and close it.

10. Format the memo document appropriately and save it.

11. Print and close the document. Close the Web toolbar if it is open.

EXERCISE 10-22

Create a document to be sent as an e-mail.

    1. Start a new document and key the text in Figure 10-24.

FIGURE 10-24

Duke City Gateway will offer a new videoconference service for our corporate travel clients beginning October 1.

In order to promote our new service, a seminar is scheduled for August 15 to explain the purpose, advantages, and concerns of videoconferencing.

Please review the following lists and insert your comments.

Reasons to consider videoconferencing:
    Reduce number of meetings
    Reduce travel expenses
    Reduce time away from the office
    Increase number of participants

Concerns related to videoconferencing:
    Coordinating multisite meetings
    Cost and operation of equipment
    Technical support
    Presenters and participants may be afraid of cameras

    2. Spell-check the document and save it as *[your initials]***10-22** in your Lesson 10 folder.
    3. Send the document for review to your instructor. Hint: File menu, Send To, Mail Recipient (for Review).
    4. Close the document.

EXERCISE 10-23 *Challenge Yourself*

Merge two revisions of the same document, create multiple versions of a document, track revisions, and accept revisions.

    1. Open the file **Rockies2**.
    2. Save the document as *[your initials]***10-23** in your Lesson 10 folder.

3. Compare and merge the document with **Rockies1**. (Turn off Legal blackline, and merge into the current document.)

4. Save the revised document as a version, with the comment:

Contains spelling and grammatical errors.

5. Accept all revisions in the document.

6. Turn on Track Changes mode, making sure the revision marks appear.

7. In the first paragraph, select the last sentence, "It is an experience you will never forget." Insert a comment with the text **This is a cliché.**

 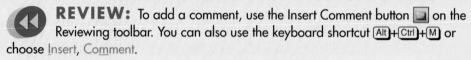

REVIEW: To add a comment, use the Insert Comment button 🔲 on the Reviewing toolbar. You can also use the keyboard shortcut Alt + Ctrl + M or choose Insert, Comment.

8. Delete the sentence you just selected and key:

It will provide you with a lifetime of memories.

9. In the sentence that begins, "Day 5 is devoted...," select the word "leisure." Insert a comment for this word with the text **Shouldn't this be "leisurely"? This is an adjective.**

10. Save the corrected document as a new version, with the comment:

Previous errors corrected; corrections made for style and grammar, with comments.

11. Turn off Track Changes mode and accept revisions in the document. Delete the comments.

 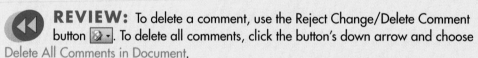

REVIEW: To delete a comment, use the Reject Change/Delete Comment button 🔲 ▾ . To delete all comments, click the button's down arrow and choose Delete All Comments in Document.

12. At the top of the document, add the following document title, formatted attractively:

Canadian Rockies
5-Day Escorted Tour

13. Save the corrected document as yet another new version, with the comment **Contains no errors. Title added.**

14. Print all three versions, making sure revision marks (and comments, if present) print in the versions that contain errors.

15. Close all versions.

On Your Own

In these exercises you work on your own, as you would in a real-life work environment. Use the skills you've learned to accomplish the task—and be creative.

EXERCISE 10-24
Key a page of text from a book or magazine you find interesting. Insert at least five comments. They can be humorous or insightful, or they can simply describe how you feel about the text. Practice editing, deleting, and navigating through the comments. Save the document as *[your initials]*10-24 in your Lesson 10 folder. Print the document with comments displayed.

EXERCISE 10-25
Compare any two similar documents. Insert a comment describing your results. Save the new document as *[your initials]*10-25 in your Lesson 10 folder and print it with revision marks displayed.

EXERCISE 10-26
Go to the Internet and research a topic that interests you. Write a report (at least two pages, double-spaced) about the topic. Insert hyperlinks to locations within the document and to Web sites. Test the hyperlinks. Use Reading Layout view to review the content. Save the document as *[your initials]*10-26 in your Lesson 10 folder and print it.

Unit 3 Applications

UNIT APPLICATION 3-1

Find and replace text and formatting, and move, copy, and paste text, using the Office Clipboard and multiple documents.

1. Open the files Albuquer, Balloon, Climate, ElMorro1, Mirage, RedRiver, and Summer.

2. Display the Office Clipboard. Clear it, if it contains any Clipboard items.

3. In Albuquer, insert a blank line after the paragraph, and then copy all the text, including the blank line, to the Clipboard.

4. Insert a blank line after the last paragraph in Balloon, and then copy the last paragraph, including the blank line, to the Clipboard.

5. Copy the first paragraph in Climate, ElMorro1, Mirage, RedRiver, and Summer. When copying, include the blank line after each paragraph. (If a document has a heading, copy the first text paragraph below the heading—do not copy the heading.)

6. Close all document files without saving changes. Start a new document and save it as *[your initials]***u3-1** in a new Unit 3 Applications folder.

 7. Redisplay the Clipboard task pane if it is no longer visible. Paste all the Clipboard entries (containing the text you just copied from the seven student files) into *[your initials]***u3-1**. Use the Paste All button . Change the document font to 11-point Arial.

8. Move the paragraph that starts "Summer is a great time" to the end of the paragraph that starts "Albuquerque is renowned." Maintain one blank line between paragraphs.

9. Move the paragraph that starts "El Morro National Monument" so it is the last paragraph in the document. Maintain one blank line between paragraphs.

10. On a new line just above the first paragraph, key **General Information** as an underlined heading. (Do not insert a blank line after the heading.)

11. As in the previous step, key **Climate** as the heading for the paragraph that starts "Albuquerque is renowned."

12. Key **Accommodations** as the heading for the paragraph that starts "La Mirage is."

13. Key **Nearby Point of Interest** as the heading for the paragraph that starts "El Morro National."

14. Edit the two paragraphs below the title "Accommodations," as shown in Figure U3-1.

FIGURE U3-1

La Mirage is a unique resort and convention center ~~located~~ in
Albuquerque / New Mexico. /Over 100 groups come ~~to La Mirage~~ each year
for ~~conventions,~~ business meetings, seminars, and training sessions.
In addition to a state-of-the-art convention center, La Mirage offers
~~a variety of~~ indoor and outdoor recreational facilities, including 3
pools, 8 tennis courts, and an 18-hole golf course.
The new Red River Hotel in Albuquerque has just been completed.
It's the perfect place to stay for seeing all the local sights or
as a stopover to other New Mexico destinatons. ~~The hotel has extended~~
~~very attractive discounts to all Gateway travel agencies, which we can~~
~~extend to our customers.~~

15. Find all underlined headings and replace them with headings that are not underlined and are uppercase, bold, and with 2 points (key **2 pt**) of spacing after paragraphs.

16. Key **Albuquerque Tourist Information** as the document title. Add 72 points of spacing before the title and 24 points of spacing after it. Change the title text to 14 point, uppercase, bold, and centered, with a text shadow.

17. Change the left and right margins for the entire document to 1 inch.

18. Spell-check the document and save it.

19. Print and close the document.

20. Clear the Office Clipboard and close the task pane.

UNIT APPLICATION 3-2

Move, copy, paste, find, and replace text.

1. Start a new document and key the text shown in Figure U3-2 on the next page.

2. Spell-check the document and save it as *[your initials]***u3-2a** in your Unit 3 Applications folder.

3. Copy the entire document.

4. Open the file **SpaLetter**.

5. Format the document as a business letter and change the font size to 11 points. Add reference initials and an enclosure notation.

FIGURE U3-2

> The Villa Rosa spa is located 25 miles from the Albuquerque
> International Airport. Set on 100 acres, its lush gardens and proximity
> to both desert and mountains make it a beautiful and relaxing vacation
> spot. Villa Rosa offers several of the spa services you enjoy, such as
> massage and sauna. It has 25 lovely rooms in the main facility and 5
> separate cottages. The restaurant offers delicious, healthy meals, and
> a gift shop showcases the work of local artists.
>
> Our manager spent a very rejuvenating weekend there last month and
> gave it the highest praise.

6. Paste the copied text (about Villa Rosa) before the paragraph that begins "Look over" in the current document. Use the Paste Options button to make sure the font size of the pasted text matches the destination formatting. Maintain a blank line between paragraphs.

7. Save this document as *[your initials]*u3-2b in your Unit 3 Applications folder.

8. Find "local artists." Key **and international** between "local" and "artists."

9. Replace each occurrence of "spa" with "resort."

10. In the second paragraph, move the sentence that begins "Set on 100 acres" so it is the last sentence in the paragraph.

11. Save, print, and close the document.

12. Make the following changes to the document *[your initials]*u3-2a:
    - Add a bold, centered title **VILLA ROSA** with a text shadow and two blank lines below it.
    - Delete the last paragraph.
    - Justify the paragraph below the title and apply 1.5-line spacing.
    - Change all the document text to Arial.
    - Set a 2-inch top margin.

13. In the third sentence, key a colon after "such as" and delete the remainder of the text in that sentence. Beginning on a new line below the colon, create a bulleted list that contains these items: Massage, Sauna, Facials, Exercise classes. (The sentence that begins "It has 25" should appear below the bulleted list.)

14. Replace the current bullets with picture bullets.

REVIEW: To add a picture bullet, open the Bullets and Numbering dialog box (Bulleted tab) and click Customize. Click Picture, select the picture bullet you want, and then click OK twice.

15. Replace "spa" with "resort" throughout the document.

16. Save, print, and close the document.

UNIT APPLICATION 3-3

Compare three documents, create three versions of a document, rewriting one in memo format.

1. Open the file Prices2.

2. Compare the document to Prices1, merging both documents into the current document, Prices2. Make a decision about each change, and accept or reject them.

3. Compare the document to Prices3, merging both documents into the current document, Prices2. Accept or reject each change.

4. Save the revised document as *[your initials]*u3-3 in your Unit 3 folder.

5. Turn on Track Changes mode to enter more revision marks.

6. Add the title **TOUR PRICES** as bold, centered, and uppercase, with two blank lines below it. Set a 2-inch top margin.

7. Save the document as a new version with the comment **Revised version**.

8. Accept all the changes and turn off Track Changes mode.

9. Add a shadow page border and justify the paragraphs under the title.

10. Save this as a version of the document with the comment **Final version with added formatting**.

11. Format the document as a memo by performing these changes:
- Remove the page border
- Left-align the paragraphs
- Delete the title and add a memo heading. The memo is to Staff, from Frank Youngblood. The subject is Tour Prices.
- Add your reference initials.

12. Turn on Track Changes mode. Read the memo carefully and edit the text so it is not written to the customer ("you," "your") but is appropriate for a memo to staff members.

13. Save the document as another version with the comment **Memo version with revisions**.

14. Print all three versions of the document.

15. Close all versions.

UNIT APPLICATION 3-4 *Using the Internet*

Copy information from three Web pages into one document, format the text consistently, and check spelling and grammar.

Locate three or more Web sites that contain information on your favorite hobby or on a topic that interests you.

Copy text from each site (make sure you select text only, no images, and press Ctrl+C to copy), and then paste it into a new Word document. Use the Keep Text Only option from the Paste Options button to eliminate Web formatting.

Create a formatted title for the document. Use paragraph and character formatting features to make the document appear consistently styled. Use the Find and Replace features to locate selected text and apply formatting for emphasis. Use the tracked changes feature as you edit the document.

Check spelling and grammar (sometimes Web sites contain misspelled words or poor grammar). Save the document as *[your initials]***u3-4** in your Unit 3 Applications folder. Print and close the document.

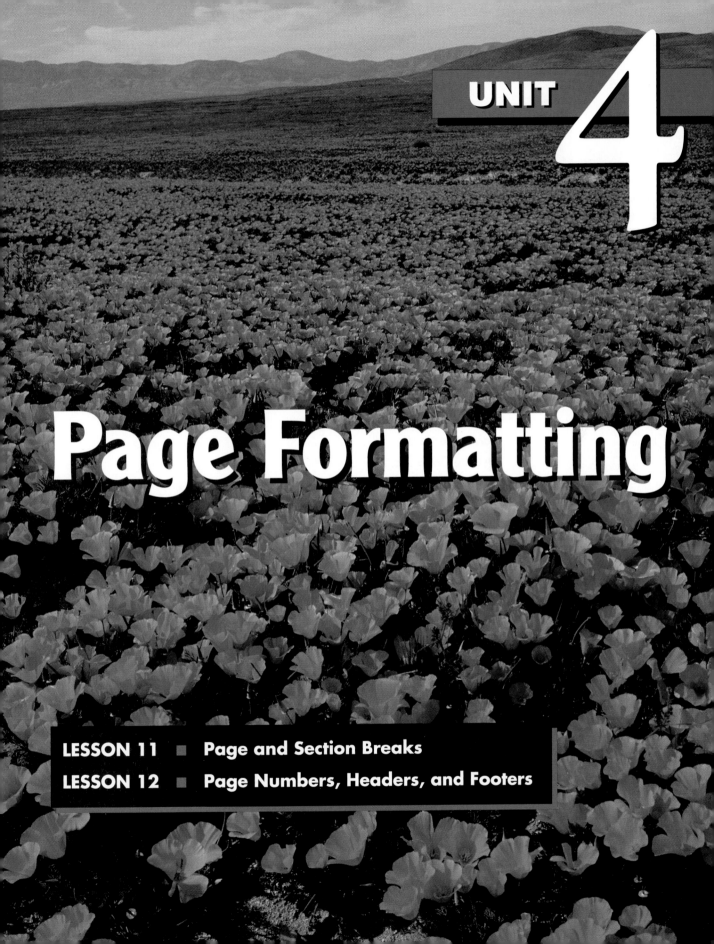

UNIT 4

Page Formatting

LESSON 11 ■ Page and Section Breaks
LESSON 12 ■ Page Numbers, Headers, and Footers

Page and Section Breaks

MICROSOFT OFFICE
SPECIALIST
ACTIVITIES
In this lesson:
WW03S-1-3
WW03S-3-5
WW03E-1-2

See Appendix C.

OBJECTIVES

After completing this lesson, you will be able to:

1. Use soft and hard page breaks.
2. Control line and page breaks.
3. Control section breaks.
4. Format sections.
5. Use the Go To feature.

 Estimated Time: 1 hour

In Word, text flows automatically from the bottom of one page to the top of the next page. This is similar to how text wraps automatically from the end of one line to the beginning of the next line. You can control and customize how and when text flows from the bottom of one page to the top of the next. This process is called *pagination*.

Sections, which were introduced in Lesson 6, are a common feature of long documents and have a significant impact on pagination. This lesson describes how to use and manage sections in greater detail.

Using Soft and Hard Page Breaks

As you work on a document, Word is constantly calculating the amount of space available on the page. Page length is determined by the size of the paper, and the top and bottom margin settings. For example, using standard-size paper and default margins, page length is nine inches. When a document exceeds this length, Word creates a *soft page break*. Word adjusts this automatic page break as you add or delete text. A soft page break appears as a horizontal dotted line on the screen in Normal view. In Print Layout view, you see the actual page break—the bottom of one page and the top of the next.

EXERCISE **11-1** **Adjust a Soft Page Break Automatically**

1. Open the file **Primer1**. Make sure the document is displayed in Normal view.

NOTE: When you first open a document, Word displays on the left side of the Status bar the total number of characters in the document. As you scroll through a document, Word displays the current page number on the left side of the Status bar.

2. Display the Reveal Formatting task pane. Make sure Show all formatting marks is checked.

3. Scroll to the bottom of page 3. Notice the soft page break separating the heading "<u>Calling Home</u>" from the paragraph below it.

NOTE: The page breaks described in this lesson might appear in slightly different locations on your screen.

4. Locate the paragraph just above the heading "<u>Calling Home</u>" (it begins "In Western Europe"). Move the insertion point to the left of "We recommend" in the eighth line and press (Enter) twice to split the paragraph. Notice the adjustment of the soft page break.

FIGURE 11-1
Adjusting the
position of a soft
page break

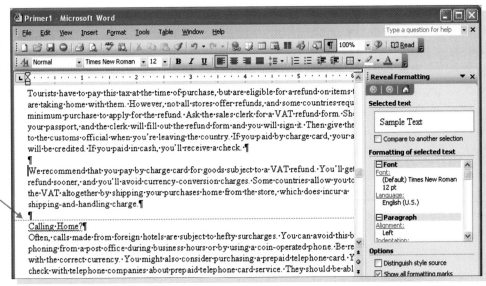

Soft page break

NOTE: The documents you create in this course relate to the Case Study about Duke City Gateway Travel, a fictional travel agency (see pages 1 through 4).

EXERCISE **11-2** **Insert a Hard Page Break**

When you want a page break to occur at a specific point, you can insert a *hard page break*. In Normal view, a hard page break appears on the screen as a dotted line with the words "Page Break."

There are two ways to insert a hard page break:

- Use the keyboard shortcut Ctrl+Enter.
- Choose <u>B</u>reak from the <u>I</u>nsert menu and click <u>P</u>age break.

1. Move the insertion point to the bottom of page 2, to the beginning of the paragraph that starts "If you're torn."

2. Press Ctrl+Enter. Word inserts a hard page break, so the paragraph is not divided between two pages.

3. Move to the middle of page 4 and place the insertion point to the left of the text that begins "For specific information."

TIP: Remember the various methods for moving within a long document. For example, you can drag the scroll box on the vertical scroll bar and use the scroll arrows to adjust the view. You can also use keyboard shortcuts: Ctrl+↑ or ↓ to move up or down one paragraph, Page Up or Page Down to move up or down one window, and Ctrl+Home or Ctrl+End to move to the beginning or end of a document.

4. Choose <u>B</u>reak from the <u>I</u>nsert menu to display the Break dialog box.

FIGURE 11-2
Break dialog box

5. Make sure <u>P</u>age break is selected and click OK. Word inserts a hard page break and adjusts pagination in the document from this point forward.

FIGURE 11-3
Inserting a hard
page break

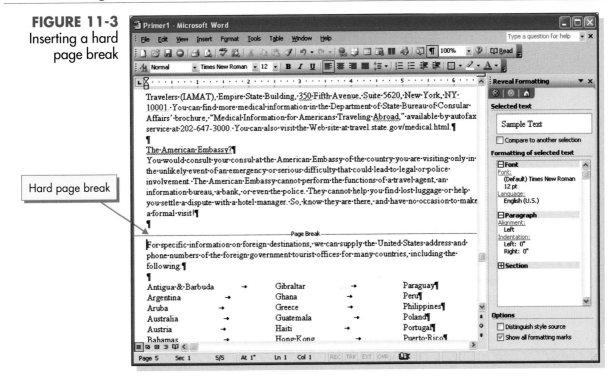

Hard page break

EXERCISE **11-3** **Delete a Hard Page Break**

You cannot delete a soft page break, but you can delete a hard page break by clicking the page break and pressing Backspace or Delete.

1. Move the insertion point to the page break you just inserted by positioning the I-beam over the page break and clicking.

2. Press Delete to delete the page break.

3. Scroll back to the hard page break you inserted at the top of page 3. Position the insertion point to the left of "If you're torn" and press Backspace. The page break is deleted and Word adjusts the pagination.

Controlling Line and Page Breaks

To control the way Word breaks paragraphs, choose one of four line- and page-break options from the Paragraph dialog box:

● Widow/Orphan control
   A *widow* is the last line of a paragraph and appears by itself at the top of a page. An *orphan* is the first line of a paragraph and appears at the bottom of a page. By default, this option is turned on to prevent widows and

orphans. Word moves an orphan forward to the next page and moves a widow back to the previous page.

- Keep lines together
  This option keeps all lines of a paragraph together on the same page rather than splitting the paragraphs between two pages.

- Keep with next
  If two paragraphs need to appear on the same page no matter where page breaks occur, use this option. The option is most commonly applied to titles that should not be separated from the first paragraph following the title.

- Page break before
  Use this option to place a paragraph at the top of a new page.

EXERCISE 11-4   Apply Line and Page Break Options to Paragraphs

1. Close **Primer1** without saving; then reopen the document.

    TIP: To reopen the file quickly, choose the filename from the bottom of the File menu.

2. At the bottom of page 3, click within the heading "Calling Home." You are going to format this heading so it will not be separated from its related paragraph.

3. Choose Paragraph from the Format menu to open the Paragraph dialog box and click the Line and Page Breaks tab.

FIGURE 11-4
Line and Page
Break options in
the Paragraph
dialog box

4. Click Keep with next to select it and click OK. Word moves the page break, keeping the two paragraphs together. Notice that this line- and page-break option appears in the Reveal Formatting task pane.

> **NOTE:** When you apply the Keep with next, Page break before, or Keep lines together option to a paragraph, Word displays a small black nonprinting square to the left of the paragraph (if the Show/Hide ¶ button ¶ is turned on).

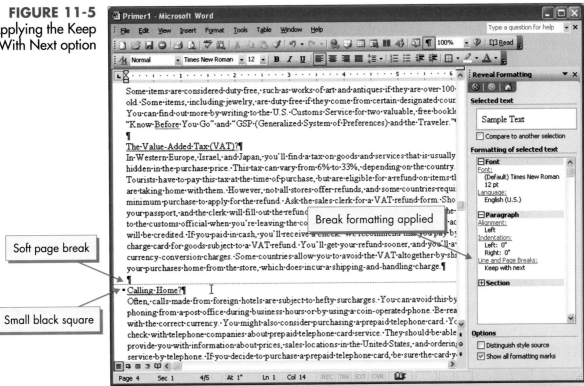

FIGURE 11-5
Applying the Keep
With Next option

5. Locate the paragraph at the bottom of page 2 that begins "If you're torn." The paragraph is divided by a soft page break.

6. Right-click the paragraph to open the shortcut menu, and then choose Paragraph.

7. Choose Keep lines together and click OK. The soft page break moves above the paragraph to keep the lines of text together.

> **TIP:** You can click the Line and Page Breaks link in the task pane to open the Paragraph dialog box, Line and Page Breaks tab, if you want to adjust the formatting.

8. Move to page 4 and place the insertion point in the paragraph that begins "For specific information." You will format this paragraph so it begins at the top of the page.

9. Open the Paragraph dialog box, click Page break before, and click OK. Word starts the paragraph at the top of page 5 with a soft page break.

10. Turn on automatic hyphenation by choosing Language from the Tools menu, and then choosing Hyphenation. Click the Automatically hyphenate document check box and click OK.

11. Click in the paragraph directly below "The American Embassy." Notice that the paragraph is hyphenated.

12. Choose Paragraph from the Format menu.

13. Click the Don't hyphenate check box and click OK. The current paragraph should no longer be hyphenated.

14. Turn off automatic hyphenation by choosing Language from the Tools menu, and then choosing Hyphenation. Click the Automatically hyphenate document check box and click OK.

15. Save the document as *[your initials]*11-4 in a new folder for Lesson 11. Leave it open for the next exercise.

Controlling Section Breaks

Section breaks separate parts of a document that have formatting different from the rest of the document. Lesson 6 explained how changing the left and right margins of selected text results in a separate section.

For better control in creating section breaks, you can insert a section break directly into a document at a specific location by using the Break dialog box. You can also specify the type of section break you want to insert.

TABLE 11-1   Types of Section Breaks

TYPE	DESCRIPTION
Next page	Section starts on a new page.
Continuous	Section follows the text before it without a page break.
Even page or Odd page	Section starts on the next even- or odd-numbered page. Useful for reports in which chapters must begin on either odd-numbered or even-numbered pages.

EXERCISE   **11-5**   **Insert Section Breaks by Using the Break Dialog Box**

1. Place the insertion point to the left of the paragraph at the top of page 5 that begins "For specific information."

2. Press Ctrl+Q. This clears the formatting for the paragraph, removing the soft page break you applied earlier.

3. Choose Break from the Insert menu. Under Section break types, choose Continuous and click OK. Word begins a new section on the same page, from the position of the insertion point.

4. Click above and below the section mark. Notice that the section number changes on the Status bar but the page number stays the same.

5. In the Reveal Formatting task pane, click the plus sign (+) to the left of Section to display the section formatting.

FIGURE 11-6
Inserting a continuous section break

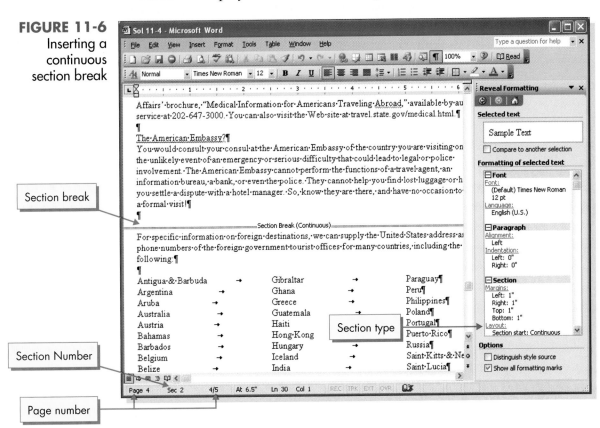

Formatting Sections

After you create a new section, you can change its formatting or specify a different type of section break. This is often useful for long documents, which sometimes contain many sections that require different page formatting, such as different margin settings or page orientation. For example, you can change a Next page section break to a Continuous section break or you can change the page orientation of a section, without affecting the rest of the document.

> **NOTE:** The formatting you apply to the section is stored in the section break. If you delete a section break, you also delete the formatting for the text above the section break. For example, if you have a two-section document and you delete the section break at the end of section 1, the document becomes one section with the formatting of the previous section 2.

EXERCISE **11-6** **Apply Formatting to Sections**

1. Use the Break dialog box to insert a Next page section break before the text "**What About**..." on page 2.

2. With the insertion point in the new section, choose Page Setup from the File menu.

3. Click the Layout tab and click to open the Section start drop-down list. From this list, you can change the section break from Next page to another type.

4. Choose Continuous, so the section does not start on a new page.

FIGURE 11-7
Using the Page Setup dialog box to modify the section

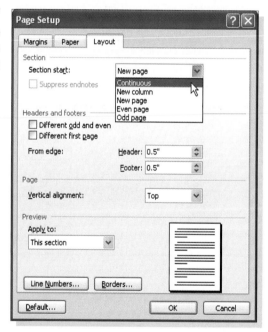

5. Click the Margins tab.

6. Set 1.25-inch left and right margins. Make sure This section appears in the Apply to box and click OK. Section 2 of the document now has new margin settings.

EXERCISE **11-7** **Change the Vertical Alignment of a Section**

Another way to format a section is to specify the vertical alignment of the section on the page. For example, you can align a title page so the text is centered between the top and bottom margins. Vertical alignment is a Layout option available in the Page Setup dialog box.

1. Move the insertion point to the last section of the document (which begins "For specific"). Notice that, because this section does not start on a new page, a page break interrupts the list of countries.

2. Open the Page Setup dialog box and click the Layout tab.

3. Use the Section start drop-down list to change the section from a Continuous to a New page section break.

4. Open the Vertical alignment drop-down list and choose Center. Click OK.

TABLE 11-2 **Vertical Alignment Options**

OPTIONS	DESCRIPTION
Top	Aligns the top line of the page with the top margin (default setting).
Center	Centers the page between the top and bottom margins with equal space above and below the text.
Justified	Aligns the top line of the page with the top margin and the bottom line with the bottom margin, with equal spacing between the lines of text (similar in principle to the way Word justifies text between the left and right margins).
Bottom	Aligns the bottom line of a partial page along the bottom margin.

FIGURE 11-8
Vertical alignment options

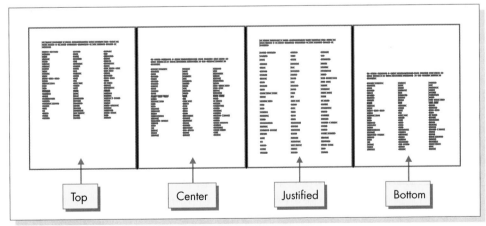

EXERCISE **11-8** **Check Pagination in Print Preview and Page Layout View**

After you apply page breaks, section breaks, or section formatting, use Print Preview or Print Layout view to check the document. Viewing the pages in relation to one another might give you ideas for improvement before printing.

Remember, you can edit and change the formatting of a document in Print Layout view or Print Preview.

1. Click the Print Preview button to preview the current section. Click the One Page button 🖻 if necessary. Notice that the text is centered between the top and bottom margins. Notice also that Print Preview doesn't show the dotted lines of the section breaks, but it does show how the page will look when you print it.

2. While still in Print Preview, open the Page Setup dialog box and change the vertical alignment to Justified. Click OK. Word justifies the last page of the document so the text extends from the top to the bottom margin.

3. Scroll back, page by page, to page 2, section 1, of the document. (Check the Status bar for location.)

4. From the Zoom box, choose Page Width. You can't see the dotted lines of the page break before the title "Primer for First-Time Overseas Travelers" or the continuous section break before "What about…" but you can check the formatting and see how the document will look when printed.

5. Click the Print Layout View button 🖻 to close Print Preview and switch to Print Layout view. Close the task pane.

6. Scroll to page 2, section 1. Notice that in Print Layout view, page breaks are indicated by the actual layout of each page as it will look when printed.

7. From the Zoom box on the Standard toolbar, choose Two Pages. This reduces the document display so you can see two pages at the same time.

8. Scroll to the end of the document. Choose Page Width from the Zoom box and click the Normal View button 🗏 to switch back to Normal view.

Using the Go To Feature

You use Go To to move through a document quickly. For example, you can go to a specific section, page number, comment, or bookmark. Go To is a convenient feature for long documents—it's faster than scrolling, and it moves the insertion point to the specified location.

There are three ways to initiate the Go To command:

- Choose Go To from the Edit menu.
- Double-click on the Status bar (anywhere to the left of "REC").
- Press Ctrl+G or F5.

EXERCISE **11-9** Go To a Specific Page or Section

> 1. With the document in Normal view, press F5. Word displays the Go To tab, located in the Find and Replace dialog box.

FIGURE 11-9
Using the Go To
feature

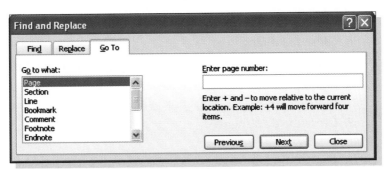

> 2. Scroll through the Go to what list to review the options. Choose Section from the list and click Previous until you reach the beginning of the document.
>
> 3. Click Next until the insertion point is located at the beginning of the last section, which is Section 3.
>
> 4. Choose Page from the Go to what list and click Previous. The insertion point moves to the top of the previous page.
>
> 5. Key **2** in the Enter page number text box and click Go To. The insertion point moves to the top of page 2.
>
> 6. Close the dialog box.

EXERCISE **11-10** Go To a Relative Destination

> You can use Go To to move to a location relative to the insertion point. For example, with Page selected in the Go to what list, you can enter **+2** in the text box to move forward two pages from the insertion point. You can move in increments of pages, lines, sections, and so on. Another option is to move by a certain percentage within the document, such as 50 percent—the document's midpoint.
>
> 1. Double-click the word "Page" on the Status bar to reopen the dialog box.
>
> 2. Choose Line from the Go to what list and enter **4** in the text box. Click Go To. The insertion point moves to the fourth line in the document.
>
> 3. Enter **+35** in the text box and click Go To. The insertion point moves forward 35 lines from the previous location.

4. Enter **-35** in the text box and click Go To. The insertion point moves back to the previous location.

5. Click Page in the Go to what list, enter **50%** in the text box, and click Go To. The insertion point moves to the midpoint of the document.

 NOTE: You must select Page in the Go to what list to use a percentage.

6. Close the dialog box.

7. Save the document as *[your initials]***11-10** in a new Lesson 11 folder.

8. Open the Print dialog box. Key **2-5** in the Pages text box and choose 4 pages in the Pages Per Sheet list box. Click OK.

9. Close the document.

USING ONLINE HELP

Working with sections can have a major impact on the overall page layout of a document.

Use Help to review section options:

1. Key **section** in the Ask a Question text box and press Enter.

2. Click the topic About sections and section breaks.

3. Review the information in the Help window. Click any of the blue text links for specific information.

4. Close the Help window when you finish.

LESSON Summary

➤ Pagination is the Word process of flowing text from line to line and from page to page. Word creates a soft page break at the end of each page. When you edit text, you adjust line and page breaks. You can adjust the way a page breaks by manually inserting a hard page break (Ctrl+Enter).

➤ Delete a hard page break by clicking it and pressing Delete or Backspace.

➤ The Paragraph dialog box contains line and page break options to control pagination. To prevent lines of a paragraph from displaying on two pages, click in the paragraph and apply the Keep lines together option. To keep two paragraphs together on the same page, click in the first paragraph and apply the Keep with ne_x_t option. To insert a page break before a paragraph, click in the paragraph and choose the Page _b_reak before option.

➤ Use section breaks to separate parts of a document that have different formatting. Apply a _N_ext page section break to start a section on a new page or a Con_t_inuous section break to continue the new section on the same page. Apply an _E_ven page or _O_dd page section break to start a section on the next even- or odd-numbered page.

➤ Change the vertical alignment of a section by clicking within the section and choosing _F_ile, Page Set_u_p. On the Layout tab, under _V_ertical alignment, choose an alignment (Top, Center, Justified, or Bottom).

➤ Check pagination in Print Preview or Print Layout view. Scroll through the document or change the zoom to display a different view.

➤ Use the Go To command to go to a specific page or section in a document. You can also go to a relative destination, such as the midpoint of the document or the 50th line.

LESSON 11   Command Summary

FEATURE	BUTTON	MENU	KEYBOARD
Insert hard page break		_I_nsert, _B_reak	Ctrl + Enter
Apply line and page break options		_F_ormat, _P_aragraph, Line and _P_age Breaks	
Insert section breaks		_I_nsert, _B_reak	
Apply formatting to sections		_F_ile, Page Set_u_p	
Go To		_E_dit, _G_o To	Ctrl + G or F5

Concepts Review

Each of the following statements is either true or false. Indicate your choice by circling T or F.

T F **1.** You can delete a hard or soft page break by pressing Delete.

T F **2.** To insert a section break, press Ctrl + Enter.

T F **3.** One way to insert a page break is to choose Break from the Insert menu.

T F **4.** Page break before is a paragraph formatting option that starts a paragraph at the top of a new page.

T F **5.** A nonprinting character appears to the left of any paragraph to which you apply the Keep with next option.

T F **6.** Section breaks appear in the Print Preview window as double dotted lines.

T F **7.** Page breaks appear in Print Layout view as single dotted lines.

T F **8.** You can use the Go To feature to move the insertion point from one section to another.

Write the correct answer in the space provided.

1. Which type of page break is automatically adjusted as you key text?

_____

2. Which type of section break does not start on a new page?

_____

3. What is the term for the last line of a paragraph that appears alone at the top of a page?

_____

4. Which option would you apply to a paragraph so it is not divided by a page break?

_____

5. Which dialog box and tab would you display to change the vertical alignment of a section?

_____

6. Which type of vertical alignment spaces text so the top line aligns with the top margin and the bottom line aligns with the bottom margin?

_____

7. In Print Layout view, which toolbar item do you use to view two pages at the same time?

_____

8. Describe the appearance of the nonprinting character Word displays next to a paragraph when you apply certain line- and page-break options.

_____

CRITICAL THINKING

Answer these questions on a separate page. There are no right or wrong answers. Support your answers with examples from your own experience, if possible.

1. In a long document that requires extensive editing, why would it be most efficient to perform all your edits before inserting hard page breaks?
2. Describe a situation where you would use a continuous section break.

Skills Review

EXERCISE 11-11

Adjust soft page breaks and insert hard page breaks.

1. Open the file **Succeed**.
2. Scroll to the bottom of page 1 to see where the soft page break occurs.
3. Change the font size for the entire document to 14 points. Notice how the change affects the soft page break.
4. Insert a page break before the text "Recreation clubs" by following these steps:
   a. Place the insertion point to the left of the text.
   b. Press Ctrl + Enter.

5. Use the menu method to insert another page break on page 2 by following these steps:

   a. Place the insertion point to the left of the paragraph that begins "<u>General</u>" on page 2.

   b. Choose <u>B</u>reak from the <u>I</u>nsert menu. Make sure <u>P</u>age break is selected and click OK.

6. View the document in Print Preview, one page at a time.

7. Use the Multiple Pages button to view three pages at once.

8. Click the Magnifier button [image], select the text for the entire document, and change the font size to 12 points.

9. Close Print Preview and remove the second page break by following these steps:

   a. Place the insertion point to the left of "<u>General</u>."

   b. Press (Backspace).

10. Save the document as *[your initials]***11-11** in your Lesson 11 folder.

11. Print page 1 and close the document.

EXERCISE 11-12

Apply line- and page-break options to paragraphs.

1. Open the file **Itin2**.

2. Change the left and right margins to 1.25 inches.

3. Format the title as 14-point centered uppercase. Add 72 points of spacing before and 24 points of spacing after the paragraph. Delete the paragraph mark following the title.

4. Delete the text from "Travel to the Southwest via bus" through "In Nevada: Las Vegas." If there is an extra blank paragraph mark, delete it.

5. Apply paragraph formatting to the heading "Sample Itinerary—The California Coast" so it begins on a new page by following these steps:

   a. Move the insertion point within the heading.

   b. Choose <u>P</u>aragraph from the F<u>o</u>rmat menu.

   c. Click the Line and Page Breaks tab.

   d. Choose Page <u>b</u>reak before and click OK.

6. Click the Undo button [image] to undo the formatting.

7. Format the same heading so it is not separated from the next paragraph by following these steps:

   a. Select the heading and the paragraph below it that begins "Day 1" (the line above the soft page break).

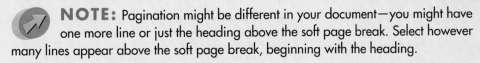

NOTE: Pagination might be different in your document—you might have one more line or just the heading above the soft page break. Select however many lines appear above the soft page break, beginning with the heading.

    b. Open the Paragraph dialog box and display the Line and Page Breaks tab.
    c. Choose Keep with next and click OK.

8. Format the text of each sample itinerary (the paragraphs under "Sample Itinerary—The Great Southwest" and "Sample Itinerary—The California Coast") with a 0.5-inch left tab and a 0.5-inch hanging indent. Replace the space that follows each Day number (i.e., "Day 1," "Day 2," etc.) with a tab character.

9. Save the document as *[your initials]*11-12 in your Lesson 11 folder.

10. Print and close the document.

EXERCISE 11-13

Specify section breaks by type and change the margin settings in sections.

1. Open the file **Duke4**.

2. Insert a continuous section break on page 2 before the heading "Corporate Travel" by following these steps:

    a. Place the insertion point to the left of the heading.
    b. Open the Break dialog box, choose Continuous, and click OK.

3. Use the Break dialog box to insert a Next page section break on page 3 before the heading "Duke City—Our Home Town."

4. At the next heading, "The Area," insert another continuous section break.

5. Use the Page Setup dialog box to format section 4 with 1.5-inch left and right margins. Make sure This section appears in the Apply to box.

6. Move the insertion point into section 2 and apply the same formatting.

TIP: Press F4 or choose Edit, Repeat Page Setup to repeat the margin formatting you previously applied.

7. View the document in Print Preview. If necessary, adjust soft page breaks to keep headings with paragraphs.

8. Save the document as *[your initials]*11-13 in your Lesson 11 folder.

9. Print and close the document.

EXERCISE 11-14

Vertically align a section and move around a document by using the Go To feature.

1. Open the file **Mexico1**.

2. Insert Next page section breaks at "The itinerary is as follows" on page 1 and at "Stateroom Locations and Rates" at the bottom of the same page.

3. Vertically align the text in section 3 by following these steps:

   a. With the insertion point in section 3, choose Page Setup from the File menu and click the Layout tab.

   b. Choose Center from the Vertical alignment drop-down list.

   c. Make sure This section appears in the Apply to box and click OK.

4. Switch to Print Preview and view only the last page (section 3).

5. From the Print Preview window, open the Page Setup dialog box again and change the vertical alignment to Justified.

6. Close Print Preview.

7. Use the Go To feature to move within the document by following these steps:

   a. Press F5.

   b. Choose Section from the Go to what list and enter **2** in the text box.

   c. Click Go To and close the dialog box.

8. With the document in Normal view, select the section break on page 1 and press Delete to delete it.

9. View the document in Print Layout view.

10. Add 72 points of spacing before the title on page 1. Delete the blank line below "Stateroom Locations and Rates" on the last page.

11. Save the document as *[your initials]***11-14** in your Lesson 11 folder.

12. Print and close the document.

Lesson Applications

EXERCISE 11-15

Insert page breaks, apply line- and page-break options, and format text as a new section.

1. Open the file **Duke4**.
2. Edit the text on page 1 as shown in Figure 11-10.

FIGURE 11-10

(Merge with previous paragraph)

Get to know us, and, ~~as we~~ we'll get to know you and your travel needs. ~~wonderful things will happen! We will plan your travel down to the~~ ~~very last detail. And you will, we hope, come to depend on us for all~~ ~~your travel needs. We built our business by building friendships, and~~ ~~the world of travel makes this possible.~~ The sky's the limit at Duke City Gateway Travel.

Gateway Travel—Your Gateway to the World

Duke City Gateway Travel is the Southwestern flagship office of Gateway Travel, an international travel agency with 50 offices throughout the United States, Canada, and Europe. All the Gateway divisions ~~work together, share their expertise,~~ (and are committed to excellence. All Gateway divisions (offer state-of-the-art service, and are accessible on the World Wide Web at www.gwaytravel.com,) Gateway Travel is your information resource for every imaginable travel query. ~~We are~~ ~~the "travel know-it-alls." What we can't answer immediately we will~~ ~~research and get back to you on within 24 hours. Our policy is "Can~~ We'll Do," and we will make your travel goals and objectives work for you.

(Ask us anything you want to know, and tell us where you want to go.)

3. Below the Gateway offices on page 1, insert a page break at the bold heading (which begins "Duke City Gateway Travel—Your Gateway to the Southwest").
4. At the beginning of the document, place the insertion point to the left of the blank paragraph mark under the boxed title and press Delete to delete it.

5. Place the insertion point to the left of "Introduction" and insert a next page section break.

6. Place the insertion point in the title section and format the title section with 2-inch left and right margins and centered vertically on the page.

 TIP: Place the insertion point to the left of "Introduction," under the title, before you insert the section break.

7. Increase the font size of the title in section 1 to 20 points and apply the shadow text effect.

8. Change the border of the title to a shadow border that is 10 points from the top and bottom of the text.

 TIP: To change the spacing between the text and the border, click the Options button in the Borders and Shading dialog box.

9. In section 2, add 6 points of spacing after each of the five bold headings. Add 3 points of spacing after each of the six underlined headings.

10. On page 3, section 2, split the paragraph below "Corporate Travel" at the third sentence (which begins "We understand that").

11. Format the heading "Automation" so it stays with the paragraph after it.

12. View the document in Print Preview to check the pagination. Shrink to fit the document on four pages.

13. Save the document as *[your initials]*11-15 in your Lesson 11 folder.

14. Print the document four sheets per page, and then close it.

EXERCISE 11-16

Add and format sections.

1. Open the file **Forman2**.

2. Set a 2-inch top margin. Add the date at the top of the document, using the correct date format and three blank lines below the date.

3. Edit the paragraph that begins "For your week" so it reads **For your week in Toronto, see the enclosed list of bed-and-breakfast recommendations.**

4. Move the three paragraphs that follow (from "The Fountaines" through "$85 per night") to the end of the document as a separate section on a new page.

5. Title the new section **TORONTO BED-AND-BREAKFAST RECOMMENDATIONS** using bold uppercase text. Center the title and increase the font size to 14 points. Add two blank lines below the title (or 24 points of paragraph spacing).

6. Format section 2 so it is vertically centered and in landscape orientation. Set a 2-inch left margin for this section and verify that there is a 2-inch right margin so the page is centered horizontally.

7. View both pages in Print Preview.

8. While in Print Preview, change the letter so it is from you as Associate Travel Counselor, and add an enclosure notation.

9. Save the document as *[your initials]***11-16** in your Lesson 11 folder.

10. Print and close the document.

EXERCISE 11-17

Apply line and page break options, add and format sections, and use the Go To feature.

1. Open the file **Europe1**.

2. Add a continuous section break before the heading "$1 Equals" and another at the bottom of the list (before the paragraph that begins "If you're wondering").

3. Format section 2 with a 2.5-inch left margin and double spacing.

4. In the paragraph that begins "If you're wondering," change the end of the sentence to read **the cities listed on the following page.**

5. Move the last paragraph of the document, which begins "The European Travel Update," so it follows the paragraph that begins "If you're wondering" (as a separate paragraph).

6. If the paragraph that begins "The European Travel" is divided by a soft page break, format it to keep the lines together.

7. Change the double-spaced text in section 2 to 1.5-line spacing.

8. Insert a Next Page section break at the heading "European Weather in June." Format this heading as bold, uppercase, centered, and followed by two blank lines.

9. Delete the date at the top of page 1 and the blank line below it.

10. Format the title at the top of page 1 as bold, uppercase, centered, and followed by two blank lines. Add 72 points of paragraph spacing before the title.

11. Use the Go To feature to go to section 4.

12. Indent the tabbed text in this section one inch from the left margin. Set two centered tab stops for the "Average High" and "Average Low" columns so the text is centered attractively below "European Weather in June."

13. Double-space the tabbed text.

14. Change the alignment in section 4 so it is centered vertically.

15. Use the Replace feature to replace "June" with **July** throughout the document.

16. Preview the document and save it as *[your initials]*11-17 in your Lesson 11 folder.

17. Print and close the document.

EXERCISE 11-18 *Challenge Yourself*

Add and delete page breaks, and add and format a new section.

1. Create a standard business memo from you to Frank Youngblood. The subject is "List of Gateway Travel Offices." Key the body text shown in Figure 11-11.

FIGURE 11-11

As you requested, attached is a separate list of the Gateway Travel offices. In addition to the U.S. offices, the list includes the offices in Canada, England, and France.

I know you wanted to include this list in the next newsletter. We can submit the list in file format to the desktop publisher. The file, "Offices," can be found in the folder C:\Gateway\Divisions.

Please note every office can also be reached at our Web site, www.gwaytravel.com.

2. At the end of the memo, add an attachment reference. After the reference, create a new section that starts on a new page.

3. In the new section, insert the file **Offices2**.

 TIP: To insert a file, place the insertion point in the new section, choose File from the Insert menu, locate the filename, and click OK.

4. Delete the hard page break in the new section.

5. Change the title at the top of section 2 to **Gateway Travel Offices in the United States.** Format the title as 14-point bold, all caps, centered, with two blank lines below it.

6. Delete the Pacific Gateway Travel office listing.

7. Insert a page break before the Ontario Gateway office listing so this address appears at the top of page 3.

8. Copy the title (including the two blank lines below it) on page 2 to the top of page 3. Modify the title to read **GATEWAY TRAVEL OFFICES OUTSIDE THE UNITED STATES.**

9. Set a 2-inch left indent for the text below the titles on pages 2 and 3.

10. If you haven't already done so, format section 1 with a 2-inch top margin. Make sure section 2 has the regular 1-inch top margin and is centered vertically.

11. Spell-check section 1 only. Check the pagination of the document in Print Preview.

12. Save the document as *[your initials]***11-18** in your Lesson 11 folder.

13. Switch to Normal view. Print and close the document.

On Your Own

In these exercises you work on your own, as you would in a real-life work environment. Use the skills you've learned to accomplish the task—and be creative.

EXERCISE 11-19

Write a short report about your ten favorite television shows or movies. Include a document title. Each show or movie should be a separate paragraph with its own heading. Adjust page breaks as needed to keep headings with their related paragraphs. Save the document as *[your initials]***11-19** and print it.

EXERCISE 11-20

Create a document that includes three poems by three different poets. Use headings to identify each poet and title. Use page breaks to start each poem on a separate page. Save the document as *[your initials]***11-20** and print it.

EXERCISE 11-21

Create a document that lists three different categories of restaurants in your area. Include descriptions of two to three restaurants per category. Make each category a separate section. Format each section differently. Save the document as *[your initials]***11-21** and print it.

Page Numbers, Headers, and Footers

OBJECTIVES

**MICROSOFT OFFICE
SPECIALIST
ACTIVITIES**
In this lesson:
WW03S-3-4

See Appendix C.

After completing this lesson, you will be able to:

1. Add page numbers.
2. Vary page numbers in Print Layout view.
3. Add headers and footers.
4. Work with headers and footers within sections.
5. Link section headers and footers.
6. Change starting page numbers.
7. Create continuation page headers.
8. Create alternate headers and footers.

 Estimated Time: 1½ hours

Page numbers, headers, and footers are useful additions to multiple-page documents. Page numbers can appear in either the top or bottom margin of a page. The text in the top margin of a page is a *header*; text in the bottom margin of a page is a *footer*. Headers and footers can also contain descriptive information about a document, such as the date, title, and author's name.

Adding Page Numbers

Word automatically keeps track of page numbers and indicates on the left side of the Status bar the current page and the total number of pages in a document. Each time you add, delete, or format text or sections, Word adjusts page breaks and page numbers. This process, called *background repagination,* occurs automatically when you pause while working on a document.

FIGURE 12-1
How Word paginates when you open a document

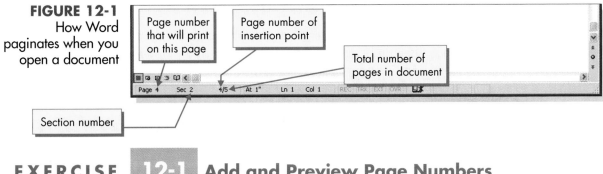

EXERCISE 12-1 Add and Preview Page Numbers

Page numbers do not appear on a printed document unless you specify that they do. The simplest way to add page numbers is to choose Page Numbers from the Insert menu and work with the Page Numbers dialog box.

1. Open the file **Mexico1**.
2. With the insertion point at the top of the document, choose Page Numbers from the Insert menu. Word displays the Page Numbers dialog box. Notice that the default settings Bottom of page (Footer) and Right alignment are chosen. Also, by default, Word shows page numbers on the first page.

FIGURE 12-2
Page Numbers dialog box

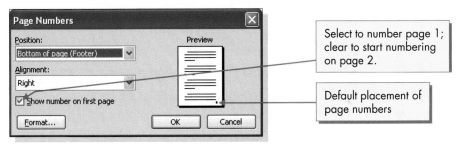

3. Click the down arrow to open the Position drop-down list. Choose Top of Page (Header). Notice the change in the Preview box.
4. Open the Alignment drop-down list and choose Center. The Preview box reflects the change.
5. Change the settings back to Bottom of Page (Footer) and Right alignment so the numbers appear on the bottom right of the page. Click OK.

 NOTE: You can see page numbers only in Print Preview, Print Layout view, or on the printed page.

6. Switch to Print Preview. Use the Multiple Pages button to view all three pages of the document.

7. If the rulers are not visible, click the View Ruler button 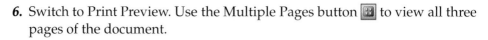 to display them.

8. Use the magnifier pointer 🔍 to click the bottom right corner of the first page. The page number appears within the 1-inch bottom margin. Specifically, the page number is positioned 0.5 inch from the bottom edge of the page at the right margin.

FIGURE 12-3
Viewing page
numbers in
Print Preview

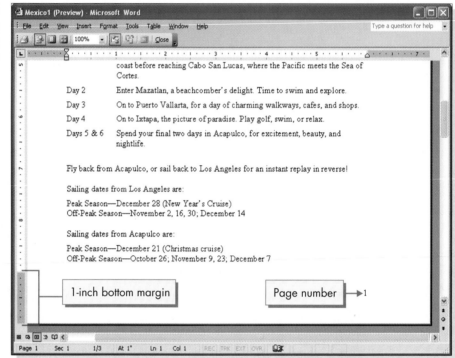

9. Scroll to see the page numbers on pages 2 and 3. Remain in Print Preview for the next exercise.

> **NOTE:** The available print area varies according to the type of printer. If your footer is not completely visible, ask your instructor about changing the footer position from 0.5 inch to 0.6 inch from the bottom edge of the page.

EXERCISE **12-2** **Change the Position and Format of Page Numbers**

Not only can you change the placement of page numbers and decide if you want to number the first page, but you can also change the format of page numbers. For example, instead of using traditional numerals such as 1, 2, and 3, you can use Roman numerals (i, ii, iii) or letters (a, b, c). You can also start page numbering of a section with a different value. For instance, you could number the first page ii, B, or 2.

1. While still in Print Preview, choose Page Numbers from the Insert menu.
2. Change the alignment to Center. Clear the Show number on first page check box so page 1 does not display a page number.
3. Click the Format button. In the Page Number Format dialog box, open the Number format drop-down list and choose uppercase Roman numerals (I, II, III…).

FIGURE 12-4
Page Number
Format dialog box

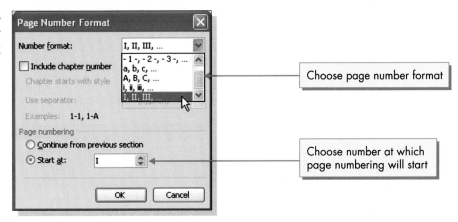

4. Click OK in the Page Number Format dialog box, and then click OK in the Page Numbers dialog box. View the document in Print Preview and note that page 1 does not display a page number. Page numbering is now centered, starting with Roman numeral II on page 2.
5. Close Print Preview and save the document as *[your initials]*12-2 in a new folder for Lesson 12.
6. Print the document four pages per sheet, and then close it.

Varying Page Numbers in Print Layout View

In addition to using the Page Numbers dialog box, you can also vary page numbers in Print Layout view. For example, you can:

- Apply character formatting such as bold or italic to page numbers.
- Add the word "Page" before the page number.
- Change the page numbering format.

EXERCISE **12-3** **Vary Page Numbers in Print Layout View**

1. Reopen the file **Mexico1**.

2. Use the Page Numbers dialog box to insert page numbers positioned at the top of the page (in the header) and right-aligned. Use the 1, 2, 3 format and do not show the page number on page 1.

3. In Print Layout view, scroll to the top of page 2. The page number appears in gray on your screen at the right margin.

4. Double-click the page number. This activates the dotted header pane (the area at the top of the page that contains the page number), displays the Header and Footer toolbar, and dims the document text.

5. In the header pane, position the I-beam just before the number 2. If the I-beam changes to a four-headed arrow, continue to move the mouse until the I-beam is just before the number. Click to position the insertion point. A shaded frame appears around the number.

FIGURE 12-5
Working with page numbers in Print Layout view

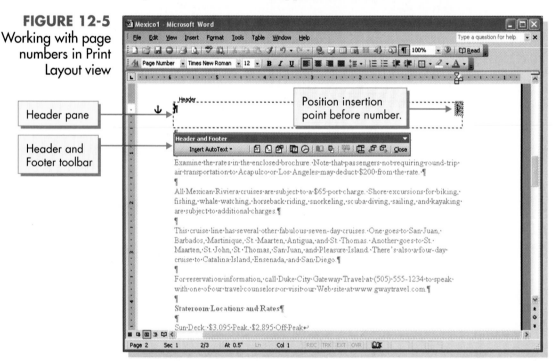

6. Key **Page** and press (Spacebar) once.

★ **TIP:** You might see the AutoText "Page X of Y (Press ENTER to Insert)." If you were to press (Enter), Word would display "Page 2 of 3" in the header. Do not press (Enter).

7. Scroll to the header pane on page 3 to view the revised header text.

NOTE: Changing one page number affects all page numbers.

8. To apply character formatting to the page number, drag the I-beam pointer over the word "Page" and the number 3 to select both.

9. Using the Formatting toolbar, change the text to Arial bold.

FIGURE 12-6
Formatted page
number with
"Page" added

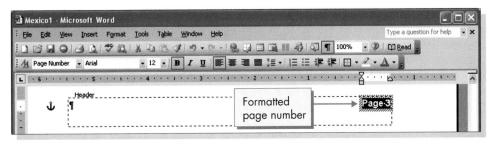

  10. Click the Format Page Number button on the Header and Footer toolbar. This opens the Page Number Format dialog box.

11. Change the number format to uppercase letters (A, B, C…) and click OK.

TIP: The page number doesn't have to be selected in the pane before you use the Format Page Number button.

12. View the page number on page 2 to see the changes. You should see "Page B" in the header.

13. Click Close on the Header and Footer toolbar to close the toolbar.

14. Save the document as *[your initials]***12-3** in your Lesson 12 folder.

15. Print the document four pages per sheet. Leave it open for the next exercise.

EXERCISE 12-4 Remove Page Numbers in Print Layout View

To remove page numbers, delete the text in the header or footer area.

1. On page 2 of the document, double-click the page number text ("Page B") to reopen the Header and Footer toolbar.

2. Use the I-beam pointer to select "Page B" and press Delete.

3. Scroll to the header area of page 3. The page number is deleted on that page as well.

4. Close the Header and Footer toolbar.

5. Close the document without saving it.

Adding Headers and Footers

Headers and footers are typically used in multiple-page documents to display descriptive information. In addition to page numbers, a header or footer can contain:

- The document name
- The date and/or the time you created or revised the document
- An author's name
- A graphic, such as a company logo
- A draft or revision number

This descriptive information can appear in many different combinations. For example, the second page of a business letter typically contains a header with the name of the addressee, the page number, and the date. A report can contain a footer with the report name and a header with the page number and chapter name. A newsletter might contain a header with a title and logo on the first page and a footer with the title and page number on the pages that follow.

FIGURE 12-7
Examples of
headers and footers

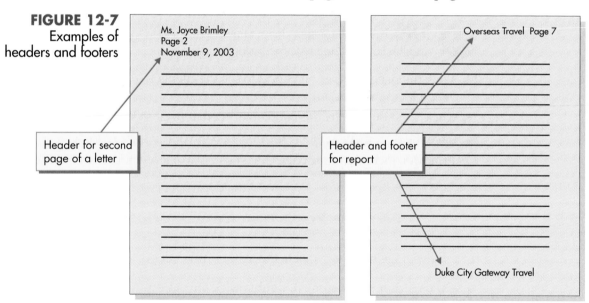

Ms. Joyce Brimley
Page 2
November 9, 2003

Header for second
page of a letter

Overseas Travel Page 7

Header and footer
for report

Duke City Gateway Travel

EXERCISE **12-5** **Add a Header to a Document**

1. Open the file **Primer1**. This five-page document includes a title page. First, you will add a header and footer to pages 2 through 5.

2. Choose Header and Footer from the View menu. Word switches temporarily to Print Layout view and displays the Header and Footer toolbar. The header pane is also visible.

3. Click the Page Setup button 📖 on the Header and Footer toolbar. The Layout tab in the Page Setup dialog box is displayed.

4. Click the Different first page check box, and then click OK. This enables you to give the document two different headers—a header for the title page, which you'll leave blank, and a header for the rest of the document, which will contain identifying text. Notice that this header pane is labeled First Page Header.

5. Click the Show Next button 🔳 on the Header and Footer toolbar to move to the header pane on the next page. Notice that this header pane is labeled Header. The Show Next and Show Previous buttons are useful when you move between different headers and footers within sections of a document, as you will see later in the lesson.

6. Key **Primer for First-Time Overseas Travelers** in the page 2 header pane. This text now appears on every page of the document except the first page.

7. Press (Tab) once. Notice that the ruler has two preset tab settings: 3-inch centered and 6-inch right-aligned. Drag the right-aligned tab marker to the right margin. Press (Tab) again to move to the right-aligned tab setting.

> **NOTE:** These preset tab settings are default settings for a document with the default 1.25-inch left and right margins. In such a document, the 3-inch tab centers text and the 6-inch tab right-aligns text. This document, however, has 1-inch left and right margins, so it's best to adjust the tabs.

8. Key **Page**, press (Spacebar), and click the Insert Page Number button 🔳 on the Header and Footer toolbar. Word inserts the page number.

9. Click the Show Previous button 🔳 and notice that the first-page header pane is still blank. Click the Show Next button 🔳 to return to the header you created.

TABLE 12-1   Buttons on the Header and Footer Toolbar

BUTTON	NAME	PURPOSE
Insert AutoText ▾	Insert AutoText	Insert common header or footer items, such as running total page numbers (Page 1 of 10), the filename, or the author's name.
🔳	Insert Page Number	Insert the page number.
🔳	Insert Number of Pages	Insert the total number of pages in the document.

continues

TABLE 12-1   Buttons on the Header and Footer Toolbar   *continued*

BUTTON	NAME	PURPOSE
	Format Page Number	Open the Format Page Number dialog box.
	Insert Date	Insert the current date.
	Insert Time	Insert the current time.
	Page Setup	Open the Page Setup dialog box.
	Show/Hide Document Text	Display or hide the document text.
	Link to Previous	Link or unlink the header or footer in one section to or from the header or footer in the previous section.
	Switch Between Header and Footer	Move back and forth between the header and footer.
	Show Previous	Show the header or footer of the previous section.
	Show Next	Show the header or footer of the next section.

EXERCISE   12-6   Add a Footer to a Document

1. With the header on page 2 displayed, click the Switch Between Header and Footer button to display the footer pane.
2. Key your name and press Tab.
3. Save the document as *[your initials]*12-6 in your Lesson 12 folder.
4. With the insertion point at the center of the footer, click Insert AutoText on the Header and Footer toolbar. Notice the types of AutoText entries you can insert into a header or footer.
5. Choose Created on from the AutoText list. This AutoText entry inserts a field that displays the text "Created on," plus the date and time the document was created, which, in this case, is the current date and time.
6. Press Tab and reopen the AutoText list.

FIGURE 12-8
Inserting AutoText
in a footer

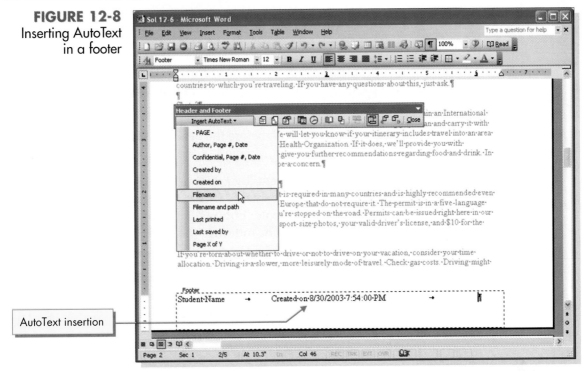

AutoText insertion

7. Choose Filename from the list of AutoText entries. The document's name is inserted. This footer information prints at the bottom of each page except the first.

8. Improve the tab positions by dragging the center tab marker to 3.25 inches and the right tab marker to 6.5 inches. (Remember, this document has 1-inch left and right margins, not the default 1.25-inch margins.)

9. Click Close to return to the document, which is again displayed in Normal view.

10. Switch to Print Preview. Check that no header or footer appears on the title page. Scroll through each page and view the header and footer.

11. Return to Normal view and save the document. Leave it open for the next exercise.

Adding Headers and Footers within Sections

Section breaks have an impact on page numbers, headers, and footers. For example, you can number each section differently or add different headers and footers.

When you add page numbers to a document, it's best to add the page numbers first, and then add the section breaks. Otherwise, you have to apply page numbering to each individual section.

EXERCISE **12-7** **Add Sections to a Document with Headers and Footers**

1. After the title page of the document, delete the hard page break and insert a Next Page section break.

2. Insert a Next Page section break before the heading "The American Embassy?" on page 3 of section 2 (page 4 of the document).

3. Return to the top of the document (by pressing Ctrl+Home) and choose Header and Footer from the View menu. If Word displays the footer, click the Switch Between Header and Footer button to switch to the header. Notice that the blank header pane indicates the section number.

4. Click the Show Next button to move to the next header, in section 2, page 1. Notice that this header is also blank, because the Page Setup option Different first page is selected for the entire document. This means the first page of each section can have a different header or footer than the rest of the pages in the section or it can have no header or footer.

5. Click the Show Next button again to move to section 2, page 2. The header and footer begin here.

6. Click the Show Next button to move to section 3, page 1. Because the Different first page option applies to the document, the first page of this section also has no header or footer.

7. Turn off the Different first page option for section 3 by clicking the Page Setup button on the Header and Footer toolbar and clearing the check box. Click OK. Now page 1 of section 3 starts with the document header and footer. Turning off this option applies only to this section, as you will see in the next step. (See Figure 12-9 on the next page.)

8. Click the Show Previous button twice to move to the header on page 1 of section 2. Notice that the header pane is still blank because the Different first page option is still checked for this section.

9. Repeat step 7 to turn off the Different first page option for this section. Now the header and footer start on page 1 of section 2.

10. View each header in the document by dragging the scroll box (on the vertical scroll bar) down one page at a time. As you display each page's header, notice the page numbering. Also notice that the text "Same as Previous" appears on the header panes.

FIGURE 12-9
The header on
page 1, section 3,
of the document

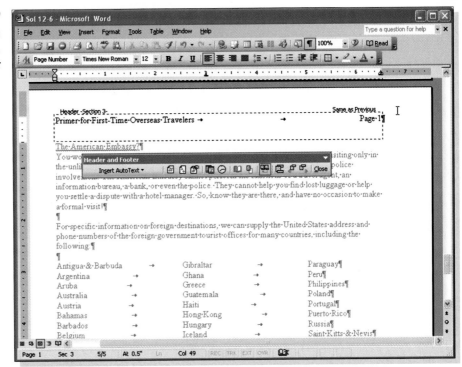

Linking Section Headers and Footers

 By default, the Link to Previous button is "on" when you work in a header or footer pane. As a result, the text you originally enter in the header (and the footer) for the document is the same from section to section. Any change you make in one section header or footer is reflected in all other sections. You can use the Link to Previous button to break the link between header/footer text from one section to another section and enter different header or footer text for a section.

 NOTE: Breaking the link for the header does not break the link for the footer. You must unlink them separately.

EXERCISE 12-8 Link and Unlink Section Headers and Footers

1. With the header for section 3, page 1, displayed, select the text "Primer for First-Time Overseas Travelers" and make it italic.

 2. Click the Show Previous button to move to the header in section 2. The header text is italic, demonstrating the link that exists between section headers and footers.

3. Click the Show Next button to return to section 3, page 1. Click the Link to Previous button to turn off this option. Now sections 2 and 3 are unlinked, and you can create a different header or footer for section 3.

4. Delete all the text in the header, including the page number.

> **TIP:** To select text in a header or footer, you can point and click from the area to the immediate left of the header or footer pane.

5. Press Tab to move to the center tab setting and key **Supplement**. Press Tab again and click the Insert Page Number button . Drag the center tab marker to 3.25 inches and the right tab marker to 6.5 inches, as needed.

6. Click the Switch Between Header and Footer button to switch to the footer. The footer text between sections 2 and 3 is still linked, so click the Link to Previous button to break the link.

7. Delete all the footer text in section 3 except your name. Click the Show Previous button to see that the original footer text is still in section 2. Click the Show Next button to return to the section 3 footer.

8. Click the Link to Previous button to restore the link between section footers. When Word asks if you want to delete the current text and connect to the text from the previous section, click Yes.

FIGURE 12-10 Restoring the link between section footers

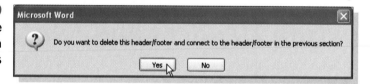

9. With the link and the original footer text now restored, close the Header and Footer toolbar.

10. Format the title page attractively. Adjust page breaks throughout the document as needed.

11. Save the document as *[your initials]***12-8**. Print the document six pages per sheet.

Changing the Starting Page Number

So far, you've seen page numbering start either with 1 on page 1 or 2 on page 2. When documents have multiple sections, you might need to change the starting page number. For example, in the current document, section 1 is the title page and the header on section 2 begins numbering with page 1. You can change this format so numbering starts in section 2, page 1, with page 2.

EXERCISE **12-9** **Change the Starting Page Number**

1. Choose <u>Header and Footer</u> from the <u>V</u>iew menu, and display the header on section 2, page 1.

2. Click the Format Page Number button 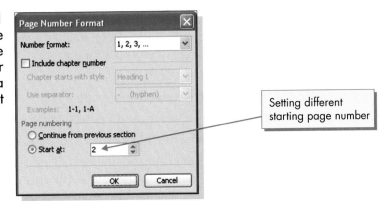 to open the Page Number Format dialog box.

3. Change the Start at number to **2**.

FIGURE 12-11
Changing the
starting page
number for
any part of a
document

4. Click OK. Section 2 now starts with Page 2.

5. Click the Show Next button to move to section 3. Click the Link to Previous button and click Yes to restore the header from the previous section. Section 3 starts numbering with Page 1.

6. Follow the same steps to open the Page Number Format dialog box. This time, choose the option <u>C</u>ontinue from previous section. Click OK. Notice that the section header begins with Page 5.

7. Save the document as *[your initials]***12-9** in your Lesson 12 folder.

8. Print the last page of the document and close it.

NOTE: You can change the starting page number for any document, with or without multiple sections. For example, you might want to number the first page of a multiple-page document "Page 2" if you plan to print a cover page as a separate file.

Creating Continuation Page Headers

It's customary to use a header on the second page of a business letter or memo. A continuation page header for a letter or memo is typically a three-line block of text that includes the addressee's name, the page number, and the date.

There are three rules for letters and memos with continuation page headers:

- Page 1 must have a 2-inch top margin.
- Continuation pages must have a 1-inch top margin.
- Two blank lines must appear between the header and the continuation page text.

EXERCISE **12-10** **Add a Continuation Page Header to a Letter**

The easiest way to create a continuation page header using the proper business format is to apply these settings to your document:

- Top margin: 2 inches
- Header position: 1 inch from edge of page
- Page Setup Layout for Headers and Footers: Different First Page
- Additional spacing: Add two blank lines to the end of the header

By default, headers and footers are positioned 0.5 inch from the top or bottom edge of the page. When you change the position of a continuation page header to 1 inch, the continuation page appears to have a 1-inch top margin, beginning with the header text. The document text begins at the page's 2-inch margin, and the two additional blank lines in the continuation header ensure correct spacing between the header text and the document text.

1. Open the file **Almy1**.
2. Add the date to the top of the letter, followed by three blank lines.
3. Open the Page Setup dialog box and display the Layout tab. Check Different first page under Headers and Footers. Set the Header to 1 inch from the edge.
4. Click the Margins tab and set a 2-inch top margin and 1.25-inch left and right margins. Click OK.
5. Display the header pane.
6. Click the Show Next button to move to the header pane on page 2.
7. Create the header in Figure 12-12, inserting the information as shown. Press (Enter) twice after the last line.

FIGURE 12-12

```
Ms. Barbara Almy
Page [Click 🔢 for the page number]
[Current date]
```

TIP: If you use the Insert Date button 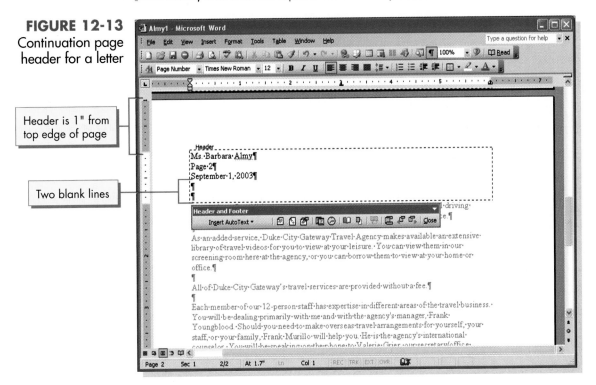 on the Header and Footer toolbar, Word inserts a date field in the format 12/25/03 by default. Letters and memos should use the spelled-out date format December 25, 2003, and the date should not be a field that updates each time you open the document. To insert the date as text, with the correct format, use the Date and Time dialog box (from the Insert menu) and clear the Update automatically check box.

FIGURE 12-13
Continuation page
header for a letter

Header is 1" from
top edge of page

Two blank lines

8. Close the header pane and view both pages in Print Preview.

9. Add **Nina Chavez** and the title **Senior Travel Counselor** to the end of the letter, followed by your reference initials.

10. Save the document as *[your initials]***12-10** in your Lesson 12 folder.

11. Print and close the document.

Creating Alternate Headers and Footers

In addition to customizing headers and footers for different sections of a document, you can also change them for odd and even pages throughout a section or document. For example, in this textbook, the header for even pages displays the unit name and the header for odd pages displays the lesson name.

EXERCISE 12-11 Create Alternate Footers in a Document

To create alternate headers or footers in a document, you use the Different odd and even check box in the Page Setup dialog box, and then create a header or footer for both even and odd pages.

1. Open the file **Cruises1**.
2. Display the Header pane.
3. Open the Page Setup dialog box. On the Layout tab, click Different odd and even, and then click OK. Notice that Odd Page Header appears at the top of the header pane on page 1.
4. Switch to the footer on page 1, which is labeled Odd Page Footer.
5. Press Tab, open the Insert AutoText drop-down list on the Header and Footer toolbar, and choose Page X of Y. This running-total page format is useful for long documents.

FIGURE 12-14
Odd page footer

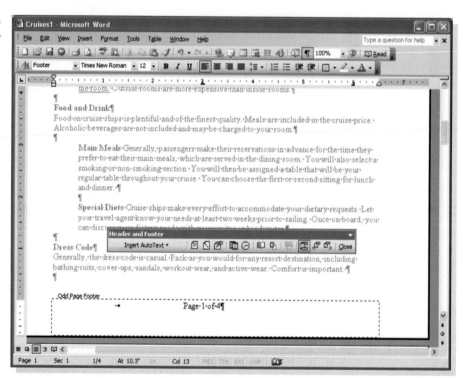

6. Adjust the center tab marker to 3.25 inches on the ruler.

7. Click the Show Next button to move to the footer on page 2. The footer pane is labeled Even Page Footer and is blank.

8. Press Tab and key **Cruise Travel Information** in italic. Adjust the center tab marker to 3.25 inches.

9. Close the Header and Footer toolbar and switch to Print Preview.

10. View each page of the document. Notice that pages 1 and 3 display page number information, and pages 2 and 4 display the document's title.

11. Return to Normal view. Add 72 points of paragraph space before the title on page 1 and adjust page breaks throughout the document as needed.

12. Save the document as *[your initials]*12-11 in your Lesson 12 folder.

13. Print the document four pages per sheet, and then close it.

NOTE: To create different odd and even headers or footers within a section, you must first break the link between that section's header or footer and the previous section's header or footer.

USING ONLINE HELP

Word provides extensive Help about headers and footers. Use Help to answer questions about creating and positioning headers and footers, creating first page headers and footers or odd and even headers and footers, and so on.

Use Help to learn more about headers and footers:

1. Key **headers and footers** in the Ask a Question text box and press Enter.

2. Click the topic Insert headers and footers.

3. Review the information in the Help window.

4. Select the link Troubleshoot headers and footers in the Search Results task pane.

5. Click the Show All link to display the topics.

6. Close the Help window when you finish.

LESSON 12 Summary

➤ A header is text that appears in the top margin of the printed page; a footer is text that appears in the bottom margin. These text areas are used for page numbers, document titles, the date, and other information.

➤ Always add page numbers to long documents. You can choose the position of page numbers (examples: bottom centered, top right) and the format (examples: 1, 2, 3 or A, B, C) in the Page Number Format dialog box. Number the first page or begin numbering on the second page.

➤ Check page numbers in Print Preview or Print Layout view (they are not visible in Normal view). In Print Layout view, you can activate the header or footer pane that contains the page number by double-clicking the text, and then modify the page number text (examples: make it bold or add the word "Page" before the number).

➤ To remove page numbers, activate the header or footer pane that contains the numbering, select the text, and then delete it.

➤ To add header or footer text to a document, choose View, Header and Footer. Use the Header and Footer toolbar buttons to insert the page number, date, and time; to insert AutoText for the filename, author, print date, or other information; and to switch between the header and footer pane and move to the next or previous pane. See Table 12-1.

➤ Adjust the tab marker positions in the header or footer pane as needed to suit the width of the text area.

➤ A document can have a different header or footer for the first page than for the rest of the pages. Apply the Different first page option in the Page Setup dialog box (Layout tab).

➤ Header and footer text is repeated from section to section because headers and footers are linked by default. To unlink section headers and footers, click the Link to Previous button 🔲. To re-link the header or footer, click the button again.

➤ Sections can have different starting page numbers. Click the Format Page Number button 🔳 to open the Page Number Format dialog box, and then set the starting page number.

➤ Memos or letters that are two pages or longer should have a continuation page header—a three-line block containing the addressee's name, page number, and date. Set the header to 1 inch from the edge, add two blank lines below the header, and use a 2-inch top margin. Apply the Different first page option in the Page Setup dialog box (Layout tab) and leave the first-page header blank.

➤ Use the Page Setup dialog box to change the position of the header or footer text from the edge of the page. The default position is 0.5 inch.

➤ A document can have different headers and footers on odd and even pages. Apply the Different odd and even option in the Page Setup dialog box (Layout tab).

LESSON 12   Command Summary

FEATURE	BUTTON	MENU	KEYBOARD
Add page numbers		Insert, Page Numbers	
Change page number format	🔳	Insert, Page Numbers, Format	
Add or edit header/footer		View, Header and Footer	
Change layout settings	📖	File, Page Setup	

Concepts Review

TRUE/FALSE QUESTIONS

Each of the following statements is either true or false. Indicate your choice by circling T or F.

T F   *1.* The simplest way to add page numbers is to choose Page Numbers from the View menu.

T F   *2.* Using the Page Numbers dialog box, you can position a page number as a header or a footer.

T F   *3.* You can change the number format (for example, from numbers to Roman numerals) by using a button on the Header and Footer toolbar.

T F   *4.* You can apply character formatting to headers and footers in Normal view.

T F   *5.* When you insert page numbers, the default position is at the bottom right of the page.

T F   *6.* Turn off ▦ on the Header and Footer toolbar to unlink a header in one section from the header in the previous section.

T F   *7.* Use ▣ to open the Page Setup dialog box from the Header and Footer toolbar.

T F   *8.* You use the Page Setup dialog box to create alternate page headers or footers.

SHORT ANSWER QUESTIONS

Write the correct answer in the space provided.

1. What is the name of the process in which Word automatically adjusts page numbers and page breaks when you edit a document?

_____

2. In addition to numbering such as 1, 2, 3 and Roman numerals such as I, II, III, what other page number formatting can you use?

_____

3. Which layout option do you use to leave the first page of a document blank and begin a header or footer on the second page?

_____

4. Which menu command displays the Header and Footer toolbar?

_____

5. ⏹ is used for what purpose?

_____

6. What three items are included in a continuation page header for a letter?

_____

7. By default, how far from the edge of the page does Word print headers and footers?

_____

8. If you set up different odd and even pages in a document, how is the header pane on page 1 labeled?

_____

CRITICAL THINKING

Answer these questions on a separate page. There are no right or wrong answers. Support your answers with examples from your own experience, if possible.

1. What information do you think most businesses would include in the header or footer for a business report? Does the information included in a business report header or footer differ from the information found in a business letter header or footer?

2. Where do you prefer to place the page number in a business report? In a business letter? Explain your answer.

Skills Review

EXERCISE 12-12

Add and modify page numbers and add a header.

1. Open the file **ElMorro1**.

2. Add page numbers to the bottom right of each page by following these steps:

    a. Choose Insert, Page Numbers.

b. In the dialog box, make sure the position is Bottom of Page, the alignment is Right, and Show number on first page is checked. Click OK.

3. Modify the page numbers in Print Layout view by following these steps:

   a. In Print Layout view, scroll to the bottom right of page 1 to see the page number (which is gray).

   b. Double-click the page number to activate the footer pane.

   c. Move the I-beam to the immediate left of the page number and click to position the insertion point.

   d. Key **Page** and press Spacebar.

   e. Drag the I-beam over the text "Page 1" and click the Italic button ⟦*I*⟧ on the Formatting toolbar to italicize the text.

   f. Scroll to the page 2 footer to see the change.

   g. Close the Header and Footer toolbar.

4. Add 72 points of spacing before the title on page 1.

5. View the document in Print Preview; then display it in Normal view.

6. Add a header to the document by following these steps:

   a. Choose Header and Footer from the View menu.

   b. In the header pane, press Tab twice to position the insertion point at the right-aligned tab setting. Drag the right tab marker to the right margin.

   c. Key your name, followed by a comma and space.

   d. Click Insert AutoText on the Header and Footer toolbar. Choose Filename from the drop-down list to insert the document filename.

   > **NOTE:** The filename ElMorro1 will change to the filename you give the document when you save it.

   e. Close the Header and Footer toolbar.

7. Save the document as *[your initials]***12-12** in your Lesson 12 folder.

8. Print and close the document.

EXERCISE 12-13

Add a footer to a document with sections, unlink the header, and change the starting page number.

1. Open the file **Tour1**.

2. Insert a Next page section break at "<u>Airfares</u>" and at "*Itinerary*."

3. On page 1, section 1, edit the text as shown in Figure 12-15 (on the next page). Use 14-point small caps and center the text vertically and horizontally on the page.

FIGURE 12-15

```
9-Day Escorted Tour of
New Mexico and Colorado
Airfares and Itinerary
```

4. Create a footer for sections 2 and 3 that is not linked to section 1 by following these steps:

   a. Move the insertion point to section 2 and choose <u>V</u>iew, <u>H</u>eader and Footer. Switch to the footer by clicking the Switch Between Header and Footer button .

   b. Click the Link to Previous button to unlink the section 2 footer from section 1 (to keep the section 1 footer blank) and key **9-Day Escorted Tour of New Mexico and Colorado**.

   c. Press (Tab) and key **Page**. Insert a space and click the Insert Page Number button .

   d. Click the Show Next button to show the footer for section 3, which should be the same as the section 2 footer (but with a different page number).

5. Change the starting page number of section 2 to Page 1 by following these steps:

   a. Click the Show Previous button to go back to section 2.

   b. Click the Format Page Number button .

   c. Click Start <u>a</u>t and make sure the number 1 appears in the text box. Click OK, and then close the Header and Footer toolbar.

6. Change the paragraph spacing for the document text in sections 2 and 3 to 6 points after. Insert a hard page break to start the Day 6 text on a new page.

7. At the top of actual pages 2 and 3 of the document, format the titles "<u>Airfares</u>" and "*Itinerary*" as centered, uppercase, and bold with 24-point paragraph spacing after. Neither should be underlined or italic.

8. Save the document as *[your initials]***12-13** in your Lesson 12 folder.

9. Print the document two pages per sheet, and then close it.

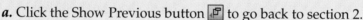

EXERCISE 12-14

Create a continuation page header for a memo.

1. Open the file **NYmemo**.

2. Change the memo so it is to Valerie Grier and from you.

3. Enter the current date.

4. Open the Page Setup dialog box and change the left and right margins to 1.25 inches. Change the top margin to 2 inches. Click the Layout tab and check Different first page. Change the Header distance to 1 inch from the edge. Click OK.

5. Add a continuation page header to page 2 of the memo by following these steps:

   a. Choose View, Header and Footer.
   b. Click the Show Next button to move to the header on page 2.
   c. Key the text in Figure 12-16, inserting the information as shown.

FIGURE 12-16

```
Ms. Valerie Grier
Page [Click ]
[Key current date or insert as text]
```

   d. To insert the current date in the correct format, choose Insert, Date and Time and select the correct format. Be sure to clear the Update automatically box so the memo date does not change.
   e. Press Enter twice after the date.

6. Click the Show Previous button to check that no header appears on page 1 and close the Header and Footer toolbar.

7. Save the document as *[your initials]*12-14 in your Lesson 12 folder.

8. Print and close the document.

EXERCISE 12-15

Add alternate footers to a document, and add a different first page footer.

1. Open the file **DukeHistory**.

2. Add page breaks before the bold headings "Company History" and "Staff."

3. Move the insertion point to the top of the document (page 1).

4. Create a footer that appears only on odd pages by following these steps:

   a. Choose View, Header and Footer.
   b. Click the Page Setup button on the Header and Footer toolbar.
   c. On the Layout tab, click Different odd and even, and then click OK.
   d. Switch to the footer, which should be the odd page footer on page 1.
   e. Key your name and press Tab.
   f. Click Insert AutoText and choose Filename from the list.
   g. Press Tab and click the Insert Date button . Key a comma, press Spacebar, and click the Insert Time button.

5. Create and format a footer that appears only on even pages by following these steps:

    a. Click the Show Next button to move to the even page footer pane on page 2.

    b. Key **Introduction to Gateway Travel** and press Tab twice.

    c. Key **Page** and press Spacebar. Click the Insert Page Number button #️⃣ and press Spacebar. Key **of** and press Spacebar. Click the Insert Number of Pages button ⬚.

    d. Select the footer text. Use the Formatting toolbar to change it to italic.

6. Close the Header and Footer toolbar.

7. On page 1, format the title in bold all caps and center it horizontally. On a new line below the title, key **An Introduction**. Format this text as bold and centered below the title. Add 144 points of paragraph spacing before the first line of the title.

8. Create a different footer for the first page by following these steps:

    a. Click File, Page Setup to open the Page Setup dialog box.

    b. On the Layout tab, check Different first page and click OK.

9. View the document in Print Preview. Check the odd and even footers and make sure the first page footer is now blank.

10. Save the document as *[your initials]***12-15** in your Lesson 12 folder.

11. Print and close the document.

Lesson Applications

EXERCISE 12-16

Add page numbers using two different formats, change the page number font, and adjust the starting page number.

1. Open the file **Primer2**.

2. Replace the page break after page 1 with a Next Page section break. Go to page 8 and insert a Next Page section break immediately preceding the bold heading "Sample Cruise Itineraries."

3. Add page numbering to section 2. (Be sure to place the insertion point in section 2 first.) Position the numbers at the bottom center and show numbering on the first page of the section.

4. Open the footer pane in section 3. Unlink the footer from the previous section.

5. Change the page number format of section 3 to uppercase Roman numerals. Instead of continuing the numbering from the previous section, start page numbering for section 3 with I.

6. In the footer of section 3, select the page number and change the format to 11-point Arial.

7. View the page numbers for sections 2 and 3 in Print Layout view.

8. On actual page 3, insert a page break at the bold heading "Category, Location, and Rates."

9. On actual page 5, before the text "Around-the-Clock Meals," key the first paragraph that appears in Figure 12-17, including the corrections. Edit the paragraph that begins "Around-the-Clock Meals," as shown in the second paragraph of the figure. Both paragraphs should have a 0.5-inch left indent. To key ê in the word "crêpes," press (Alt) and key **0234** on the numeric keypad (make sure (Num Lock) is on).

FIGURE 12-17

World-Class Cuisine Cruise ship chefs are often specialists in cuisine from around the world. Depending on where you are traveling and the whim of the chef, cruises typically feature food ~~of~~ with regional ~~origin~~ themes. For example, one night may be Italian Night, serving ~~up~~ pastas, pizzas, and veal scaloppini. Another night may be French Night, with escargots, crêpes, and an endless assortment of French pastries. A Mexican buffet may offer a savory spread of tacos, ~~and~~ tamales, and enchiladas.

continues

FIGURE 12-17 *continued*

> non-stop serving of food. In addition to the regular three-meal-a-day schedule, you'll find announcements for midday snack times, afternoon tea times, and

Around-the-Clock Meals Cruise ships are famous for ~~the many other meals that make up the day, from midday snack times to~~ midnight gourmet buffets. These meals are open seating and are served in different parts of the ship, including poolside.

10. Scroll through section 2 and insert a page break before "Ship Terminology" and "Types of Cruises." In section 3, insert page breaks where necessary to keep itinerary text together.

11. Format the text on page 1 (the title page) attractively, centering it vertically on the page.

12. Spell-check the document. Do not spell-check the itineraries in the last section.

13. Save the document as *[your initials]***12-16** in your Lesson 12 folder.

14. Print the document as four pages per sheet, and close the document.

EXERCISE 12-17

Create and unlink headers and footers within sections.

1. Open the file **Duke4**.

2. Insert a Next Page section break at the bold heading "Introduction." Copy the boxed title, including the blank line below it, to the top of section 2. On the title page, center the boxed title vertically on the page and reduce the width of the border by formatting the text in the box with 1-inch left and right indents.

3. Format the table on page 2 (which starts with the text "Golden Gateway Travel") as a separate section by placing a Continuous section break before it and a Next Page section break after it. Format this new section (section 3) with a 3-inch left margin.

4. If section 4 (actual page 3) starts with a blank line, delete the blank line.

5. Create a header that starts on the first page of section 4 and is not linked to section 3. In the heading, key **Duke City Gateway Travel Agency**. At the right-aligned tab, key **Page** followed by a space and include the page number. Drag the tab marker to right-align the page number at the right margin.

6. Italicize the header text and insert two blank lines below it.

7. On the first page of section 4, create a footer that is not linked to the previous section and includes your name aligned at the left. Insert the "Created on" AutoText entry so it is aligned at the right margin.

NOTE: Remember that the "Created on" AutoText entry will be updated when you save and print the document.

8. Decrease the footer font size by 1 point (remember the keyboard shortcut Ctrl+Shift+<).

9. Copy the footer on page 3 to page 2, so all the pages except page 1 have the footer. (Unlink the footer in section 2 before pasting the footer text.) Check that page 1 contains no header or footer.

10. Add Duke City Gateway's address (15 Montgomery Boulevard, Albuquerque, NM 87111) to the title page, within the border. The address text should be 14-point (not bold). Change the border to a 2 1/4-point shadow border that has 10-point spacing between the top and bottom of the text.

11. View the document in Print Preview. If the last page is very short, shrink the document to fit on one less page. Adjust page breaks as needed.

12. Save the document as *[your initials]*12-17 in your Lesson 12 folder.

13. Print and close the document.

EXERCISE 12-18

Create a continuation page header for a business letter.

1. Open the files **Wolinsky** and **Special2**.

2. Format Wolinsky as a business letter. Use the address shown in Figure 12-18. The letter will be from Tom Carey, Senior Travel Counselor.

FIGURE 12-18

```
Ms. Rachel Wolinsky

32 Tramway Boulevard

Albuquerque, NM 87123
```

3. After the paragraph that begins "Choose your," start a new paragraph by keying **For a completely different trip, consider a special-interest tour:**.

4. Switch to the Special2 window. Copy paragraphs 4 through 8 (from "Our local" through "Salzburg") and paste them below the text you keyed in the letter document. There should be one blank line before and after the pasted text.

5. Format the copied paragraphs with bullets (use the standard bullet style) and a 0.5-inch right indent.

6. Adjust page setup options for a continuation page header by choosing the Different first page option, changing the header to 1 inch from the edge, and setting a 2-inch top margin and 1.25-inch left and right margins.

7. Create a three-line continuation page header that prints on page 2. (Use the correct date format.)

8. Switch to Print Preview to view the document.

9. Add your reference initials and spell-check the document.

10. Save the document as *[your initials]*12-18 in your Lesson 12 folder.

11. Print the document and close both documents. (If asked, do not save changes to Special2.)

EXERCISE 12-19 *Challenge Yourself*

Create alternate footers, unlink and format section footers, change starting page numbers, and change page formats.

1. Open the file **Duke2**.

2. Replace the page break after page 1 with a Next Page section break. On page 7, insert a Next Page section break at the bold heading "Duke City Gateway Travel Itineraries."

3. Format the entire document for different odd and even headers and footers.

4. Display the section 2 odd page footer. Unlink it from section 1, center-align the insertion point, and key **Page** followed by the page number. Italicize the footer. Change the starting page number to 1. (Word adjusts pagination so the odd page footer appears on the appropriate page. Word also adds a forced blank page after page 1, which you'll see when you preview the document.)

5. Display the section 2 even page footer. Key **An Introduction to Duke City Gateway Travel Agency**. Italicize and center the footer.

6. Go to the section 3 odd page footer. Unlink it from the previous footer, delete "Page," and change the page number format to lowercase Roman numerals, starting with "i."

7. Go to the section 3 even page footer. Unlink it from the previous footer and change the text to **Duke City Gateway Travel Itineraries**.

8. Go to the odd page footer of section 1, which should be blank. Insert the Filename AutoText entry. Center and italicize the footer.

9. In sections 2 and 3, format the bold titles that are not indented as small caps with 6 points of spacing after the paragraph. (There are six unindented titles.)

10. In the section 3 sample itinerary paragraphs (which start "Day 1:"), replace the space following each colon with a tab character and format the paragraphs with a 0.75-inch hanging indent.

11. On the first page of section 2, apply a 0.5-inch left indent to the three Gateway Travel offices.

12. Find the text "All Gateway divisions" in the document. Delete text in this paragraph, starting with the found text through the end of the paragraph ("work for you"), but excluding the paragraph mark.

13. Find the paragraph that starts "There is a cruise for everyone" and delete the paragraph.

14. Adjust page breaks that separate paragraphs from headings or that look unattractive. (Insert hard page breaks or apply text flow options.)

15. Make the following changes to the title page:
   - Center the text horizontally and vertically.
   - Replace the date with the current date, using the format June 6, 2004 and change the spacing before and after to 24 points.
   - Insert the text **Prepared by** on a new line above "Susan Allen."
   - Format the first two lines as small caps and 16 points.

16. Format the entire document with a 1/2-point shadow page border.

17. Preview the document; then save it as *[your initials]***12-19** in your Lesson 12 folder.

18. Print the document four pages per sheet, and then close it.

On Your Own

In these exercises you work on your own, as you would in a real-life work environment. Use the skills you've learned to accomplish the task—and be creative.

EXERCISE 12-20
Write a two-page letter (to a prospective employer, a friend, or whomever). Create your letterhead in the first-page header pane and create a continuation header in the second-page header pane. Save the document as *[your initials]***12-20** and print it.

EXERCISE 12-21
Research a topic on the Internet that interests you. Copy several paragraphs and assemble the text into a new document at least two pages long. Check pagination. Add appropriate headers and footers and include page numbering. Create a title page as a separate section without a header or footer. Save the document as *[your initials]***12-21** and print it.

EXERCISE 12-22
Write a short report about ten places you'd like to visit. Each place should be a separate paragraph with its own heading. Include a title page. Adjust page breaks as needed. Format the document for odd and even headers and footers, and then insert different identifying information in the headers or footers. Save the document as *[your initials]***12-22** and print it.

Unit 4 Applications

UNIT APPLICATION 4-1

Create a cover letter for a document as a separate section; insert a page break and a continuation page header.

1. Open the file **RockArt**.
2. Add a cover letter to the beginning of the document (use a section break to divide the cover letter and the itinerary). To create the cover letter, key the text shown in Figure U4-1, including the corrections. Use the standard letter format. The letter should be from Alice Fung, Senior Travel Counselor. Include your reference initials and an Enclosure notation.

FIGURE U4-1

Ms. Florence Gibbons

City University of New York

Brooklyn College

Department of Archaeology

James Hall

2900 Bedford Avenue

Brooklyn, NY 11210

Dear Ms. Gibbons:

~~As~~ you requested, enclosed is the information about the Rock Art tour of New Mexico, Utah, and Arizona.

The tour is an excellent way to teach your students about petroglyphs, the ancient rock carvings and drawings of the Southwest. The tour is led by expert archaeologists, who offer insights into the mysterious rock art. Each evening there will be opportunity^ies for discussion at informal get togethers. The tour also ~~builds in~~ *includes* plenty of time to shop or relax.

A maximum of 30 people can be scheduled on each tour, so please keep this in mind when you are organizing your group.

^limit

add paragraph

Ground transportation is via air-conditioned buses with restrooms.

continues

Figure U4-1 *continued*

note

In addition, please ~~keep in mind~~ the following important information:

6 pt spacing after

The tour is for people who can handle a moderate amount of strenuous physical activity.

The tour will be at altitudes above 4000 feet. People unaccustomed to higher altitudes might experience fatigue or dizziness.

Bulleted paragraphs with 0.5" left and right indents and 6 pt spacing after; no blank lines

Some cliff sites are accessible only by ladders, most between six and ten feet in height.

You might want to advise people with heart or respiratory problems to consult their physicians before the tour.

Let me know if you have any questions. I look forward to speaking with you soon.

3. Insert a page break in the letter at the paragraph that begins "You might want to."

4. Add a continuation page header to page 2 of the letter. (Remember to set the margin and page layout settings.) Do not include a header in section 2.

5. In section 2, center the text vertically.

6. In the itinerary text, replace the space after each colon with a tab character. Set a 0.75-inch hanging indent for the itinerary text.

7. Spell-check the letter only.

8. In Print Preview, change the bold heading in section 2 to all caps.

9. Save the document as *[your initials]***u4-1** in a new folder for Unit 4 Applications.

10. Print and close the document.

UNIT APPLICATION 4-2

Create a cover page for a report; add and format headers and footers; apply section formatting.

1. Open the file **CostaRica**.

2. Add a cover page using the text shown in Figure U4-2 (on the next page). Separate the cover page and the remaining text with a Next Page section break.

FIGURE U4-2

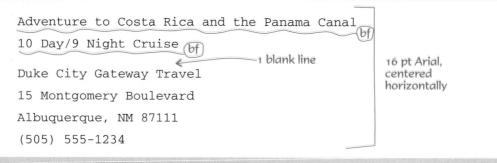

Adventure to Costa Rica and the Panama Canal (bf)

10 Day/9 Night Cruise (bf)

← 1 blank line

Duke City Gateway Travel

15 Montgomery Boulevard

Albuquerque, NM 87111

(505) 555-1234

16 pt Arial, centered horizontally

3. Insert a page break at the end of the document and key the itinerary shown in Figure U4-3. To key the ç character in "Curaçao," press [Alt] and key **0231** on the numeric keypad. Start the second tabbed column at 1 inch with a hanging indent.

FIGURE U4-3

Barbados to Costa Rica ←

Bold, centered, all caps
24 pt spacing after

Date	Port
Day 1	Bridgetown, Barbados
Day 2	Bequia, Grenadines
Day 3	At sea
Day 4	Curaçao, Netherlands Antilles
Day 5	At sea
Day 6	Cartagena, Columbia
Day 7	San Blas Islands, Panama
Day 8	Panama Canal, Panama
Day 9	At sea
Day 10	Puerto Caldera, Costa Rica. Disembark and enjoy a day of leisure. Overnight in Costa Rica.
Day 11	San Jose, Costa Rica. Full-day tour of Poas Volcano. Overnight in Costa Rica.
Day 12	San Jose, Costa Rica. Transfer to airport and return flight home.

24 pt spacing after

4. Begin section 2 with the title **BEAUTY, ADVENTURE, AND FRIENDLY PEOPLE** as bold, uppercase, and centered, with 24 points of spacing after the paragraph.

5. Add a header to section 2 with the text **Cruise to Costa Rica and the Panama Canal** at the left and the page number (with the word "Page") at the right. Start page numbering of section 2 with Page 1. Format the header in Arial italic.

6. Add a footer to section 2 with the centered text **Duke City Gateway Travel**. Format the footer as Arial italic.

7. Add a footer to the title page that contains the Filename AutoText entry. Center the footer and format it as Arial italic. Be sure section 1 contains no header.

8. Scroll through the document in Print Layout view, checking the headers and footers.

9. Center all pages of the document vertically.

10. Spell-check pages 1 and 2 of the document, but do not spell-check the itinerary.

11. Preview the document, and then save it as *[your initials]*u4-2 in your Unit 4 Applications folder.

12. Print and close the document.

UNIT APPLICATION 4-3

Write a cover memo for a document, and add alternate headers and footers.

1. Assume that Frank Youngblood (Duke City Gateway Travel's manager) asked you to research the historic points of interest in downtown Santa Fe for a walking tour he is planning. Open the file **SantaFe**. Assume you wrote this document based on Frank's request. (Review the document content.)

2. As a separate section at the beginning of the document, add a cover memo to Frank. In the memo, tell Frank you are submitting the information he requested, and describe which historic points you think would interest tour participants who like art objects and crafts.

3. Because section 2 will be distributed to tour participants, you don't need to create a continuation page header for the memo. Format section 2 with Different odd and even headers and footers and the header to print one inch from the edge. Format the Even Page Header for section 2, so that it contains a right-aligned page number preceded by the word "Page."

4. Add an odd page footer to section 2 that reads **Walking Tours Sponsored by Duke City Gateway Travel**. Use 10-point Arial italic and center the text.

5. Make sure no header or footer appears in section 1.

6. Add a page break at the end of section 2, and then key the text shown in Figure U4-4. For the tabbed text, set an appropriate indent or tab stop before the first column and set an appropriate tab stop before the second column. Both columns should be left-aligned.

FIGURE U4-4

```
Learning Spanish in Santa Fe ←  Bold, centered, all caps
                                   72 pt spacing before, 24 pt spacing after

Common Spanish terms you may encounter during your walking tour of
Santa Fe are listed below.

Spanish Word(s)          Translation
1 blank line →

Plaza Vieja              Old Plaza

esperanza                hope

bultos                   hand-carved images of saints

retablos                 painted images of saints

Dia de los Muertos       Day of the Dead

santeros                 traditional Hispanic folk artists

mercado                  market
```

6 pt spacing after

7. View the document in Print Preview.

8. Format the title at the beginning of section 2 with 72 points of spacing before paragraphs.

9. In section 2, start the paragraph "Loretto Chapel" on the next page.

10. Spell-check the document (except for the last page).

11. Save the document as *[your initials]*u4-3 in your Unit 4 Applications folder.

12. Print and close the document.

UNIT APPLICATION 4-4 *Using the Internet*

Work with sections, page numbers, and headers and footers.

Using the Internet, research Santa Fe (or another city) and design for yourself a five-day detailed itinerary that highlights your personal interests.

Organize the document into sections, with each day as a separate section. Include in each of these sections:

● A descriptive list of the day's activities. For each site you will visit, include its hours and entry fees (if applicable).

● If available, include a map or a written route for reaching each site.

Using the skills you learned in this unit, prepare an attractive finished product. Make sure you do the following:

● Create a title page for the document as a separate section.

● Title each subsequent section of the document.

● Check pagination, and apply line and page break options where needed.

● Include appropriate headers/footers and page numbering on all pages except the title page.

Save the document as *[your initials]*u4-4 in your Unit 4 Applications folder and print it.

Tables and Columns

LESSON 13

Tables

OBJECTIVES

**MICROSOFT OFFICE
SPECIALIST
ACTIVITIES**
In this lesson:
   WW03S-2-1

See Appendix C.

After completing this lesson, you will be able to:

1. Create a table.
2. Key and edit text in tables.
3. Select cells, rows, and columns.
4. Edit table structures.
5. Format tables and cell contents.
6. Convert tables and text.

Estimated Time: 1½ hours

A table is a grid of rows and columns that intersect to form *cells*. The lines that mark the cell boundaries are called *gridlines*. It's often easier to read or present information in table format than in paragraph format. Using Word's table feature, you can create a table, and insert text, pictures, and other types of data into the table's cells.

Creating a Table

There are two ways to create a table:

- Insert a table by using the Insert Table dialog box or the Insert Table button on the Standard toolbar.
- Draw a table by using the Tables and Borders toolbar.

FIGURE 13-1
Columns, rows, and
cells in a table

HOTELS FOR LARGE GROUP EVENTS			
Hotel	**Location**	**Amenities**	**Banquet Rooms**
Crown West	San Diego	✓ Marina Setting ✓ Tennis Court	3
Rosedale Suites	Chicago	✓ Athletic Club ✓ Shopping Arcade	4
Amandsons	Atlanta	✓ Four Star Restaurant ✓ Golf Course ✓ Health Spa	3

Row

Column

Cell

NOTE: You can apply formatting options, such as borders and shading, to tables. You can also display tables without gridlines.

EXERCISE 13-1 Insert a Table

Use the Insert Table command to create a table with the number of rows and columns you specify. There are two ways to insert a table:

● Use the Insert Table button 🔲 on the Standard toolbar to provide an adjustable grid.

● Use the Insert Table dialog box, which is accessed by pressing the Insert Table button 🔲▾ on the Tables and Borders toolbar or by choosing Insert, and then Table from the Table menu.

NOTE: The button for Insert Table functions differently on the Tables and Borders toolbar than on the Standard toolbar. The button on the Tables and Borders toolbar opens the Insert Table dialog box. The button on the Standard toolbar displays the table grid.

1. Open the file **Rates**. Enter today's date in the memo date line.

2. Position the insertion point at the second paragraph mark below the paragraph that begins "Each week."

3. Click the Insert Table button 🔲 on the Standard toolbar. A grid containing five columns and four rows appears below the button.

4. Position the pointer in the upper-left cell of the grid. Drag the pointer across to highlight four columns, and then drag the pointer down past the bottom of the grid to highlight five rows. The table dimensions appear below the grid. (See Figure 13-2 on the next page.)

FIGURE 13-2
Insert Table
dialog box

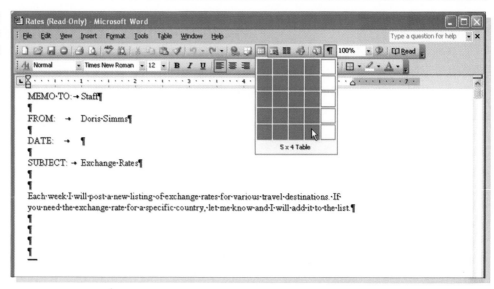

5. Release the mouse button. A five-row by four-column table appears with the insertion point in the first cell. Each column is the same width and the table extends from the left to the right margin. Notice the markers on the ruler that indicate the column widths.

NOTE: When you create a table, each gridline automatically has a 0.5-point black border that is printed. You can remove borders by opening the Borders and Shading dialog box, displaying the Borders tab, and choosing None under Setting. Gray table gridlines appear on-screen only and not in the printed document.

6. Click the Undo button to undo the table. You will now try the other method of inserting a table.

FIGURE 13-3
Specifying table
dimensions with the
table grid

7. From the Table menu, choose Insert, and then choose Table. The Insert Table dialog box appears.

8. Key **3** in the Number of columns text box and **6** in the Number of rows text box. You can also click the up or down arrows to the right of these boxes.

9. In the Fixed column width text box, change the default (Auto) to **1.5"** and click OK. Word inserts a 6 × 3 table with 1.5-inch columns. The insertion point is positioned in the first cell and the table is left-aligned.

EXERCISE 13-2 Draw a Table

You can draw a table in your document by using the Tables and Borders toolbar, which you open by using the Tables and Borders button on the Standard toolbar.

TABLE 13-1 Tables and Borders Toolbar Buttons

BUTTON	PURPOSE
Draw Table	Draw a freehand table
Eraser	Erase table lines
Line Style	Choose a border style
Line Weight	Choose a border weight
Border Color	Choose a border color
Outside Border	Choose or remove a border
Shading Color	Choose a shading color
Insert Table	Choose to insert a cell, row, column, or table
Merge Cells	Merge two or more cells
Split Cells	Split cells
Align Top Left	Choose to align cell text left, right, top, bottom, and/or centered
Distribute Rows Evenly	Make rows the same height
Distribute Columns Evenly	Make columns the same width

continues

TABLE 13-1   Tables and Borders Toolbar Buttons *continued*

BUTTON	PURPOSE
Table AutoFormat	Choose a predefined table format
Change Text Direction	Orient text horizontally
Sort Ascending	Sort in ascending order
Sort Descending	Sort in descending order
Σ AutoSum	Calculate the sum of values in cells

1. Click the Tables and Borders button on the Standard toolbar. Word changes to Print Layout view, displays the Tables and Borders toolbar, and changes the pointer to a pencil shape *✎* .

2. Position the pencil pointer at the third paragraph mark below the current table. Drag diagonally down to draw a rectangle about the same size as the current table. As you drag, the pointer creates a dotted rectangle. Release the mouse button.

> **NOTE:** You can click the Undo button at any time to undo a drawing action.

3. After drawing the outside border of the table, draw two vertical lines in the table to create three columns (just as in the other table). (See Figure 13-4 on the next page.)

4. Draw two horizontal lines, creating three rows. Don't worry about creating perfectly spaced rows or columns—you'll space them evenly later in the lesson. (See Figure 13-5 on the next page.)

5. Click the Eraser button on the Tables and Borders toolbar and drag the eraser pointer across one of the row lines you drew. Release the mouse button.

6. Click the Undo button to restore the line.

7. Press (Esc) or click the Eraser button to restore the normal pointer.

8. Close the Tables and Borders toolbar and switch back to Normal view.

FIGURE 13-4
Drawing a table

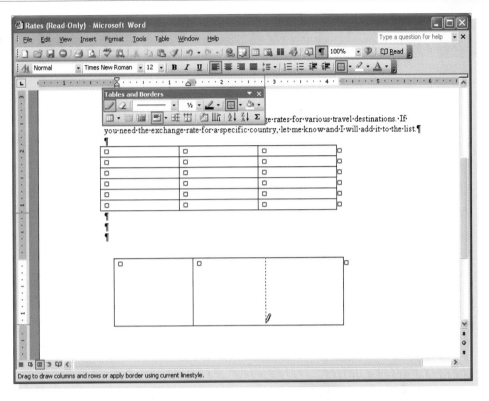

FIGURE 13-5
Adding rows

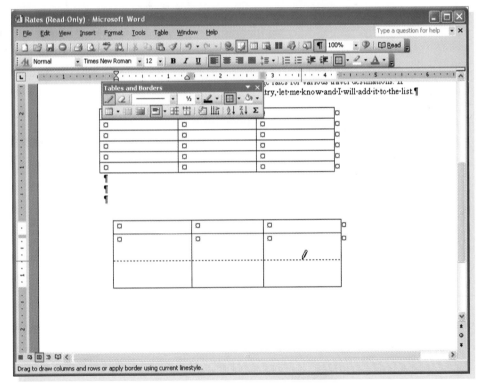

Keying and Editing Text in Tables

Keying and editing text in tables is similar to working with text in paragraphs. (Insert) and (Delete) work the same way. But if you key text in a cell and press (Enter), a new paragraph is created within the same cell.

To move the insertion point to different cells in a table, use the mouse, the arrow keys, or the following keyboard shortcuts:

TABLE 13-2

Shortcuts for Moving Between Cells

TO MOVE TO	PRESS
Next cell	(Tab)
Previous cell	(Shift)+(Tab)
First cell in the current row	(Alt)+(Home)
Last cell in the current row	(Alt)+(End)
Top cell in the current column	(Alt)+(Page Up)
Last cell in the current column	(Alt)+(Page Down)

EXERCISE **13-3** **Key and Edit Text in a Table**

1. In the first table, position the insertion point in the first cell and key **Country**.
2. Press (Tab) and key **Currency** in the next cell. Press (Tab) and key **$ Equals**. This is the *header row* for the table, the first row of a table (or the second row, if the table has a title row), in which each cell contains a heading for the column of text beneath it.
3. Press (Tab) to go to the first cell of the second row and key **Britain**.
4. Press (Tab) or (→) to go to the next cell and key **Pound**. Go to the next cell and key **0.632**.
5. Key the text shown in Figure 13-6 in the remaining rows of the table, inserting each word or number into a different cell. Remember not to press (Enter).

FIGURE 13-6

Canada	Dollar	1.37
Brazil	Reais	2.927
Japan	Yen	106.47
Mexico	Pesos	10.83

6. Press Alt + Home to move to the first cell in the last row and Shift + Tab to move to the third cell in the previous row. Change "106.74" to **116.74**.

7. In the third cell in the first row, key **1** after "$."

Selecting Cells, Rows, and Columns

There are several ways to select the contents of cells, rows, and columns. With them selected, you can delete, copy, or move their contents or change the format.

To help with selection, *end-of-cell markers* (◻) indicate the end of each cell. In addition, *end-of-row markers* (◻) to the right of the gridline of each row indicate the end of each row.

EXERCISE   13-4   Select Cells

1. Click the Show/Hide ¶ button ¶ to display your end-of-cell markers, if hidden characters are not visible.

2. To select the first cell in the first table, position the pointer just inside the left edge of the cell (between the cell's left border and the letter "C"). When the pointer becomes a solid black right-pointing arrow ◢, click to select the cell.

FIGURE 13-7
Selecting a cell

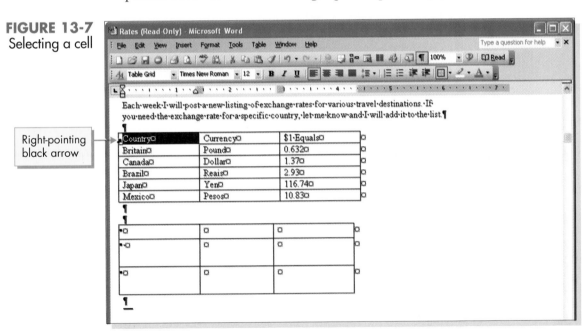

Right-pointing black arrow

3. Next, using the I-beam pointer, drag over the text "1.37" in the third row. Notice that the text is highlighted but the end-of-cell marker is not.

4. Press Tab to move to the next cell. Again, the text in the cell is highlighted, but the end-of-cell marker is not.

5. Using the I-beam pointer, triple-click within any cell to select the entire cell.

EXERCISE **13-5** **Select Rows, Columns, and Tables**

1. In the first table, point to the left of the fourth row. When you see the white right-pointing arrow ◁, click to select the row.

FIGURE 13-8
Selecting a row

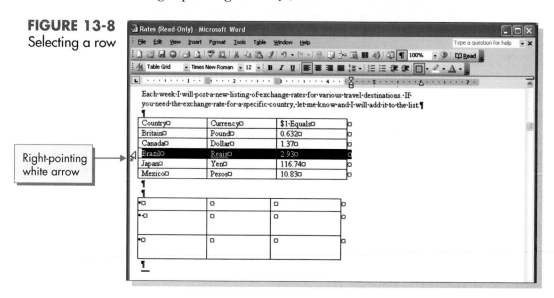

Right-pointing white arrow

2. Point just inside the left border of any cell in the previous row and double-click. That row is selected.

3. Position the insertion point anywhere in "Reais" in the fourth row. From the Table menu, choose Select, and then choose Row. The fourth row is selected again.

4. Point to the left of the third row and drag the pointer down one row to select both the third and fourth rows.

5. Point to the top border of the third column. When the pointer changes to a solid black down arrow ↓, click to select the column. (See Figure 13-9 on the next page.)

6. Position the insertion point before "Currency" in the second column. Drag down through "Pesos" to select column 2.

7. Point to the top of column 2 and drag the black arrow pointer across to select columns 2 and 3. Click anywhere in the table to deselect the columns.

8. With the insertion point anywhere in the table, choose Select, and then choose Table from the Table menu to select the entire table. Click anywhere in the table to deselect it.

FIGURE 13-9
Selecting a column

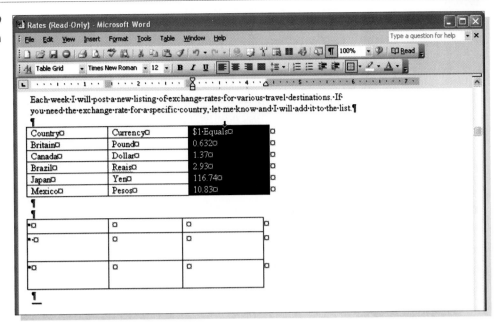

9. Change to Print Layout view. (Click the Print Layout View button ▣ at the bottom of the screen.)

10. Scroll until you can see the top of the first table.

11. Move the I-beam over the table until the table move handle ⊞ appears at the top left of the table.

12. Move the I-beam over the handle. When the I-beam changes to a four-headed arrow pointer ⊞, click to select the table.

13. Click anywhere in the table to deselect it and return to Normal view.

NOTE: Another way to select a table is to position the insertion point anywhere in the table, make sure Num Lock is turned off, and press (Alt)+(5) on the numeric keypad.

TABLE 13-3   Selecting Table Elements

TO SELECT	MOUSE	MENU	KEYBOARD
Cell	Click left inside edge of cell	Table, Select, Cell	(Shift)+(End)
Row	Click to the left of the row *or* double-click left inside edge of a cell	Table, Select, Row	
Column	Click column's top border	Table, Select, Column	(Alt)+click in column
Table	Click table move handle in Print Layout view	Table, Select, Table	(Alt)+(5) (numeric keypad)

Editing Table Structures

In addition to editing the contents of a table, you can edit a table's structure. You can add, delete, move, and copy cells, rows, and columns. You can also merge and split cells, or change a table's position or dimensions.

To modify tables, you can use the Table menu, Standard toolbar, the Tables and Borders toolbar, or the shortcut menu.

E X E R C I S E   13-6   Insert Cells, Rows, and Columns

1. Display the Tables and Borders toolbar by opening the View menu and choosing Toolbars, Tables and Borders. Notice that Word remains in Normal view when you open the toolbar this way.

2. Position the insertion point anywhere in the first table. Notice that both toolbars show a button for inserting tables.

3. In the first table, select the first cell in the first row. Notice that the Insert Table button 📖 on the Standard toolbar changes to the Insert Cells button .

> **NOTE:** The Insert Table button 📖 on the Standard toolbar changes to an Insert button for cells, rows, or columns, depending on what you select in the table.

FIGURE 13-10
Insert Cells
dialog box

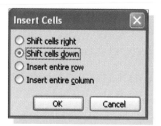

4. On the Tables and Borders toolbar, click the down arrow to the right of the Insert Table button 📖▾ and choose Insert Cells from the drop-down list. The Insert Cells dialog box opens.

5. Click Shift cells right and click OK. Word inserts a new cell and shifts the other cells in row 1 to the right. Also notice that the Insert Cells button 📋▾ replaced the Insert Table button 📖▾ on the Tables and Borders toolbar.

6. Click the Undo button ↺▾.

> **NOTE:** The Insert Table button 📖▾ on the Tables and Borders toolbar changes to reflect the last item you inserted from the button's drop-down list.

7. Drag the insertion point from "Britain" through "Dollar" in the second column. Click the Insert Cells button on the Standard toolbar.

8. Click Shift cells down, if it is not already selected. Click OK. Four new cells appear above the selected cells.

9. Select the first row. Click the Insert Rows button 🔳 on the Standard toolbar. A new row appears at the top of the table.

10. Select the second row. Click the down arrow to the right of the Insert Cells button 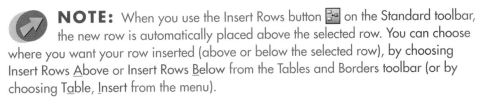 on the Tables and Borders toolbar and choose Insert Rows <u>B</u>elow from the drop-down list. A new row appears below the selected row.

11. Click the Undo button 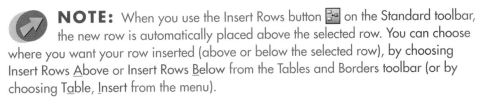.

> **NOTE:** When you use the Insert Rows button 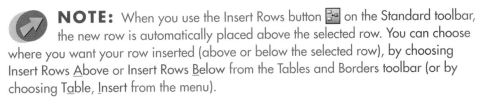 on the Standard toolbar, the new row is automatically placed above the selected row. You can choose where you want your row inserted (above or below the selected row), by choosing Insert Rows <u>A</u>bove or Insert Rows <u>B</u>elow from the Tables and Borders toolbar (or by choosing T<u>a</u>ble, <u>I</u>nsert from the menu).

12. Move the insertion point to the left of the paragraph mark below the first table. Click the Insert Rows button 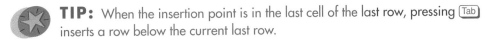 on the Standard toolbar.

13. In the Insert Rows dialog box, key **2** in the <u>N</u>umber of rows text box and click OK. Two more rows are added to the table.

> **TIP:** When the insertion point is in the last cell of the last row, pressing (Tab) inserts a row below the current last row.

14. Select the second column of the first table. Click the Insert Columns button on the Standard toolbar. The new column appears to the left of the selected column.

15. Select the third column. Click the down arrow to the right of the Insert Rows Below button 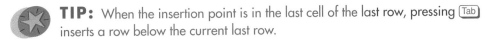 on the Tables and Borders toolbar. Choose Insert Columns to the <u>R</u>ight from the drop-down list. A new column appears to the right of the selected column.

16. Click the Undo button 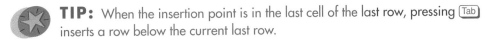.

> **NOTE:** When you use the Insert Columns button on the Standard toolbar, the new column is automatically placed to the left of the selected column. You can insert a column to the right by choosing Insert Columns to the <u>R</u>ight from the Tables and Borders toolbar (or by choosing T<u>a</u>ble, <u>I</u>nsert from the menu).

17. Close the Tables and Borders toolbar.

> **TIP:** When just the end-of-row markers are selected, you can click the Insert Columns button from the Standard toolbar to extend the table to the right.

EXERCISE **Delete Cells, Rows, and Columns**

Deleting cells, rows, and columns is different from deleting text (selecting text and pressing (Delete)). You must first select the table structure you want to delete, and then choose <u>D</u>elete from the T<u>a</u>ble menu.

1. In the first table, select the blank cells in the third and fourth rows. From the Table menu, choose Delete, and then choose Cells. The Delete Cells dialog box opens.

FIGURE 13-11
Delete Cells
dialog box

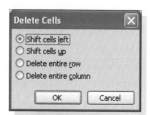

2. Click Shift cells up. Click OK and the blank cells disappear.

3. Select the blank rows at the bottom of the first table. From the Table menu, choose Delete, and then choose Rows. The blank rows are deleted.

4. Select the second column (which is blank). Right-click the column and choose Delete Columns from the shortcut menu. Only the top row of your table is now blank.

EXERCISE   13-8   Move and Copy Cells, Rows, and Columns

In addition to using toolbar buttons and the shortcut menu, you can also use keyboard shortcuts to cut, copy, and paste.

1. Select the row that begins "Country."

2. Point to the first selected cell in the row. When you see a left-pointing arrow, drag and drop the selection into the empty row above. The text now appears in the first row.

3. Select the row that begins "Canada," but not its end-of-row marker.

 4. Click the Cut button ✂ The text is deleted, but the empty row remains.

5. Position the insertion point in the first cell of the second row. Press Ctrl+V to paste the text.

6. Select the bottom three cells in the third column and copy them to the Clipboard.

7. Click within the cell containing "Reais" and paste the text. The pasted cells overwrite the previous text.

8. Undo the Paste command.

9. Select only the text in the first three rows (not the end-of-row markers) and cut the text.

10. Click in the first cell in the second row and paste the text. Your table should have the top row blank and the "Britain" and "Canada" rows reversed.

 NOTE: If you paste text somewhere on the table where there is not enough room for all the cells, Word adds additional columns or rows to accommodate the text.

EXERCISE **Merge and Split Cells**

1. Select the first row. From the T<u>a</u>ble menu, choose <u>M</u>erge Cells. The cells in the first row merge into a single cell.

> ⭐ **TIP:** If the Clipboard task pane is open, you can close it now.

2. In the first row, key the table title **Sample Exchange Rates**.

3. With the insertion point in the merged cell, right-click the first row and choose S<u>p</u>lit Cells from the shortcut menu. The Split Cells dialog box appears.

4. Key **3** in the Number of <u>c</u>olumns text box and click OK. The row is split into three cells again and the title text wraps in the first cell.

  > ⭐ **TIP:** You can also merge and split cells by using the Merge Cells button and the Split Cells button 🔲 on the Tables and Borders toolbar.

5. Display the Tables and Borders toolbar by using the button on the Standard toolbar. Change to Normal view.

6. With the first row selected in the table, click the Merge Cells button 🔲 on the toolbar to merge back to a single cell.

EXERCISE **13-10** **Change Table Dimensions and Position Tables**

You can adjust and position a table in the following ways:

- Change the width of columns, the space between columns, and the height of rows.
- Use AutoFit to change the width of a column to fit the longest text.
- Indent a table or center it horizontally on the page.

1. Select the second table (position the insertion point in any cell and choose T<u>a</u>ble, Sele<u>c</u>t, <u>T</u>able, or position the insertion point in any cell and press <kbd>Alt</kbd>+<kbd>5</kbd> on the numeric keypad with <kbd>NumLock</kbd> turned off).

2. From the T<u>a</u>ble menu, choose Table P<u>r</u>operties. Click the Col<u>u</u>mn tab in the Table Properties dialog box. (See Figure 13-12 on the next page.)

3. Click Preferred <u>w</u>idth and key **2** in the text box. Click OK. The columns are now 2 inches wide.

4. Click the Undo button 🔽.

  5. With the second table selected and using the Tables and Borders toolbar, click the Distribute Rows Evenly button ⊞. Then click the Distribute Columns Evenly button ⊞. The hand-drawn rows and columns are now evenly spaced.

FIGURE 13-12
Table Properties
dialog box

6. Select the first table. From the T<u>a</u>ble menu, choose Table P<u>r</u>operties and click the <u>R</u>ow tab.

7. Click <u>S</u>pecify height and key **0.4** in the text box. Choose At least from the Row height <u>i</u>s drop-down list, if it is not already selected.

 NOTE: Like column width, row height is expressed in inches, not points.

8. Click the <u>T</u>able tab in the Table Properties dialog box. Under Alignment, choose <u>C</u>enter and click OK. The table is centered horizontally on the page and the row heights are taller.

TIP: When a table appears on a page by itself, center it vertically and horizontally. Use the <u>L</u>ayout tab in the Page Setup dialog box to choose vertical alignment.

9. Deselect the table and position the pointer on the right border of the last column until it changes to a vertical double bar for resizing.

10. Drag the border a half inch to the right to widen the column. (Hold down ⟨Alt⟩ as you drag to see the exact ruler measurements.) (See Figure 13-13 on the next page.)

11. Click the Undo button 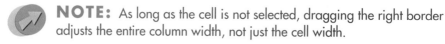 and position the pointer on the right border of the cell containing "Pound." Drag a half inch to the right. The second column is wider, but the last column is now narrower.

NOTE: As long as the cell is not selected, dragging the right border adjusts the entire column width, not just the cell width.

12. Click the Undo button ⟨↩ ▾⟩. Hold down ⟨Shift⟩ and drag the second column border a half inch to the right again. The second column is now wider and the last column remains the same width.

FIGURE 13-13
Dragging a
table border

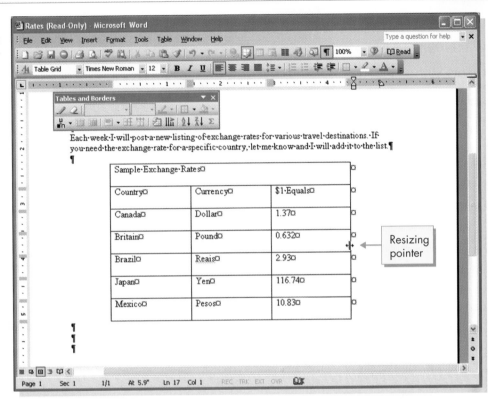

Resizing pointer

13. Click the Undo button 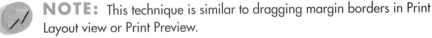. Double-click the right border of the second column. The column is adjusted to the width of the widest cell entry.

> **TIP:** You can simultaneously adjust all the columns to fit the width of each column's widest cell entry by choosing AutoFit, and then AutoFit to Contents from the Table menu.

14. Click anywhere in rows two through seven of the first table. Move the Tables and Borders toolbar to the bottom of the screen, if it is in your way.

15. On the ruler, point to the right column marker for the second column. When you see the ScreenTip "Move Table Column," drag the marker a short distance to the right, and then release the mouse button. (See Figure 13-14 on the next page.)

16. Hold down [Alt] and drag the marker until the ruler measurement for the second column is 1 inch.

> **NOTE:** This technique is similar to dragging margin borders in Print Layout view or Print Preview.

17. Use the same technique to adjust the first and third column widths to 1 inch. Each table column is now 1-inch wide and the merged row adjusts to the new table width.

FIGURE 13-14
Dragging a column
marker

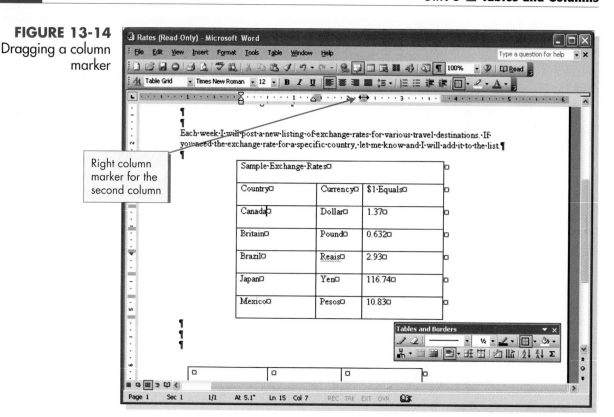

> **TIP:** The ruler measures column width by the actual space for text within the cells. The Table Properties dialog box (Column tab, Preferred width text box) includes the left and right cell margins, which are .08 inch by default. This means 1-inch wide columns set by using the ruler appear as 1.16-inch wide in the dialog box.

Formatting Tables and Cell Contents

There are many ways to make a table more attractive and easier to read. For example, you can:

- Format table text.
- Align text horizontally within columns.
- Align text vertically within cells or rows.
- Apply borders and shading.
- Use Table AutoFormat to apply a predesigned table style.
- Rotate the direction of text from horizontal to vertical.
- Sort table text.

E X E R C I S E 13-11 Format and Sort Table Text

1. Select the table title in the first row of the first table and format it as centered, bold, 14 points, and small caps.

2. Select row 2 and make the text bold.

3. Select "Currency" and the cells below it. Click the down arrow next to the Align Top Left button on the Tables and Borders toolbar to display a menu of alignment buttons.

⭐ **TIP:** If the Align Top Left button does not appear on the toolbar, click the down arrow next to the active Align button to display the menu of alignment buttons.

4. Click the Align Center Left button to left-align and center the text vertically in the selected cells. Notice that this button now appears on the Tables and Borders toolbar.

5. Click within the title in the first row, and then click the Align Center button on the Tables and Borders toolbar to center the text vertically.

6. Select "$1 Equals" and the cells below it. Click the down arrow next to the Align Center button .

7. Click the Align Center Right button to right-align the text and vertically center it within the selected cells.

⭐ **TIP:** When a column contains all text, it should be left-aligned. When it contains all numbers, it should be right-aligned.

8. Format all the text below "Currency" as italic.

9. Select "Country" and the cells below it. Click the down arrow next to the Align Center Right button .

10. Click the Align Center Left button to left-align the text and vertically center it within the selected cells.

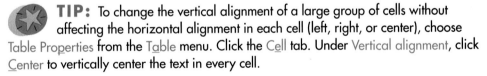

⭐ **TIP:** To change the vertical alignment of a large group of cells without affecting the horizontal alignment in each cell (left, right, or center), choose Table Properties from the Table menu. Click the Cell tab. Under Vertical alignment, click Center to vertically center the text in every cell.

11. Select rows 3 (the text below the header row) through 7 of the first table. In the next step, you sort these rows.

🔷 **NOTE:** Select only the cells you want sorted when using the Sort command.

12. From the Table menu, choose Sort. Choose Column 1 from the Sort by drop-down list, choose Text from the Type drop-down list, and click Ascending if it is not already selected.

13. Click OK to sort the rows alphabetically in ascending order.

EXERCISE 13-12 Apply Borders and Shading

You can increase the attractiveness of your table by adding borders and shading. It is important to choose appropriate line color, line style, and shading options so that text is easy to read and to ensure an attractive document.

1. Select the first row (the title). On the Tables and Borders toolbar, click the arrow next to the Shading Color button and choose Gray-10% from the drop-down palette (the third color in the first row).

2. Select the entire table.

3. Use the Line Style button ——— and the Line Weight button ½ drop-down lists on the Tables and Borders toolbar to set a double-line border that is ½-point weight.

4. Click the Outside Border button to make the line changes to the outside border. Click within the table to deselect the table and view the borders. Notice that Word has changed to Print Layout view.

5. Click the arrow beside the Border Color button and choose Pale Blue from the palette.

6. With the pencil pointer activated, drag it over each outside border of the table, one at a time, and then over the bottom border of the first row.

FIGURE 13-15
Drawing a border

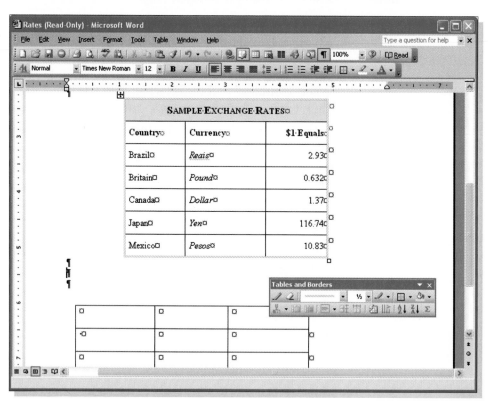

7. Click the Draw Table button 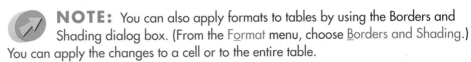 or press (Esc) to restore the normal pointer.

NOTE: You can also apply formats to tables by using the Borders and Shading dialog box. (From the Format menu, choose Borders and Shading.) You can apply the changes to a cell or to the entire table.

EXERCISE 13-13 Use Table AutoFormat

Word can format your table attractively when you use the Table AutoFormat button ⊞.

1. Select the text in rows 2 through 4 (from "Country" to "10.83"), but not the end-of-row markers. Copy the selected text to the Clipboard.

2. Click the first cell in the blank table and paste the copied text.

3. On the Tables and Borders toolbar, click the Table AutoFormat button ⊞.

4. Click the various AutoFormats in the Table styles list box and view the effect in the Preview box. (See Figure 13-16 on the next page.)

5. Choose the Table Grid 3 format. Under Apply special formats to, choose Heading rows.

6. Click Apply to apply the format.

7. Click the Table AutoFormat button ⊞.

8. Click the Modify button.

9. Next to Apply formatting to, choose Header row from the drop-down list.

10. Click the Italic button **I**, click OK, and click Apply to apply the adjusted formatting.

TIP: Using the AutoFormat dialog box to adjust formatting after a format has been applied is not optimal. You can simply select the header text and click the Italic button **I** on the Formatting toolbar or press (Ctrl)+(I). The best time to use the Modify button in the AutoFormat dialog box is when you are applying a predefined Word format.

11. Select the second table, including the blank paragraph marks above and below it, and press (Delete).

12. Add an extra paragraph mark above the table and add your reference initials on a second blank line below the table.

13. Save the document as [your initials]13-13 in a new Lesson 13 folder.

14. Print and close the document.

FIGURE 13-16
Applying a Table
AutoFormat

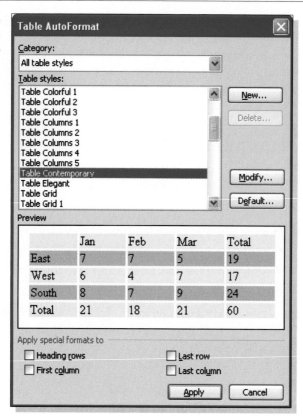

EXERCISE 13-14 Rotate Text

1. Start a new document. Display the Tables and Borders toolbar, if it is not already open.

2. Click the Insert Table button ▦ ▾ on the Tables and Borders toolbar. (If this button is not displayed, click the down arrow next to the Insert button that is displayed, and then choose Insert Table.)

3. In the Insert Table dialog box, set the Number of columns to **2** and the Number of rows to **8**. Click OK.

4. In the first cell of the first row, key **Customary Units**. Press Tab and key **Metric Equivalent**. Format both headings as bold.

5. Key **1 ounce** in the first cell of the second row. Press Tab and key **23.3495 grams**. Key the data shown in Figure 13-17 (on next page) for the remaining table cells.

6. Select the first column and click the Insert Columns button on the Standard toolbar.

7. Select the first row and click the Insert Rows button on the Standard toolbar.

FIGURE 13-17

1 pound	0.4536 kilograms
1 ton	0.907 tonnes
1 inch	2.54 centimeters
1 foot	30.48 centimeters
1 yard	0.9144 meters
1 mile	1.6093 kilometers

8. Position the insertion point in the first cell of the second row. Key **Weight** in bold. Position the insertion point in the first cell of the sixth row. Key **Length** and format it as bold.

9. Select the table and format all text as 14 points.

10. Select the cell containing "Weight" and the three blank cells below it. From the Table menu, choose Merge Cells.

 11. With the insertion point in the merged cell, click the Change Text Direction button on the Tables and Borders toolbar to rotate "Weight" until it reads from top to bottom. Click again so "Weight" reads from bottom to top.

> **NOTE:** If you are in Normal view when you rotate text, Word changes to Print Layout view.

12. Select the cell containing "Length" through the last cell of the first column. Click the Merge Cells button on the Tables and Borders toolbar.

13. With the merged cell selected, click the Change Text Direction button on the Tables and Borders toolbar twice to rotate "Length" until it reads from bottom to top. (See Figure 13-18 on the next page.)

14. Merge the cells in the first row and key the title **Metric Conversion Table** in the merged row. Format the title as 16-point bold uppercase.

15. Select the entire table. From the Table menu, choose Table Properties and click the Row tab. Set the row height to exactly **0.5"**.

16. Click the Table tab and, as the alignment, choose Center to horizontally center the table on the document page. Click OK.

17. With the table still selected, choose AutoFit, and then AutoFit to Contents from the Table menu to change all column widths to be as wide as the longest text in the column.

> **NOTE:** When a table has a title, AutoFit changes the title's column to fit the length of the title. This might make the column much wider than is required for the widest text in the remaining columns. You can insert a line break to create a two-line title.

18. With the table still selected, click the down arrow next to the button for alignment on the Tables and Borders toolbar. Click the Align Center Left button ▣ ▾ to vertically center all text.

19. Apply Gray-20% shading to the first row and to the cells that contain "Customary Units," "Metric Equivalent," "Weight," and "Length."

20. Select the table again. Using the Tables and Borders toolbar, change the border color to black and the line weight to 3/4 point. Apply a double-line border to the outside of the table.

21. Draw the same style border to separate the "Weight" and "Length" sections of the table.

22. Open the Page Setup dialog box. On the Layout tab, set the Vertical alignment to Center and click OK.

23. Save the document as *[your initials]***13-14** in your Lesson 13 folder. Print and close the document.

FIGURE 13-18
Table with
rotated text

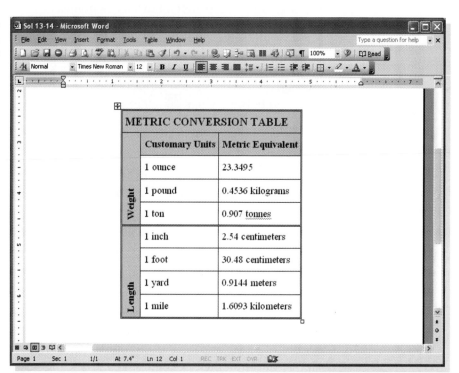

Converting Tables and Text

You can convert existing tabbed text to a table, which can be faster than keying text into an existing table. You can also convert an existing table to paragraphs of text, and you can choose to separate the converted text by paragraph marks, commas, or tabs. This might be useful when converting Word data for use in a database.

EXERCISE 13-15 Convert Text into Tables

1. In a new document, key the text shown in Figure 13-19, with one tab character between each entry. Press ⌈Enter⌋ at the end of each line as usual. Use single-spacing. (Your text might line up differently than in the figure, but don't worry about that. You will correct the document in the following steps.)

FIGURE 13-19

```
Quarters     2001       2002       2003

Q1           42.8       40.8       41.3

Q2           33.5       39.0       39.1

Q3           34.6       38.2       37.9

Q4           41.6       39.2       40.4
```

2. Select all the text. From the Table menu, choose Convert, and then choose Text to Table. The Convert Text to Table dialog box appears.

3. Make sure the Number of columns text entry is **4**. In addition, check that Tabs is chosen under the Separate text at section. Click OK. The selected text appears in a 5 × 4 table.

4. Choose Table AutoFormat from the Table or shortcut menu. Choose the Table Columns 4 format. Under Apply special formats to, deselect Last row and Last column. Click Apply.

5. Select the table and copy it to the Clipboard. You will need this table for the next exercise.

6. With the table still selected, open the Table Properties dialog box. Click the Row tab and set the row height to at least **0.3"**. Click OK.

7. With the table still selected, click the Align Center button 🖻 on the Tables and Borders toolbar to vertically and horizontally center the text.

8. Use the Page Setup dialog box to center the table vertically on the page.

9. Insert a new row at the top of the table and merge the cells in the row. In the first row, key the title **GATEWAY NATIONAL SALES** as 16-point bold type.

10. Save the document as *[your initials]***13-15** in your Lesson 13 folder.

11. Print and close the document.

EXERCISE 13-16 Convert Tables to Text

1. Start a new document. Paste the Clipboard contents that contain the copied table from step 5 of the previous exercise.

2. Select the last four rows of the table.

3. From the Table menu, choose Convert, and then choose Table to Text. The Convert Table to Text dialog box opens.

FIGURE 13-20
Convert Table to Text dialog box

4. Choose Tabs, if it is not already selected, to separate the contents of each cell with tabs.

5. Click OK. The text appears outside the table, with a tab replacing each end-of-cell marker and a paragraph mark replacing each end-of-row marker.

6. Close the document without saving it.

7. Close the Tables and Borders toolbar if it is still open.

USING ONLINE HELP

FIGURE 13-21
Help window about resizing tables

Not only can you resize the cells, rows, and columns of a table, but you can also resize the entire table. Help can show you how.

Ask Word Help how to resize an entire table:

1. Key **resize tables** in the Ask a Question text box and press Enter.

2. Click the topic Resize all or part of a table.

3. Click the topic Resize an entire table to expand it.

4. Read the information in the Help window. Then close the window.

LESSON 13 Summary

➤ A table is a grid of rows and columns that intersect to form cells. The lines that mark the cell boundaries are called gridlines.

➤ There are two ways to create a table: use the Insert Table dialog box or draw a table by using the Tables and Borders toolbar.

➤ Keying and editing text in tables is similar to working with text in paragraphs. [Insert] and [Delete] work the same way, but if you key text in a cell and press [Enter], a new paragraph is created within the same cell.

➤ To move the insertion point to different cells in a table, use the mouse, the Arrow keys, or keyboard shortcuts.

➤ Select cells by using the black right-pointing arrow or the shortcut keys, or by dragging over text. With the cells selected, you can delete, copy, or move their contents or change the format.

➤ To help with selection, end-of-cell markers (◻) indicate the end of each cell. In addition, end-of-row markers (◻) to the right of the gridline of each row indicate the end of each row.

➤ Select rows, columns, and tables by using the white right-pointing arrow or the Table menu or by dragging.

➤ The header row is the first row of a table (or second row if the table has a title row), in which each cell contains a heading for the column of text beneath it.

➤ In addition to editing the contents of a table, you can edit a table's structure by adding, deleting, moving, and copying cells, rows, and columns. You can also merge and split cells or change a table's position or dimensions.

➤ In addition to using toolbar buttons and the shortcut menu, you can also use keyboard shortcuts to cut, copy, and paste.

➤ There are many ways to make a table more attractive and easier to read: format table text, align text horizontally and vertically, apply borders and shading, use Table AutoFormat, rotate the direction of text, and sort table text.

➤ You can convert existing tabbed text to a table, which can be faster than keying text into an existing table. You can also convert an existing table to paragraphs of text, and you can choose to separate the converted text by paragraph marks, commas, or tabs.

LESSON 13 Command Summary

FEATURE	BUTTON	MENU	KEYBOARD
Insert table		Table, Insert, Table	
Draw table		Table, Draw Table	
Insert cells		Table, Insert, Cells	
Insert rows above		Table, Insert, Rows Above	
Insert rows below		Table, Insert, Rows Below	
Insert columns to left		Table, Insert, Columns to the Left	
Insert columns to right		Table, Insert, Columns to the Right	
Select a column		Table, Select, Column	
Select a table		Table, Select, Table	Alt+5 (Numeric keypad)
Delete selected table		Table, Delete, Table	Backspace
Merge cells		Table, Merge Cells	
Split cells		Table, Split Cells	
Distribute rows evenly		Table, AutoFit, Distribute Rows Evenly	
Distribute columns evenly		Table, AutoFit, Distribute Columns Evenly	
Sort text	or	Table, Sort	
Table AutoFormat		Table, Table AutoFormat	
Change text direction		Format, Text Direction	

Concepts Review

TRUE/FALSE QUESTIONS

Each of the following statements is either true or false. Indicate your choice by circling T or F.

T　F　**1.** You can insert a table by using the Standard toolbar.

T　F　**2.** To move to the top cell in the current column, you press `Alt`+`Page Down`.

T　F　**3.** You can use the TableAutoFormat dialog box to apply formats to existing tables.

T　F　**4.** You press `Alt`+`7` to select a table.

T　F　**5.** After inserting a cell, row, or column in a table, the Insert Table button 　 on the Tables and Borders toolbar becomes a button for the last inserted item.

T　F　**6.** To split a merged cell, first delete all paragraph marks in the merged cell.

T　F　**7.** By default, when you insert a table, it has borders that will be printed.

T　F　**8.** You can apply shading by using the Tables and Borders toolbar.

SHORT ANSWER QUESTIONS

Write the correct answer in the space provided.

1. Which dialog box do you use to change column width?

_____

2. Which keyboard shortcut moves the insertion point to the previous cell?

_____

3. Which nonprinting character marks the end of a cell?

_____

4. What are the nonprinting lines called that mark the boundaries of cells?

_____

5. What is the button 　 used for?

_____

6. Which feature on the Tables and Borders toolbar do you use to delete a line you drew?

_____

7. Which table element can you select when the pointer appears as ◢?

_____

8. What would the button 🔲 do to selected cells in a table?

_____

CRITICAL THINKING

Answer these questions on a separate page. There are no right or wrong answers. Support your answers with examples from your own experience, if possible.

1. You learned how to insert a table automatically and draw a table manually. What are the benefits of each method? Which do you prefer?

2. Review the table formats in the TableAutoFormat dialog box. Which table formats would you be most likely to use on a day-to-day basis? Can you think of any formats that are not particularly useful?

Skills Review

EXERCISE 13-17

Create a table. Key and edit text in the table.

1. Open the file **CarRent**.

2. Delete the paragraph that begins "In the compact category."

3. Add another paragraph mark (there should be two) above the paragraph that begins "All Rates."

4. At the paragraph mark just created, insert a table by following these steps:

   a. Click the Insert Table button 🔲 ▾ on the Standard toolbar.
   b. Drag in the grid to create a 4-row by 3-column table.
   c. Release the mouse button. Make sure there is one blank line above and below the table.

5. Key text in the first row of the table by following these steps:

   a. With the insertion point in the first cell, key **Category**.
   b. Press Tab and key **Car**.
   c. Press Tab and key **Price**.

6. Key the text shown in Figure 13-22 in the remaining cells of the table.

FIGURE 13-22

Compact	Ford Focus	$101
Intermediate	Ford Taurus	$110
Full-size	Ford Crown Victoria	$115

7. Move around and edit text in the table, using keyboard shortcuts, by following these steps:

   a. With the insertion point in the last cell, press Alt + Page Up and then Alt + Home. Change "Category" to **Class**.

   b. Press Tab twice and change "Price" to **Weekend Rate**.

   c. Press Alt + Page Down and change "$115" to **$140**.

8. Select the text in the first row and apply bold formatting.

9. Format the document as a memo to Doris Simms, from you, with the subject "Weekend Car Rentals."

10. Save the document as *[your initials]*13-17 in your Lesson 13 folder.

11. Print and close the document.

EXERCISE 13-18

Select, move, and copy cells, rows, and columns. Edit table structures.

1. Open the file **Frequent**.

2. Cut the text in the third row by following these steps:

   a. Make sure the Show/Hide ¶ button ¶ is turned on.

   b. Drag to select only the text in the row—not the end-of-row marker.

   c. Cut the text.

3. Paste the text in a new row at the bottom of the table by following these steps:

   a. Position the pointer in the last cell of the last column and press Tab to create a new row.

   b. Paste the text.

4. Delete the third row and insert a column by following these steps:

   a. Point just outside the left edge of the third row and click to select the row. From the Table menu, choose Delete, and then choose Rows.

   b. Point to the top border of the second column and click to select the column. Click the Insert Columns button on the Standard toolbar.

5. Copy and paste the text from the fourth column into the second column and delete column 4 by following these steps:

    a. Select the text "California" through "Florida" in column 4 and copy the text.

    b. Click the second cell of the second column and paste the text.

    c. Select the fourth column, right-click, and choose Delete Columns from the shortcut menu.

6. Change the column width and row height, and center the table by following these steps:

    a. Select the entire table by opening the Table menu, choosing Select, and then choosing Table.

    b. From the Table menu, choose Table Properties.

    c. Click the Column tab and set the Preferred width text box to **1.75"**.

    d. Click the Row tab. Select Specify height and set the height to at least **0.3"**.

    e. Click the Table tab. Under Alignment, choose Center. Click OK.

7. Merge the cells in the first row by following these steps:

    a. Select the first row.

    b. Display the Tables and Borders toolbar by clicking the Tables and Borders button on the Standard toolbar.

    c. Click the Merge Cells button on the Tables and Borders toolbar.

8. Center the text in row 1 and format it as 14-point bold all caps.

9. Insert a line break in the title before "CITIES" by positioning the insertion point and pressing `Shift`+`Enter`.

10. Center text vertically in each cell of the table without changing the horizontal alignment by following these steps:

    a. Select the entire table.

    b. From the Table menu, choose Table Properties and click the Cell tab. Under Vertical alignment, choose Center. Click OK.

11. Use the Page Setup dialog box to center the table vertically on the page.

12. Save the document as *[your initials]***13-18** in your Lesson 13 folder.

13. Print and close the document.

EXERCISE 13-19

Convert text to a table and format the table.

1. Open the document **Tour1**.

2. Convert tabbed text to a table by following these steps:

    a. Select the tabbed text from "Departing From" through the end of the line that begins "Arkansas."

    b. From the Table menu, choose Convert, and then choose Text to Table.

    c. Make sure **4** appears in the Number of columns text box and that Tabs is selected. Click OK.

3. Copy the text "Departing From" and "Rate" to the corresponding blank cells in row 1. Remove the underline formatting.

4. Using the Tables and Borders toolbar, click Table AutoFormat.

5. From the Table styles list, choose Table Columns 2. Under Apply special formats to, deselect every option except Heading rows. Click Apply.

6. Delete the blank line above "Airfares." Format "Airfares" and the line above it as 16-point bold, centered, with no underline. Add 24 points of paragraph spacing after "Airfares."

7. Use the Table Properties dialog box to increase the height of every row in the table to at least **0.3"**.

8. Using the Tables and Borders toolbar, make the text in every cell left-aligned and vertically centered.

9. Delete all the text below the table through the end of the document.

10. Center the document vertically on the page.

11. Save the document as *[your initials]***13-19** in your Lesson 13 folder.

12. Print and close the document.

EXERCISE 13-20

Convert text to a table and format the table.

1. Open the file **Revenue2**.

2. Convert the text to a 5-row by 4-column table. Without closing the Convert Text to Table dialog box, click AutoFormat and apply the Table Professional format.

3. Select the first row and increase the font to 18 points.

4. Select the table and double-click the right border of the last column with the vertical double bar to AutoFit each column.

TIP: When a table is selected, you can double-click the right border of the last column to AutoFit each column in one quick step.

5. Change row height for the entire table to exactly **0.5"**.

6. Right-align columns 2 through 4.

7. Vertically center the text in every cell without affecting the horizontal alignment in the cell.

8. Center the table on the page vertically and horizontally.

9. Sort the table alphabetically in ascending order.

10. Save the document as *[your initials]***13-20** in your Lesson 13 folder.

11. Print and close the document.

Lesson Applications

EXERCISE 13-21

Create a table, enter text in the table, and edit and format the table.

1. Start a new document.
2. Create a table with nine rows and four columns.
3. Key the text shown in Figure 13-23 in the table, leaving the first cell blank.

 TIP: For convenience, after you key the first set of months, copy the text to the cells at the bottom of the column.

FIGURE 13-23

	Month	High	Low
Santa Fe	January	40	18
	April	59	35
	July	82	56
	October	62	38
Taos	January	40	10
	April	53	21
	July	87	50
	October	67	32

4. Merge the cell that contains "Santa Fe" with the three blank cells below it. Do the same for the cell that contains "Taos" and the blank cells below it.
5. Change the entire table to 16-point Arial and the row height to at least **0.45"**.
6. Center the table text vertically within the cells.
7. Use the column markers or column borders to adjust the second column to 1.25 inches wide, and the third and fourth columns to 1-inch wide.
8. Rotate the text "Santa Fe" until it reads from the bottom to the top of the cell. Format the text as 26-point bold. Format "Taos" the same way.
9. Change the first column to 0.75 inch wide.
10. Center-align the text in columns one, three, and four.

11. Insert a new row at the top of the table, merge the cells, and key the title **Average Temperatures** in 24-point bold small caps.

12. Change the font color of the second row to white. Apply black shading to the row.

13. Apply 20% gray shading to the first row.

14. Add a 1/2-point double-line outside border to the table.

15. Center the table vertically and horizontally on the page.

16. Save the document as *[your initials]***13-21** in your Lesson 13 folder.

17. Print and close the document.

EXERCISE 13-22

Convert text to a table, and edit and format the table.

1. Open the file **Itin2**.

2. On page 1, convert the text from "Day 1" through "Hoover Dam" (under "Sample Itinerary—The Great Southwest") to a one-column table separated at paragraphs.

3. Set the column width to 3 inches.

4. Insert a 1-inch wide column to the left of the 3-inch column.

5. Move the text "Day 1" to the first cell in the new column. Move all the remaining day numbers to the appropriate cell in the new column.

6. Delete all the text below the table. Above the table, delete the text from "Travel to the Southwest via bus" through "Las Vegas."

7. Insert a new row at the top of the table and merge the cells in the row.

8. Cut the bold title that is just above the table (but not its paragraph mark) and paste it into the first row. There should be a total of two blank lines above the table.

9. Apply the table AutoFormat Table Grid 4 to the table, removing the special formats for the last row and the last column.

10. Adjust the second column to be 5 inches wide.

11. In each cell in the second column, start each second sentence as a separate paragraph.

12. Format the text in each cell in column 2 as a bulleted list, using the arrowhead bullet.

13. Format the bulleted text paragraphs with no left indent, no left tab setting, and a 0.25-inch hanging indent.

14. Format the table title with 6 points of spacing before and after it.

15. Set a 2-inch top margin for the document, and format the document title as 14-point bold small caps, centered, with two blank lines below it.

16. Find and replace straight apostrophes with smart apostrophes.

17. Save the document as *[your initials]***13-22** in your Lesson 13 folder.

18. Print and close the document.

EXERCISE 13-23

Create a table, and edit and format the table.

1. Start a new document and change the page to landscape orientation.

2. Insert a table with five rows and six columns, and key the text shown in Figure 13-24 in the table.

FIGURE 13-24

Monday	Tuesday	Wednesday	Thursday	Friday	sat./sun.
			1	2	3/4
5	6	7	8	9	10/11
12	13	14	15	16	17/18
19	20	21	22	23	24/25

REVIEW: You might need to use the AutoCorrect Options button to undo the fraction ¾ after you key "3/4."

3. Add a sixth row to the table and key **26**, **27**, **28**, **29**, and **30**, beginning with the first cell.

4. Change each column to 1.2 inches wide.

5. Change the weekday headings in the top row to 16-point bold italic and centered. Change the top row height to at least **0.4"**, and vertically center the text.

6. In the cell for the 6th, key **RSVP for Red Rock Ranch Trip** as a new paragraph below the date.

7. In the cell for the 1st, key **Mail travel documents for Vegas** as a new paragraph below the date.

8. In the cell for the 19th, key **Noon visit with regional managers** as a new paragraph below the date.

9. Key **Staff meeting** below the date in the cells for the 16th and the 30th.

10. Change the row height for rows 2 through 6 to exactly **0.9"**.

11. Insert a new first column, merge the cells in the column, and key **JUNE**.

12. Rotate the text "JUNE" until it reads from the bottom to the top of the column. Change the font size to 60 points and the column width to 1 inch. The text should be bold but not italic.

13. Center the text in column 1 vertically and horizontally within the column.

14. Change the zoom to Page Width to view the entire table.

15. Apply a 1/2-point triple-line outside border to columns 2 through 7.

16. Apply a 3-point solid-line outside border and 20% gray shading to the first column.

17. Draw a 1/2-point double-line border at the bottom of row 1.

18. Center the table vertically and horizontally on the page.

19. Save the document as *[your initials]*13-23 in your Lesson 13 folder.

20. Print and close the document.

EXERCISE 13-24 *Challenge Yourself*

Draw a table, enter text in the table, and then edit and format the table.

1. Start a new document and change the page orientation to landscape.

2. Using the Draw Table feature and the default 1/2-point single-line style, draw a rectangle large enough to fit the page (in Print Layout view), leaving about 1-inch left and right margins. (The exact dimensions are unimportant; just draw a rectangle that you can see clearly on the page.)

3. Draw five vertical lines to create six columns. The lines don't have to be evenly spaced.

4. Draw four horizontal lines to create five rows.

5. Merge the cells in row 1.

6. Select the table and use the Tables and Borders toolbar to distribute the columns and rows evenly.

7. Key the text shown in Figure 13-25 into the table.

FIGURE 13-25

3-Year Computer-Related Office Expenditures

Year	Paper	Toner	Machines	Maintenance	Total
2001	288.50	468.00	9,767.00	108.00	$10,631.50
2002	367.00	389.00	3,568.00	127.50	$4,451.50
2003	402.00	332.00	850.00	235.00	$1,819.00

8. Apply the Table Contemporary Table AutoFormat.

9. Format the title row as 26-point small caps, centered, and bold.

10. Format the text in row 2 as 20-point bold. Increase the font size for the remaining rows to 18 points.

11. Change the row height for all rows to 1 inch.

12. Horizontally center the column headers from "Paper" through "Total." Right-align the numeric text under these headers.

13. Center all text vertically within the cells.

14. Center the table vertically and horizontally on the page.

15. Sort the bottom three rows in descending order by column 1.

16. Add a dollar sign to each of the amounts in row 3.

17. Apply a 2 1/4-point solid-line outside border to the table.

18. Check the document in Print Preview. Save it as *[your initials]*13-24 in your Lesson 13 folder.

19. Print and close the document.

20. Close the Tables and Borders toolbar, if it is still open.

On Your Own

In these exercises, you work on your own, as you would in a real-life work environment. Use the skills you've learned to accomplish the task—and be creative.

EXERCISE 13-25

Create a table of monthly expenses. Include an expense name column, an average amount column, and a due date column. Adjust the row height and column width, and format the table, so it is readable and attractive. Spell- and grammar-check your document and save it as *[your initials]*13-25. Print the document.

EXERCISE 13-26

Log on to the Internet and find statistics on five stocks in which you are interested. Create a table showing the history of the stocks. (You might want to copy and paste information into your document and then convert the text to a table). Give the document a title, and spell- and grammar-check the document. Save the document as *[your initials]*13-26 and print it.

EXERCISE 13-27

Using the table feature, create a calendar for one of the months of the year, similar to the one you created in Exercise 13-23. Format the calendar attractively and add information you want to remember for particular days. Save the document as *[your initials]*13-27 and print it.

Columns

OBJECTIVES

**MICROSOFT OFFICE
SPECIALIST
ACTIVITIES**

In this lesson:
WW03S-3-3

See Appendix C.

After completing this lesson, you will be able to:

1. **Create multiple-column layouts.**
2. **Key and edit text in columns.**
3. **Format columns and column text.**
4. **Control column breaks.**

Estimated Time: 1 hour

Word can arrange document text in multiple columns on a single page like those used in newspapers and magazines. The continuous flow of text from one column to another can make a document more attractive and easier to read.

Documents commonly have between one and three columns. A single document can also use different column layouts. For example, a document might use a standard one-column layout in one section and a three-column layout in another section.

Creating Multiple-Column Layouts

There are two ways to create multiple-column layouts:

- Use the Columns button on the Standard toolbar.
- Use the Columns dialog box.

You can change the column format for the whole document, for the section containing the insertion point, or for selected text.

EXERCISE 14-1 **Use the Toolbar to Create Columns**

1. Open the file **Primer1** and switch to Print Layout view. If nonprinting characters are not displayed, click the Show/Hide ¶ button ¶ to display them.

NOTE: Normal view displays columnar text in a single, continuous, narrow column. Use Print Layout view or Print Preview to see how columns are arranged on a page.

2. Position the insertion point at the beginning of the bold heading "What about..." on the second page of the document.

3. Insert a continuous section break before the heading. (Insert menu, Break, Continuous.)

4. Click the Columns button ▦ on the Standard toolbar. A palette containing four columns appears below the button. The number of columns you highlight determines the number of columns that will appear in your document.

FIGURE 14-1
Choosing a column
layout

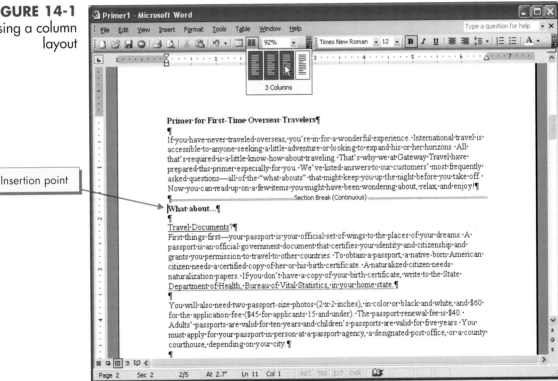

5. Point to the third column in the palette and click to choose a three-column layout. The new section is now in three columns. (See Figure 14-2 on the next page.)

FIGURE 14-2
One- and three-
column layouts in
Print Layout view

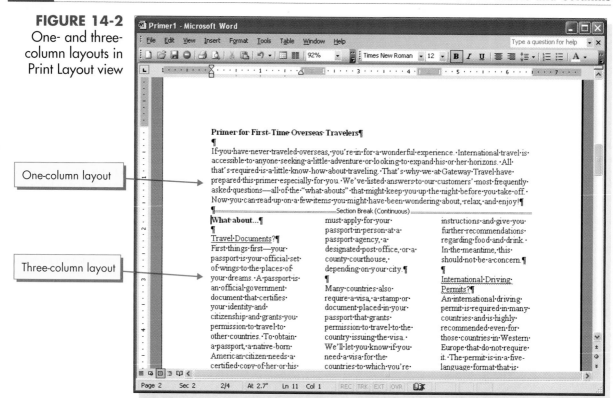

One-column layout

Three-column layout

EXERCISE | **14-2** | **Use the Columns Dialog Box to Set Up Columns**

When you use the Columns button 🔲, Word applies default settings for column width, and starts a new column. You can change these settings in the Columns dialog box.

1. While still in Print Layout view, go to the last page and position the insertion point at the beginning of the paragraph that starts "For specific information." (The paragraph is near the bottom of the second column.)

2. From the F*o*rmat menu, choose *C*olumns to open the Columns dialog box. (See Figure 14-3 on the next page.)

3. Key **1** in the *N*umber of columns text box or click *O*ne in the Presets section.

4. Choose This point forward from the *A*pply to drop-down list. The Preview box reflects the options you specify.

5. Click OK. A continuous section break is automatically inserted at the location of the insertion point, and the new section is formatted as one column.

6. Press (Enter) to insert an extra blank line between the sections.

FIGURE 14-3
Columns dialog box

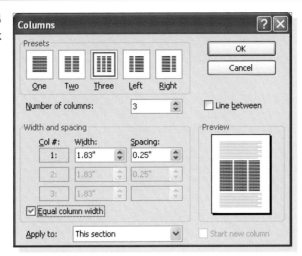

7. Position the insertion point at the beginning of the heading "What about…" on the second page, and open the Columns dialog box.

8. Set the Spacing text box for Col #1 to **0.25"** to specify spacing between columns. Click the Equal column width check box to make all the columns equal in width.

9. Choose This Section from the Apply to drop-down list to apply these settings to only this section; then click the Line between check box. Notice that the Preview box reflects these options.

10. Click OK. The columns in this section are spaced 0.25 inch apart with a line between them.

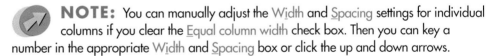 **NOTE:** You can manually adjust the Width and Spacing settings for individual columns if you clear the Equal column width check box. Then you can key a number in the appropriate Width and Spacing box or click the up and down arrows.

11. Choose Options from the Tools menu and click the View tab in the Options dialog box.

12. Under Print and Web Layout options, click Text boundaries to select this option. Click OK. A single dotted line displays column boundaries and page margins. (See Figure 14-4 on the next page.)

 NOTE: Text boundary lines do not appear when a document is printed.

13. Scroll through the document and review all boundary lines.

14. Reopen the Options dialog box. On the View tab under Print and Web Layout options, click Text boundaries to deselect this option.

15. Click OK. The boundary lines are hidden.

FIGURE 14-4
Display column
boundaries

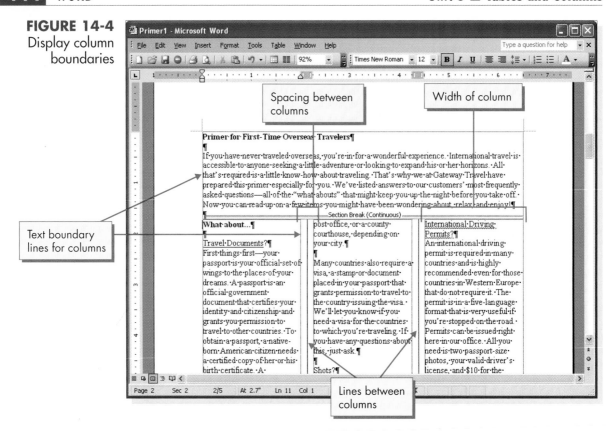

Keying and Editing Text in Columns

You can key and edit text in multiple-column layouts the same way you do in standard one-column layouts. Moving the insertion point around is a little different, however.

EXERCISE 14-3 Key and Edit Column Text

Keying and editing text in columns is very similar to keying and editing text in a standard document.

To move the insertion point from the bottom of a column to the top of the next column, you press →. You press ← to move from the top of a column to the bottom of the previous column. To move from the bottom line of a column to the first line of the same column on the next page, press ↓.

> **TIP:** If a document has a page that contains blank lines between the end of the text in a column and the page break, press ↓ more than once to move to the top line of the column on the next page.

1. Scroll down to the bottom of page 2, to the paragraph above the heading "<u>The Language Barrier</u>" in the third column. Position the insertion point before the text "Check gas costs."

2. Key the text shown in Figure 14-5. Use single spacing. Notice that the text is inserted into the column paragraph just as it would be in a standard one-column paragraph.

FIGURE 14-5

```
Driving is an excellent way to tour an area—providing you are not in a
hurry. It allows you to explore freely and to mingle with local people
at your own pace.
```

 REVIEW: The keyboard combination for an em dash (—) is `Ctrl`+`Alt`+the minus sign from the numeric keypad.

3. Position the insertion point at the end of the text in the third column and press →. The insertion point moves to the first column at the top of the next page.

 NOTE: When you use arrow keys to navigate in columns, you might need to press an arrow key twice to move past a space.

4. Press ← to return to the previous location.

5. Position the insertion point at the end of the first column on page 2 and press → again. The insertion point moves to the top of the second column.

6. Press ← to return to the end of the first column. Position the insertion point anywhere in the middle of the last line of the first column.

7. Press ↓ and the insertion point moves to the same position in the same column at the top of the next page. If the first line is a blank line, you might have to press ↓ again.

8. Move the insertion point to the end of the paragraph above the heading "<u>The Language Barrier</u>." The insertion point should be located after "the road." Press the spacebar, and then key the text shown in Figure 14-6. Use single spacing.

FIGURE 14-6

```
This may take a little getting used to, so you may want to be a
passenger first before trying your hand at driving.
```

Formatting Columns and Column Text

You can create and change font and paragraph formats for text in columns the same way you create and change them for standard layouts.

There are two ways to change the width of a column and the amount of space between columns:

- Use the ruler.
- Use the Columns dialog box.

EXERCISE **14-4** **Format Column Text**

1. Just below the section break on the second page, select the text "What about...."

2. Format the text as 14-point bold italic, with 18 points of spacing after the paragraph.

3. Delete the blank line above "Travel Documents?"

4. Select the text "Travel Documents?" and the paragraph below it.

5. Click the Justify button on the Formatting toolbar to justify the text. Notice the additional white space between words as a result of the justified text.

6. Click the Undo button ⟲▾.

TIP: When you justify column text, it looks like text in a newspaper. Left-align column text to create more of a newsletter effect. It's rarely appropriate to center column material.

EXERCISE **14-5** **Change Column Width and Spacing**

1. Display the ruler, if it is not displayed.

2. On page 2, position the insertion point in the first column, at the bottom of the page. Notice that the column has its own indent markers and margins.

3. Point to the column's right margin on the ruler. When you see the ScreenTip "Right Margin" and the two-headed arrow ↔, drag the right margin 0.5 inch to the left. Hold down Alt while dragging to see the exact measurement.

FIGURE 14-7
Using the ruler to
adjust column width

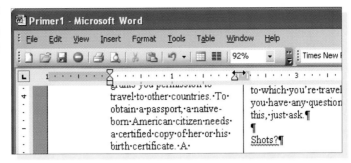

4. Release the mouse button. Because these columns were set to be of equal width, they all become narrower and the spacing between them increases. Notice that the page breaks change because the text reflows.

 NOTE: When adjusting columns of unequal width, only the width of the adjusted column changes.

5. Open the Columns dialog box. Click Equal column width to deselect this option.

6. Click OK. The ruler now contains column markers in the blue areas. You can drag these markers to adjust the width of individual columns.

7. Position the pointer over the first column marker. A ScreenTip displays "Move Column."

FIGURE 14-8
Dragging a
column marker

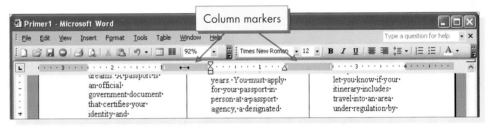

8. Drag the column marker 0.5 inch to the right. The width of the first column increases as it makes the second column narrower.

9. Move the pointer to the left margin area of the second column. The ScreenTip "Left Margin" appears.

FIGURE 14-9
Adjusting left
margin of
second column

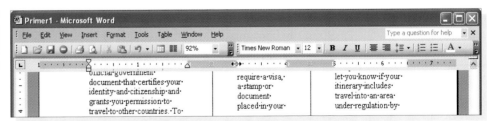

10. Drag the two-headed arrow pointer ↔ 0.5 inch to the left. This decreases the space between the first two columns.

11. Open the Columns dialog box. Notice the different settings for columns 1 and 2.

12. Set the spacing for column 1 to **0.4"** and check Equal column width.

13. Click OK. Word changes the width for all three columns to 1.9 inches and the space between columns to 0.4 inch.

Controlling Column Breaks

Word automatically breaks columns at the end of the page or section, so column lengths are often uneven. However, you can insert a column break manually, so the columns end where you want them to end. You can also balance the columns by adjusting the column breaks to make them even.

EXERCISE **14-6** **Insert Column Breaks**

There are three ways to insert column breaks:

- Use the Break dialog box.
- Press Ctrl+Shift+Enter.
- Use the Columns dialog box.

In this exercise, you practice inserting column breaks by using each of the three methods.

1. On the second page of the document, position the insertion point before "Shots."

2. From the Insert menu, choose Break to open the Break dialog box.

3. Choose Column break and click OK. A column break is inserted and "Shots" appears at the top of the third column.

4. Press Backspace to delete the column break.

5. Press Ctrl+Shift+Enter to insert a column break again.

6. Press Backspace to remove it.

7. Open the Columns dialog box.

8. Choose This point forward from the Apply to drop-down list, click Start new column, and click OK. A new section is inserted. Notice that the section break has the same effect as the previous column breaks.

9. Click the Undo button.

TIP: You can also use the Paragraph dialog box to control column breaks. For example, if a column break separates a paragraph from its heading, you can use the Keep with ne<u>x</u>t pagination option.

EXERCISE **14-7** **Balance the Length of Columns**

Many people feel that balancing the length of your columns on a partial page adds a professional appearance to a document.

1. At the end of the three-column section, select the paragraph mark after "make a formal visit!" and all of the text following it, up to—but not including—the last paragraph mark after "Zambia."

2. Press (Delete). Section 2 becomes one column again. (The formatting in the last paragraph mark now applies to section 2.)

3. Format the first line on page 2 (the title) as uppercase, bold, and centered.

4. Add a blank line after the title. Two blank lines should now separate the title from the text.

5. Add 72 points of paragraph spacing before the title.

6. Click within section 2, open the Columns dialog box, and reapply the formatting for this section: three columns with equal column width, spacing between the columns of 0.25 inch, and a vertical line between the columns.

7. Open Print Preview and notice that the columns are not balanced.

8. Close Print Preview and position the insertion point at the end of the three-column section, after "formal visit!"

9. Insert a Continuous section break. Word creates columns of equal length on the last page.

10. Switch to Print Preview to view the document. (See Figure 14-10 on the next page.)

11. Return to Print Layout view.

12. At the end of page 1, replace the page break with a <u>N</u>ext page section break.

13. Apply page numbering, beginning in the second section, with numbers centered at the bottom of the page. Start numbering at 1 and make sure there is no page number in section 1.

14. Save the document as *[your initials]***14-7** in a new Lesson 14 folder.

15. Open the Print dialog box. In the Pages text box, key **s2-s3** to print sections 2 and 3 of the document. Click OK.

16. Close the document.

FIGURE 14-10
Document with columns of equal length

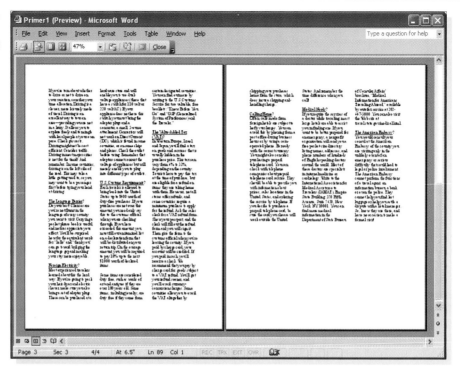

USING ONLINE HELP

FIGURE 14-11
Help about columns

You can use Word's column feature to display text in many ways. Help is available to provide you with more information and to answer questions.

Explore Help topics about columns:

1. Key **newspaper columns** in the Ask a Question text box and press Enter.

2. Click one of the topics to open a Help window.

3. Explore the topics in the Help window.

4. Close the Help window when you have finished.

LESSON 14 Summary

➤ Word can arrange document text in columns like those used in newspapers and magazines. There are two ways to create multiple-column layouts: by using the Columns button ▦ and by using the Columns dialog box.

➤ You can key and edit text in multiple-column layouts the same way as in standard one-column layouts. Use the arrow keys to move between columns.

➤ You can create and change font and paragraph formats for text in columns the same way you create and change them for standard layouts.

➤ Adjust column width and spacing by dragging column markers and column margins on the ruler or by using the Columns dialog box. Create equal column widths by clicking the Equal column width check box in the Columns dialog box.

➤ Word automatically breaks columns at the end of the page or section. However, you can insert a column break manually, so the columns end where you want them to end.

➤ Column lengths are often unequal at the ends of sections or pages. Insert a Continuous section break after the unbalanced columns and Word will balance the columns.

LESSON 14   Command Summary

FEATURE	BUTTON	MENU	KEYBOARD
Insert columns	▦	Format, Columns	
Insert column break		Insert, Break	Ctrl + Shift + Enter

Concepts Review

Each of the following statements is either true or false. Indicate your choice by circling T or F.

T F   **1.** You can open the Columns dialog box by using the Insert menu.

T F   **2.** A continuous section break is used to balance columns.

T F   **3.** Formatting text in multiple columns is very different from formatting text in a typical document.

T F   **4.** Pressing ⏎ at the end of any column moves the insertion point to the top of the next column.

T F   **5.** You use the Table Properties dialog box to change the document's column width and spacing.

T F   **6.** You can use the ruler to change column width.

T F   **7.** To quickly choose a column layout for the current section, click ▦.

T F   **8.** When working with columns, use Normal view to see the columns as they are arranged on the page.

Write the correct answer in the space provided.

1. What is the keyboard shortcut to insert a column break?

_____

2. Which dialog box do you use to add a vertical line between columns?

_____

3. Which arrow key do you press to move from the top of any column to the bottom of the previous column?

_____

4. Which toolbar includes the Columns button ▦?

_____

5. Which key do you press while adjusting column widths on the ruler to see an exact measurement display?

6. What shape is the pointer when you use the ruler to adjust column widths?

7. What type of line can you apply from the Columns dialog box?

8. What option in the Columns dialog box do you select to make all columns the same width?

CRITICAL THINKING

Answer these questions on a separate page. There are no right or wrong answers. Support your answers with examples from your own experience, if possible.

1. Imagine a newspaper that doesn't use columns. Then imagine an 8½ × 11-inch newsletter, also without columns. What problems would these publications present to readers?

2. Find examples in magazines, newspapers, and newsletters of columns of different widths. Which width do you feel is most readable?

Skills Review

EXERCISE 14-8

Create multiple-column layouts.

1. Open the file **Succeed**.
2. Change to Print Layout view.
3. Format the text from "Churches and religious organizations" to the end of the document into two equal-width columns with a vertical rule between columns by following these steps:
   a. Position the insertion point to the left of "Churches."
   b. From the Format menu, choose Columns.
   c. Specify **2** for Number of columns and **0.25"** for the Spacing of column 1.
   d. Choose Equal column width.
   e. Choose This point forward from the Apply to drop-down list and check the Line between box.
   f. Click OK.

4. Locate the paragraph that begins "General" toward the end of the document (after "Businesses").

5. Format this text through the end of the document as one column by following these steps:

   a. Insert a continuous section break before "General."

   b. With the insertion point in the new section, click the Columns button and click one column.

6. Create a title on the first page, HOW TO SUCCEED IN THE TRAVEL AGENCY BUSINESS.

7. Format the title as uppercase, bold, and centered. Position it 2 inches from the top of the page and two line spaces (24 points) from the first paragraph. Remember that the second page top margin should only be 1 inch.

8. Insert page numbering at the bottom center of the second page only. The number "2" should display at the bottom of the second page.

9. Insert a blank line above the paragraph beginning "General."

10. Replace all straight quotes with smart quotes.

11. Save the document as *[your initials]*14-8 in your Lesson 14 folder.

12. Print and close the document.

EXERCISE 14-9

Key and edit text in columns.

1. Open the file **Special3**.

2. Insert column text by following these steps:

   a. Position the insertion point at the end of the paragraph that begins "Our local area."

   b. Press Enter twice; then key the text shown in Figure 14-12.

FIGURE 14-12

Our ski tours have included the Rockies in Colorado, Mt. Hood in Oregon, and the Berkshires in Vermont, to name some of our domestic adventures. However, the list would not be complete if we did not include our international ski trips to France, Italy, and the magnificent Swiss Alps.

3. Edit the text at the top of the next column by following these steps:

   a. Press → to move to the next column.

b. Select "Greece" and key **Portugal**.

c. Select "They" and key **These tours**.

4. Change the document top margin to 2 inches. Apply this change to the whole document.

5. Add a box page border, using the diagonal line style that appears fifth from the bottom in the Style list.

6. Save the document as *[your initials]***14-9** in your Lesson 14 folder.

7. Print and close the document.

EXERCISE 14-10

Format columns. Adjust column width and spacing.

1. Open the file **RockArt**.

2. Create columns of unequal width by following these steps:

   a. Position the insertion point before the text "June 4."

   b. Open the Columns dialog box.

   c. Choose the Left option in the Presets section, choose Line between, and choose This point forward from the Apply to list.

   d. Click OK.

3. Because this format is not appropriate for the document, use the ruler to change column width and spacing by following these steps:

   a. Place the insertion point in column one and point to the column marker (a ScreenTip displays "Move Column").

   b. Hold down [Alt] and drag the marker to the right to change column 1 to 2.5 inches wide.

   c. Point to the left margin area on the ruler for column 2 (a ScreenTip displays "Left Margin").

   d. Drag the margin 0.5 inch to the right, making the second column 2.5 inches wide and increasing the space between columns to 1 inch.

4. Open the Columns dialog box. Change columns 1 and 2 to **2.75** inches wide with **0.5**-inch spacing.

5. Balance the columns by following these steps:

   a. Move the insertion point to the end of the document, after "Depart for home."

   b. Open the Break dialog box, choose Continuous, and click OK.

6. Change the top margin of the whole document to 2 inches.

7. Change the entire document font to Arial.

8. Format the first line of the heading as 14-point, uppercase.

9. Save the document as *[your initials]***14-10** in your Lesson 14 folder.

10. Print and close the document.

EXERCISE 14-11

Control column breaks.

1. Open the file **SantaFe**.

2. Set a 2-inch top margin for the document.

3. Format the document from "The Plaza" forward as two columns of equal width, with no line between columns.

 4. Switch to Print Preview and click the Shrink to Fit button so the document fits on one page.

5. Close Print Preview and insert a column break before the text "Palace of the Governors" by following these steps:

   a. Position the insertion point to the left of the text.

   b. From the Insert menu, choose Break.

   c. Choose Column break and click OK.

6. Press Backspace to remove the column break.

7. Balance the columns by following these steps:

   a. Move the insertion point to the end of the document, after "North America."

   b. Open the Break dialog box, choose Continuous, and click OK.

8. Save the document as *[your initials]*14-11 in your Lesson 14 folder.

9. Print and close the document.

Lesson Applications

Create a three-column layout, key text in columns, and balance the columns.

1. Open the file **Foreign** and switch to Print Layout view.
2. Use the Columns dialog box to create a three-column layout for the list of countries, with 1 inch between the columns and no vertical lines. Keep the introductory paragraph in one-column format.
3. Insert the countries shown in Figure 14-13 at the end of the list.

FIGURE 14-13

| Latvia | Bulgaria | Romania |
| Finland | Estonia | Lithuania |

4. Select the entire list of countries and sort by paragraphs in ascending order.
5. Balance the columns on the page.
6. Key the title **FOREIGN TOURIST OFFICES IN THE UNITED STATES** in all caps followed by two blank lines at the top of the document. Center the heading and change it to 14-point bold.
7. Change the top margin to 2 inches.
8. Spell-check the document and save it as *[your initials]***14-12** in your Lesson 14 folder.
9. Print and close the document.

Create a three-column layout, key text in columns, change column spacing, and balance columns.

1. Open the files **Offices2** and **Duke2**.
2. Copy the second heading on the second page of **Duke2** ("**Gateway Travel—Your Gateway to the World**") and the paragraph that follows it.

3. Paste the selected text at the beginning of **Offices2**, and insert two blank lines after the pasted text.

4. Delete the hard page break before "Ontario Gateway Travel" and insert a continuous section break before "Golden Gateway Travel."

 5. Use the Columns button ▦ to create a three-column layout for the list of travel offices.

6. Change the spacing between columns to 0.2 inch. Keep the columns equal width (with no line between columns).

7. Add the Duke City Gateway Travel office (shown in Figure 14-14) as the first listing in section 2. Use single spacing and include a blank line below the new listing.

FIGURE 14-14

```
Duke City Gateway Travel
15 Montgomery Boulevard
Albuquerque, NM 87111
Tel. (505) 555-1234
Fax (505) 555-1244
```

8. Insert a column break at the start of the listing for "New England Gateway Travel" in the first column using Ctrl+Shift+Enter.

9. Insert a column break at the start of the listing for "Ontario Gateway Travel" using the Break dialog box.

10. Insert two blank lines below the document title and format the title as all caps, 14-point bold italic, centered.

11. Format the heading "Gateway Travel Offices" as small caps, centered.

12. Change the top margin of the whole document to 2 inches.

13. Save the document as *[your initials]*14-13 in your Lesson 14 folder.

14. Print the document. Close both documents.

EXERCISE 14-14

Create a two-column layout, key text in columns, and format and balance the columns.

1. Open the file **Turquoise**.

2. Change the page orientation to landscape.

3. Format the entire document as two columns.

4. At the top of the first column, key the heading **New Mexico/Santa Fe Sightseeing: The Turquoise Trail Scenic and Historic Area**.

5. Insert two blank lines after the new heading.

6. Format the heading as 24-point italic.

7. At the end of the document, key the new paragraph shown in Figure 14-15. Use single spacing and insert a blank line before the paragraph.

FIGURE 14-15

No tour of New Mexico is complete without a visit to one of the 18 timeless pueblos. Circle off Route 14 going toward Interstate 25 and visit the Santo Domingo Pueblo. The pueblo was settled about 700 years ago and today is the home of 3,500 Native Americans. Don't miss Trader Fred's Trading Post, world-famous for its rugs, jewelry, and crafts for over 100 years. But keep a lookout along the way because there are many beautiful sights to see and other places to stop for local goods.

8. Change the space between columns to 1 inch and keep the columns equal width.

9. At the beginning of the text under the heading, insert a continuous section break.

10. Format the heading section as one column.

11. In the heading, delete the colon and start the heading text beginning "*The Turquoise Trail*" on a new line.

12. Change the text in the new line to small caps and center both lines of the heading. Make sure there are two blank lines after the heading.

13. Balance the length of the columns in section 2.

14. Center the entire document vertically on the page.

15. Switch to Print Preview to view the document.

16. Save the document as *[your initials]***14-14** in your Lesson 14 folder.

17. Print and close the document.

EXERCISE 14-15   *Challenge Yourself*

Create a three-column layout, format columns and column text, and balance columns.

1. Open the file **Bike2**.
2. Key the title **Big Turnout for the Bike-a-Thon** as 24-point Arial bold. Center the title and insert a blank line below it.
3. Insert a continuous section break at the first paragraph below the title.
4. Format the second section as three equal columns.
5. Balance the columns.
6. Change the font in section 2 to Arial.
7. Format all the paragraphs in section 2 with a 0.25-inch first-line indent.
8. Change the space between the columns to 0.4 inch and make sure there's a line between columns.
9. Change the document left and right margins to 1.25 inches. Center the whole document vertically on the page.
10. In section 2, delete the blank lines between paragraphs and change the spacing to 6 points after paragraphs.
11. Hyphenate the document manually. Avoid hyphenating words that are six characters or less or any capitalized words.
12. Apply a triple-line 1½-point page border to the document.
13. Save the document as *[your initials]*14-15 in your Lesson 14 folder.
14. Print and close the document.

On Your Own

In these exercises, you work on your own, as you would in a real-life work environment. Use the skills you've learned to accomplish the task—and be creative.

EXERCISE 14-16

Create a three-column newsletter about a neighborhood activity. Left-align the text and format the newsletter attractively. Add a title to the newsletter and balance the columns. Spell- and grammar-check your document. Save the document as *[your initials]*14-16. Print the document.

EXERCISE 14-17

Create a newspaper article about a topic you would enjoy reporting. Use landscape orientation and format the article with four columns and justified alignment. Apply very simple formatting that might appear in a newspaper. Balance the columns. Add a title to the article and spell- and grammar-check the document. Save the document as *[your initials]*14-17 and print it.

EXERCISE 14-18

Log on to the Internet. Find Web sites about a health issue that interests you. Create a magazine article using the information and apply a two-column format. Left-align the text and format the article attractively. Add a title to the article and balance the columns. Spell- and grammar-check the document. Save the document as *[your initials]*14-18 and print it.

Unit 5 Applications

UNIT APPLICATION 5-1

Create a memo that includes a table. Edit and format table structures and text.

1. Start a new document and format it as a standard business memo to Frank Murillo from Anna Svenkova. As the subject, key **Updated Cruise Packages**.

2. As the body of the memo, key **The table below includes the updated cruise information you requested.**

    NOTE: Refer to Appendix B: "Standard Forms for Business Documents" if you need help with the memo.

3. Insert two blank lines and create a 5-row by 3-column table.

4. Key the text shown in Figure U5-1 in the table.

FIGURE U5-1

```
Tropical        South American      8 days/7 nights
Vacation        Mexican             5 days/4 nights
Northern        Canadian            12 days/11 nights
Exotica         Alaskan             11 days/10 nights
Navigator       Virgin Islands      8 days/7 nights
```

5. Copy "Alaskan." Insert a cell above "Canadian" and paste the copied text into the new cell.

6. Delete the cell that contains "Canadian" (use Shift Cells Up).

7. Delete the last row in the table.

8. Go to the fourth cell in the second column and change "Alaskan" to **Hawaiian**.

9. Insert a new row after the second row in the table. Cut and paste only the text (not the row itself) from the first row to the new row.

10. Key the column heads **Cruise Line**, **Cruise Type**, and **Cruise Length** into the cells in the first row. Format the text as bold.

11. Apply a 1½-point single-line outside border to the table by using the Tables and Borders toolbar.

12. Apply 20% gray shading to the first row.

13. AutoFit the column widths and center the table horizontally on the page.

14. Change the row height in the table to **0.33** inch.

15. Center-align the column headings in row 1 vertically and horizontally.

16. Align rows 2 through 6 as Align Center Left.

17. Sort the table alphabetically by cruise type.

18. Add your reference initials.

19. Spell-check the document.

20. Save the document as *[your initials]***u5-1** in a new Unit 5 Applications folder.

21. Print and close the document.

UNIT APPLICATION 5-2

Create column layouts, change column width and spacing, add a line between columns, and balance columns.

1. Open the file **Memo2**.

2. Delete the following text:

   - The memo heading
   - The first paragraph
   - The paragraph that begins "Write these deadlines"
   - The entire section called "Record Keeping"
   - The text at the end of the document beginning with "For more detailed information"

3. Format the document as two columns.

4. Delete all blank paragraph marks between paragraphs. Change the spacing for text to 6 points after paragraphs, but do not change the spacing for the list at the end of the document.

5. Edit the text, as shown in Figure U5-2.

FIGURE U5-2

July 1 — Give a preliminary count to ~~suppliers (including~~ airlines, hotels, and tour operators. Let participants know you will be billing them and that final payment is due by July 15.

July 15 — This is the due date for all land and air travel. We would like payment received at the agency by this time.

continues

FIGURE U5-2 *continued*

July 17 ¹/ₘ Start calling or e-mailing participants who have not yet
~~remitted~~. ^payments Valerie can help with this, ^task as can Darrell. The idea is not
to pressure people, but ^to let them know we ⟨simply⟩ want to make sure
they are included in the group.

August 1 ¹/ₘ Issue the air tickets and mail final payment to all
suppliers. Darrell is most familiar with this procedure. In addition,
send suppliers the participation ~~and rooming~~ lists.

August 15 ¹/ₘ Mail all travel documents to the participants. Some of our
local customers may prefer to stop by and pick up their documents.
Give them ^this ~~the~~ option.

6. Change the column spacing to 0.4 inch. Keep the column width equal and add a vertical line between columns.

7. Key the title **PROCEDURES FOR GROUP TRAVEL** in 18-point bold italic, no underline, above "Deadlines."

8. Format the title as one column, center the text, and add 72 points of spacing before and 24 points of spacing after.

9. Start the text "Finalizing the Arrangements" as a new section on the same page, with a one-column format. Add a blank line before the text.

10. Sort the list at the end of the document alphabetically.

11. Format the list as a bulleted list, using the check-mark bullet. The bulleted text should have no left indent, no left tab, and a 0.25-inch hanging indent.

12. Shrink the document to fit on one page if it does not already fit.

13. Add a footer that contains the date at the left margin, the filename in the center, and the page number (preceded by the word "Page") at the right margin. Adjust the footer's tab settings by moving the right tab marker to the right margin and moving the center tab marker to 3.25 inches.

14. Format the footer text as 10-point italic.

15. Save the document as *[your initials]***u5-2** in your Unit 5 Applications folder.

16. Print and close the document.

UNIT APPLICATION 5-3

Create column layouts, format column text, balance columns, convert text to a table, use Table AutoFormat, change column width, and compose a document.

1. Open the files **Mexico2** and **Flights2**.
2. Copy the text from **Flights2** and paste it at the end of **Mexico2**. Add two blank lines between the documents, if they don't already appear.
3. Change the font of the pasted text to 12 points.
4. Format the text from "Day 1" through "nightlife" as two equal columns without a vertical line between columns.
5. Add to the end of the itinerary: **Day 7 Sail back to Los Angeles.**
6. Format the paragraph titles "Day 1" through "Day 7" as Arial bold.
7. Insert a column break before "Day 3."
8. Change the spacing for the itinerary text ("Day 1" through "Day 7") to 6 points after paragraphs.
9. Change the spacing between the columns to 0.4 inch.
10. Format the title as 14-point Arial bold small caps. Center the title and insert a line break before "FROM."
11. Convert the text from "<u>Depart</u>" through "Flight 968" into a 7-row by 3-column table.
12. Use Table AutoFormat and apply the Table Professional format. Deselect the first column, last row, and last column options.
13. Change the column width of the first two columns to 1.5 inches.
14. Change the row height of the entire table to 0.35 inch and center the text vertically within the rows, but not horizontally.
15. Remove the underlines from the headings in the first row.
16. Center the table horizontally on the page.
17. Above the table, change the heading that begins "Flights from Tucson" to center alignment, 14-point Arial bold small caps. Insert two more blank lines above it and one blank line below it.
18. As a separate section on a new page at the beginning of the document, create a business letter from Alexis Johnson (Travel Counselor) to Mr. Andreas Smith at 45 Main Street in Tucson, AZ 87743. Use 12-point Times New Roman. In the letter, tell Mr. Smith that the Riviera Cruise itinerary is set. Describe his flight options to Los Angeles and let him know he has 30 days to decide which flight he wants to take. Include an enclosure notation and your reference initials. Change the top margin for the whole document to two inches.

NOTE: Refer to Appendix B: "Standard Forms for Business Documents" if you need help setting up the letter.

19. Save the document as *[your initials]*u5-3 in your Unit 5 Applications folder.

20. Print the document. Close all documents.

UNIT APPLICATION 5-4 *Using the Internet*

Create a newsletter by using tables and columns.

Using the Internet, create a simple newsletter that is one to two pages in length and contains the following:

- A newsletter title. Use appropriate spacing and character formatting.
- Four articles, which can be about school, your town, your family, a hobby, or any topic that interests you. Research the content of the articles on-line. You can write the articles from information you obtain from various Web sites or copy and paste the content from the Web sites. (Paste the text without its formatting.) Make sure the information contained in the newsletter is based directly on information you find.
- Each article should have a title.
- All paragraphs within an article should be separated by the appropriate paragraph spacing.
- One of the articles should contain a bulleted list, using any bullet style.
- Use column breaks, continuous section breaks, and page breaks to arrange the columns in a visually suitable layout.
- Include one or two formatted tables with rotated text and shading.
- Determine if you need to horizontally center your table(s) within the boundaries of a column or the margins of the newsletter.
- Each page should have a header and footer.
- Include any other formatting to make the newsletter attractive.

On a new page in the document, create a bulleted list containing the list of references or Web sites you used. Save the document as *[your initials]*u5-4 in your Unit 5 Applications folder.

Advanced Topics

Styles

After completing this lesson, you will be able to:

1. **Apply styles.**
2. **Create new styles.**
3. **Redefine, modify, and rename styles.**
4. **Use style options.**
5. **Use AutoFormat and the Style Gallery.**

 Estimated Time: 1¼ hours

A *style* is a set of formatting instructions you can apply to text. Styles make it easier to apply formatting and ensure consistency throughout a document. In every document, Word maintains a *style sheet*—a list of style names and their formatting specifications. A style sheet, which is stored with a document, includes standard styles for body text and headings. You can use these styles, modify them, or create your own styles.

Applying Styles

The default style for text is called *Normal* style. Unless you change your system's default style, Normal is a paragraph style with the formatting specifications 12-point Times New Roman, English language, left-aligned, single-spaced, and widow/orphan control.

To change the appearance of text in a document, you can apply four types of styles:

- A *character style* is formatting applied to selected text, such as font, font size, and font style.
- A *paragraph style* is formatting applied to an entire paragraph, such as alignment, line and paragraph spacing, indents, tab settings, borders and shading, and character formatting.
- A *table style* is formatting applied to a table, such as borders, shading, alignment, and fonts.
- A *list style* is formatting applied to a list, such as numbering or bullet characters, alignment, and fonts.

EXERCISE **15-1** **Apply Styles**

There are three ways to apply styles:

- Open the Styles and Formatting task pane and select a style to apply. To open the task pane, choose Styles and Formatting from the Format menu.
- Click the Styles and Formatting button on the Formatting toolbar.
- Click in the Style box on the Formatting toolbar to select the style to apply.

1. Open the file **Budget**.

   NOTE: The documents you create in this course relate to the Case Study about Duke City Gateway Travel, a fictional travel agency (see pages 1 through 4).

2. Click the Styles and Formatting button on the Formatting toolbar to open the Styles and Formatting task pane. The task pane lists formatting currently used in the document (such as for bold and bulleted text) and includes some of Word's built-in heading styles. (See Figure 15-1 on the next page.)

3. In the first line of text, edit the date to the current date. Replace the tab character with two space characters.

4. With the insertion point still in the first line of text, place the mouse pointer (without clicking it) over the Heading 1 style in the task pane. A ScreenTip displays the style's attributes.

5. Click the Heading 1 style to apply it to the paragraph. The style includes 16-point Arial bold formatting.

NOTE: To apply a style to a paragraph, you can simply click anywhere in the paragraph without selecting the text. Remember, this is also true for applying a paragraph format (such as line spacing or alignment) to a paragraph.

FIGURE 15-1
Using the Styles and
Formatting task pane
to apply a style

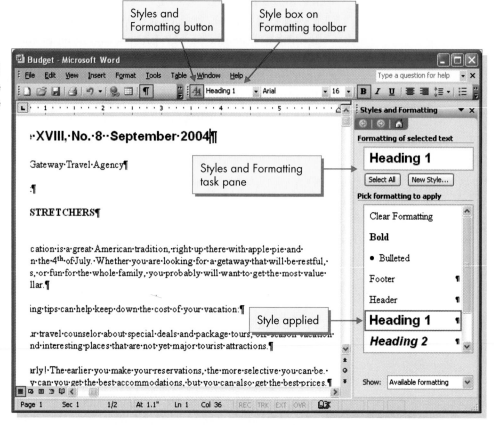

6. Close the Styles and Formatting task pane.
7. Position the insertion point in the text "HOME ON THE RANGE" at the bottom of the first page.

8. On the Formatting toolbar, click the down arrow to the right of the Style box to open the Style drop-down list. (See Figure 15-2 on the next page.)
9. Choose Heading 2. Notice the applied formatting.
10. Position the insertion point in the heading "CIBOLA NATIONAL FOREST" on page 2. Press F4 to repeat the Heading 2 style.

TIP: To remove a style from text and restore the Normal style, choose Clear Formatting from the Styles and Formatting task pane or the Style drop-down list.

FIGURE 15-2
Using the Style
drop-down list

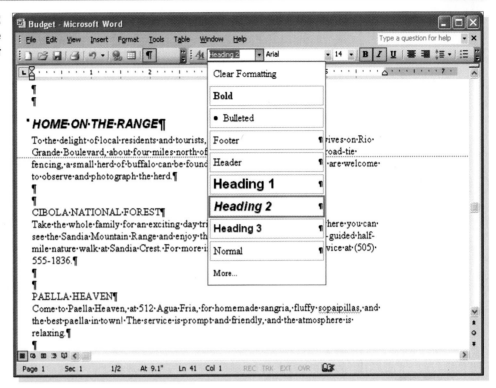

Creating New Styles

Creating styles is as easy as formatting text, and then giving the set of formatting instructions a style name. Each new style name must be different from the other style names already in the document.

Word saves the styles you create for a document when you save the document.

EXERCISE **15-2** **Create a Paragraph Style**

There are two ways to create a new paragraph style:

- Use the Style box on the Formatting toolbar.
- Click the New Style button in the Styles and Formatting task pane.

1. Reopen the Styles and Formatting task pane by clicking the Styles and Formatting button ⚌ or by choosing Format, Styles and Formatting from the menu.

2. Select the heading "BUDGET STRETCHERS" near the top of page 1.

TIP: Click to the left of the text to select the entire paragraph. (Be sure your mouse is in the margin area and the mouse pointer changes to a white arrow.)

3. Increase the font size to 18 points. The formatting of the selected text appears in the task pane.

4. With the heading still selected, click in the Style box on the Formatting toolbar. The current style name is selected.

5. Key **Headline** and press (Enter). You created and named the paragraph style "Headline." The new style appears in the task pane and in the Styles box on the Formatting toolbar.

FIGURE 15-3
Creating a new style

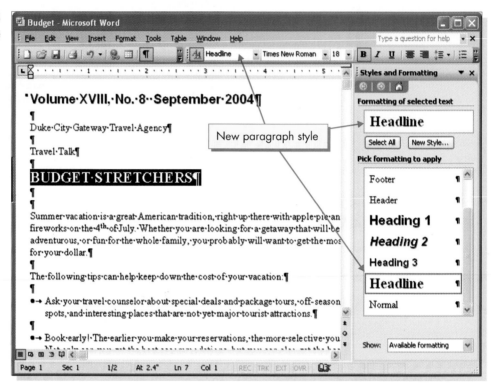

NOTE: If you key a style name that already exists in the Style box, you apply the existing style; you do not create a new one.

6. Click within the paragraph below "BUDGET STRETCHERS," which begins "Summer vacation is."

7. Click the New Style button in the task pane. The New Style dialog box appears. (See Figure 15-4 on the next page.)

8. Key **Body** in the Name text box. The Style type is Paragraph, because you're creating a paragraph style.

9. Under Formatting in the New Style dialog box, notice two rows of buttons: the first row for font formatting and the second row for paragraph formatting. Point to the buttons on the second row that are unfamiliar to you—a ScreenTip will identify each button.

TIP: Use the Formatting buttons in the New Style dialog box to apply basic font and paragraph formatting. For more formatting options, click the Format button and choose Font, Paragraph, Tabs, Border, or Numbering to open the corresponding dialog boxes.

10. Open the Font Size drop-down list in the New Style dialog box and change the font to 16 points. Make sure the font is Times New Roman.

FIGURE 15-4
New Style
dialog box

Font and paragraph formatting options

Click for more formatting options

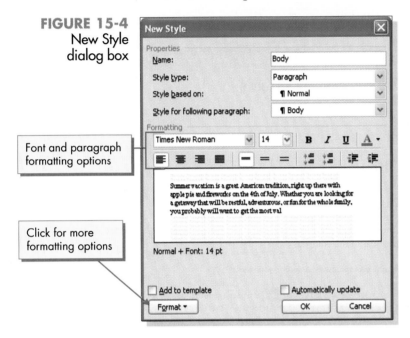

11. Click OK. Notice that the Body style is listed in the task pane, but it has not been applied to the text.

12. With the insertion point still in the paragraph that begins "Summer vacation is," click the Body style in the Styles and Formatting task pane. The style is applied to the paragraph.

EXERCISE 15-3 Create a Character Style

You use the New Style dialog box to create a new character style.

1. Open the New Style dialog box and key **Prominent** in the Name box.

2. Choose Character from the Style type drop-down list box.

3. Click the Format button, choose Font, and set the formatting to 11-point Arial Black, italic.

4. Click OK to close the Font dialog box. Click OK to close the New Style dialog box. The Prominent style appears in the task pane.

NOTE: In a list of styles, paragraph styles display a paragraph symbol (¶) and character styles display a text symbol (a) to the right of the style name.

5. Select the text "special deals" in the first bulleted paragraph and apply the Prominent style from the task pane. Note that a character style is applied to selected text, not the entire paragraph.

FIGURE 15-5
Applying the
character style

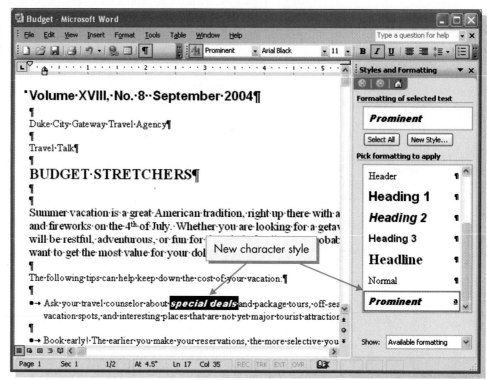

6. Save the document as *[your initials]*15-3 in a new folder for Lesson 15. Do not print the document; leave it open for the next exercise.

TIP: At the bottom of the Styles and Formatting task pane, you can open the Show drop-down list to choose which types of styles are displayed. The default setting, "Available formatting," lists styles and unnamed formats available to the current document. "Formatting in use" lists styles and unnamed formats applied in the current document. "All styles" lists styles in the current document and all of Word's built-in styles.

Modifying and Renaming Styles

After creating a style, you can modify it by changing the formatting specifications or renaming the style. When you modify a style, the changes you make affect each instance of that style. You can quickly replace one style with another by using the Replace dialog box.

NOTE: You can modify any of Word's built-in styles as well as your own styles. However, you cannot rename Word's standard heading styles.

EXERCISE   **15-4**   **Modify and Rename Styles**

To modify a style, right-click the style in the Styles and Formatting task pane, and then choose Modify. Or, select the styled text, modify the formatting, right-click the style name in the Styles and Formatting task pane, and choose Update to Match Selection.

1. In the Styles and Formatting task pane, right-click the style name Heading 2. Choose Modify from the drop-down list.

 TIP: Instead of using the right mouse button to open this drop-down list of style options, you can use the left mouse button to click the down arrow next to a style name. Remember, if you click a style name (not its down arrow) with the left mouse button, you'll apply the style to the text containing the insertion point.

2. In the Modify Style dialog box, change the point size to 12 and click OK to update the style.

3. Right-click the style name Heading 2 in the task pane. Choose Select All 2 Instance(s). Notice the changes in "HOME ON THE RANGE" and "CIBOLA NATIONAL FOREST."

 TIP: You can also click the Select All button in the Styles and Formatting task pane to select all instances of a style.

4. Deselect the text. Place the insertion point in the first line of text at the top of the document (the text with Heading 1 style).

5. Center the text. Note that the task pane now lists the original Heading 1 style as well as Heading 1+Centered.

6. Right-click the original Heading 1 style in the task pane. Choose Update to Match Selection. The style now includes center alignment (and the style Heading 1+Centered disappears). (See Figure 15-6 on the next page.)

7. Position the insertion point in the text "BUDGET STRETCHERS." Right-click the Headline style in the task pane and choose Modify.

8. Rename the Headline style by keying **First Head** in the Name text box. Click OK. The style First Head appears in the Style box and in the task pane, replacing the style name Headline.

 NOTE: After modifying or renaming a style, you can undo your action (for example, choose Undo Style from the Edit menu).

FIGURE 15-6
Modifying a style
by updating it

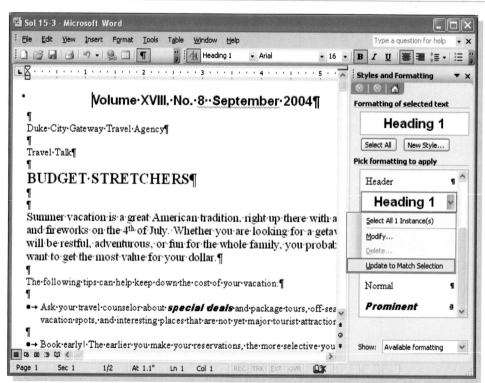

EXERCISE 15-5 Replace a Style

1. Choose Replace from the Edit menu. Click the More button, if needed, to expand the dialog box. Clear any text or formatting from a previous search.

2. Click the Format button and choose Style.

3. In the Find Style dialog box, choose Heading 2 from the Find what style list and click OK.

FIGURE 15-7
Find Style dialog box

4. Click in the Replace with text box, click Format, and choose Style.

5. Choose Heading 3 from the Replace With Style list and click OK.

6. Click Replace All. After Word replaces all occurrences of the Heading 2 style with the Heading 3 style, click OK in the notification dialog box, and then click Close. Notice the changes in the headings "HOME ON THE RANGE" and "CIBOLA NATIONAL FOREST."

TIP: You can also use the Styles and Formatting task pane to replace one style with another style: right-click a style name in the task pane, choose Select All Instances, and then click another style name in the task pane.

E X E R C I S E   15-6 Delete a Style

---

1. Right-click the style named Body in the Styles and Formatting task pane. Click Delete.

2. Click Yes when prompted to verify the deletion. The Body style is deleted and the paragraph below "BUDGET STRETCHERS" returns to Normal, the default style.

NOTE: When you delete a style from the style sheet, any paragraph that contained the formatting for the style returns to the Normal style. You cannot delete the standard styles (Word's built-in styles) from the style sheet.

Using Style Options

Word offers two options in the Style dialog box to make formatting with styles easier:

- Style based on
  This option helps you format a document consistently by creating different styles in a document based on the same underlying style. For example, in a long document, you can create several different heading styles that are based on one heading style and several different body text styles that are based on one body text style. Then, if you decide to change the formatting, you can do so quickly and easily by changing just the base styles.
- Style for following paragraph
  This option helps you automate the formatting of your document by applying a style to a paragraph and specifying the style that should follow immediately after the paragraph. For example, you can create a style for a heading and specify a body text style for the next paragraph.

NOTE: The standard styles available with each new Word document are all based on the Normal style.

E X E R C I S E   15-7 Use the Style for Following Paragraph Option

---

1. Click in the paragraph that begins "Summer vacation is a great American tradition," and then click the New Style button in the Styles and Formatting task pane.

2. Key **Bodystyle** in the Name text box.

3. Click Format and choose Font. Set the font to 12-point Arial Regular and click OK.

4. In the New Style dialog box, check that the Align Left button ▤ is selected. Click OK. You're going to assign this new style (Bodystyle) as the style to follow the First Head style.

5. Right-click the First Head style in the task pane and choose Modify.

6. Choose Bodystyle from the Style for following paragraph drop-down list. Click OK.

FIGURE 15-8
Choosing a style for the following paragraph

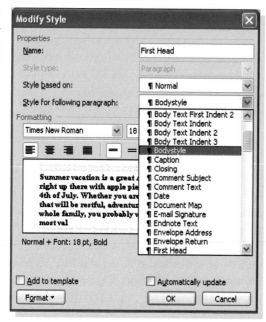

7. Move the insertion point to the end of "ARTS & CRAFTS FAIR" and click First Head in the task pane to apply the style.

8. Press Enter. Note that Bodystyle is indicated as the style for the next paragraph.

9. Key the text shown in Figure 15-9. The text appears in Bodystyle (12-point Arial).

NOTE: When you key a telephone number, it's a good idea to use a nonbreaking space (Ctrl+Shift+Spacebar) after the area code and a nonbreaking hyphen (Ctrl+Shift+Hyphen) after the first three digits to prevent a line break within the number.

FIGURE 15-9

```
June 20-22
10 a.m. to 6 p.m.
State Fairgrounds

The annual Arts & Crafts Fair is a much-anticipated event in
Albuquerque. More than 200 artists and craftspeople will gather to
sell their wares and enjoy the cloudless weather. Last year over 5,000
attendees appreciated the high-quality goods, the fine food, and the
local musicians. This year, sound stages will be set up at both ends
of the exhibit area for more elaborate music and entertainment.

Admission is free. For more information, contact the Albuquerque
Convention and Visitors Bureau at (505) 555-3696.
```

EXERCISE   15-8   Use the Based On Option

1. Select the three lines of text near the bottom of page 2, from "Travel Talk" through "Franchise." Change the font size to 11 points. Click in the Style box on the toolbar. Key the name **Basebody** and press Enter.

2. Select the text from "Editorial/Advertising Office" through "87111" and apply the Basebody style.

3. With the Basebody text still selected, click the New Style button in the task pane.

4. Key **Body2** as the name of the new style. Click the Italic button *I* in the New Style dialog box to change the font to italic. Check that Basebody appears in the Style based on box. Click OK.

   TIP: When you want to use an existing style as the based-on style, select the text with that style before opening the New Style dialog box. The style will automatically appear in the Style based on box.

5. In the task pane, place the mouse pointer over (without clicking) the Body2 style. The ScreenTip indicates that the Body2 style is based on the Basebody style.

6. Click the New Style button in the task pane to create another style.

7. Key **Body3** as the name of the new style. Choose Basebody from the Style based on list, if it is not already selected. Change the font size to 10 points. Click OK.

8. Select the text from "President" through the end of the document. Apply the Body3 style.

9. Right-click the style Basebody in the task pane and choose <u>M</u>odify.

10. Change the font to Arial Narrow. Click OK. All the text using or based on the Basebody style changes to Arial Narrow.

EXERCISE 15-9 **Display and Print Styles**

To make working with styles easier, you can display a document's styles on the screen and print the style sheet.

1. Choose <u>O</u>ptions from the <u>T</u>ools menu and click the View tab.

2. Set the Style ar<u>e</u>a width box to 0.5" and click OK. The style area appears in the left margin.

NOTE: The style area is intended for onscreen purposes in Normal view only (it is not available in Print Layout view). If you switch to Print Preview, this area does not display or appear on the printed document. Additionally, when you display the style area, it will be displayed for any document you open unless you reduce the view to 0 inches. The default Style area width is 0".

FIGURE 15-10
Styles shown in style area

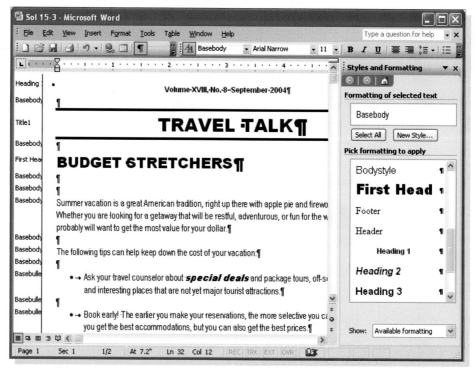

3. Choose Replace from the Edit menu to replace all occurrences of the Normal style with Basebody style. Next, replace all occurrences of Bodystyle with Basebody style.

4. Because the four paragraphs starting with "Ask your travel counselor" are no longer bulleted, format them as bulleted, creating a new style called Basebullet that is based on Basebody.

5. Change the First Head style so it is based on Basebody. Change the font of the First Head style to Arial Black and remove the bold attribute.

6. Delete the text "Duke City Gateway Travel Agency" at the top of the document, along with the blank line below it.

7. Format the text "Travel Talk" at the top of the document as 22-point Arial Black, all caps, centered, with a 3-point top and bottom border. Use the Style box on the toolbar to name this style Title1.

8. Change the font size of the Heading 1 style to 9 points.

9. Wherever two consecutive blank lines occur, delete one line (except before "Travel Talk" on page 2).

10. Apply the Body 2 style to the text "36th Annual" and apply the Heading 3 style to "Paella Heaven."

11. Choose Options from the Tools menu and click the View tab. Set the Style area width box to 0" and click OK.

12. Save the document as *[your initials]*15-9 in your Lesson 15 folder.

13. Open the Print dialog box. Choose Styles from the Print what drop-down list and click OK. Word prints the styles for your active document.

14. Print and close the document.

> **NOTE:** You can assign a shortcut key to a style you use frequently. In the Styles and Formatting task pane, right-click the style to which you want to assign a shortcut key. Choose Modify. Click Format, and then click Shortcut key. In the Customize Keyboard dialog box, press an unassigned keyboard combination, such as Alt+B for a body text style. The shortcut key is saved with the document.

Using AutoFormat and the Style Gallery

You can use AutoFormat to format an entire document quickly. When you use this feature, Word evaluates each paragraph and applies one of the standard styles, using heading styles and body text styles where appropriate.

EXERCISE 15-10 Use AutoFormat

1. Reopen the file **Budget**. Make sure the Styles and Formatting task pane is open.

2. Choose AutoFormat from the Format menu. (Expand the menu, if necessary.)

FIGURE 15-11
AutoFormat
dialog box

3. Check that General document is set as the document type. Click OK to verify that you want to proceed with AutoFormat now.

4. Scroll through the document and notice the changes made. Note that Word assigns styles based on the document's structure; for example, short lines are treated as headings.

5. View the new styles in the Styles and Formatting task pane.

6. Using the new styles, change the text "Duke City Gateway Travel Agency" at the beginning of the document to the Body Text style and change "BUDGET STRETCHERS" to the Heading 1 style.

EXERCISE 15-11 Use the Style Gallery

All style sheets are contained within a template. Word provides several templates that contain specific styles, such as templates for letters, reports, and publications. You can use the Style Gallery to apply styles to your document that are assigned to specific templates.

 NOTE: You learn about templates in the next lesson.

1. Choose Theme from the Format menu and click the Style Gallery button. The Style Gallery dialog box appears, displaying a list of templates and a view of the current document. (See Figure 15-12 on the next page.)

2. Choose Elegant Report from the Template list box. Under Preview, choose Example. You should see an example of the Elegant Report template. Under Preview, choose Document. This option displays your document as it would appear with the Elegant Report styles applied.

 NOTE: Some templates might not be installed on your computer. Check with your instructor if you see the message "Template not installed yet" above the Preview window.

3. Click Style samples under Preview to display a list of the template's styles. Scroll through the Preview of box to view the various style names and formatting.

4. Choose Contemporary Report from the Template list box. Click Example to see an example of the template styles; then click OK to apply the template. Notice the new styles in the Styles and Formatting task pane.

FIGURE 15-12
Style Gallery
dialog box

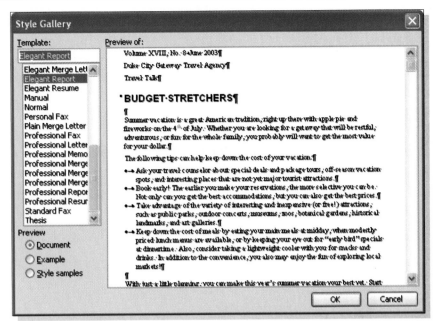

5. Locate the Normal style in the task pane and modify the paragraph formatting by setting the left indent to 0 inches and the font size to 12 points. Notice the change to the text with Body Text style, because this style is based on Normal style.

6. In the first line of text, update the date to the current month and year. Apply the Header style to the first two lines; then right-align the date at the right margin.

7. Apply Title style to the text "Travel Talk."

8. Delete the blank line below "BUDGET STRETCHERS."

9. In the first bulleted paragraph, apply the character style Emphasis to the text "special deals." Delete the blank paragraph mark below the last bulleted paragraph.

10. At the bottom of page 1, delete the line of text that begins "36th Annual" along with the line of text and the two blank lines below it.

11. Change the Heading 3 headings to Heading 2 style. Delete the blank paragraph mark above each of these headings.

12. Below "PAELLA HEAVEN" on page 2, format the text from "Travel Talk" through the end of the document as Block Quotation style. Modify the style by changing the left indent to 0.15 inches.

13. Adjust the soft page break at the bottom of page 1 if it interrupts a paragraph.

14. Save the document as *[your initials]***15-11** in your Lesson 15 folder.

15. Print and close the document.

USING ONLINE HELP

Styles offer many options for document formatting. Word provides extensive Help about styles options and troubleshooting.

Use Help to troubleshoot styles:

1. In the Ask a Question box, key **troubleshoot styles** and press Enter.
2. Click the topic Troubleshoot using styles and applying formatting.
3. Choose a troubleshooting topic in the Help window by clicking one of the links. Continue exploring the topics.
4. Close the Help window when you finish.

LESSON Summary

➤ A style is a set of formatting instructions you can apply to text to give your document a unified look. The four types of styles are character (applied to selected text), paragraph, table, and list.

➤ Word's default style for all text is called Normal style, which is 12-point Times New Roman, left-aligned, single-spaced, with widow-orphan control. Word provides built-in heading styles (for example, Heading 1, Heading 2).

➤ To apply a style, select the text you want to style (or click in a paragraph to apply a paragraph style). Then choose the style from the Style box on the Formatting toolbar or the Styles and Formatting task pane.

➤ View the attributes of a style by placing the mouse pointer over the style name in the Styles and Formatting task pane, and reading the text in the ScreenTip.

➤ Select all instances of a style by clicking the Select All button in the Styles and Formatting task pane or by right-clicking a style name in the task pane and choosing Select All Instance(s).

➤ To create a new paragraph style: select text, modify the text, click in the Style box, key a new style name, and press Enter. Or, click the New Style button in the Styles and Formatting task pane and set the style's attributes in the New Style dialog box. You must use the New Style dialog box to create a character style and specify Character as the style type.

➤ To modify or rename a style, right-click the style name in the Styles and Formatting task pane (or point to the style name and click the down arrow), choose Modify, and then change the attributes. Or, select text that uses the style, change the format, right-click the style name in the task pane, and choose Update to Match Selection.

➤ After applying a style throughout a document, you can replace it with another style. Choose Replace from the Edit menu (in the dialog box, click Format, choose Style, and select the style name in both the Find what and Replace with boxes).

➤ You can also replace styles by using the Styles and Formatting task pane (select all instances of a style and then choose another style).

➤ To delete a style, right-click the style name in the Styles and Formatting task pane (or point to the style name and click the down arrow), choose Delete, and click Yes.

➤ When creating new styles, you can specify that they be based on an existing style. You can also specify that one style follows another style automatically. Both of these options are offered in the New Style dialog box.

➤ Display styles down the left margin of a document in Normal view by choosing Options from the Tools menu and setting the Style area width box to 0.5". Do the reverse to hide the styles.

➤ Print a style sheet by choosing Print from the File menu and choosing Styles from the Print what drop-down list.

➤ The AutoFormat feature applies Word's built-in styles to each paragraph, automatically. The Style Gallery makes styles available in your document from a specific template.

LESSON 15   Command Summary

FEATURE	BUTTON	MENU	KEYBOARD
Styles and Formatting task pane		Format, Styles and Formatting	
Apply styles	Normal ▾	Format, Styles and Formatting	
View style area		Tools, Options	
AutoFormat		Format, AutoFormat	
Style Gallery		Format, Theme, Style Gallery	

Concepts Review

TRUE/FALSE QUESTIONS

Each of the following statements is either true or false. Indicate your choice by circling T or F.

T F 1. Paragraph styles can include both paragraph- and character-formatting instructions.

T F 2. You can apply character styles to selected text within a paragraph.

T F 3. You can use either the Styles and Formatting task pane or the Style box on the Formatting toolbar to apply a paragraph style.

T F 4. If you select only part of a paragraph to change the paragraph style, only the selected portion is reformatted.

T F 5. You can create a character style by changing the formatting of selected text, and then entering a new style name in the Style box on the Formatting toolbar.

T F 6. A style named and created for a specific document cannot be modified for that document.

T F 7. When you delete a style from the style sheet, any paragraph or text containing that style returns to the Normal style.

T F 8. You save the styles created for a document by saving the document.

SHORT ANSWER QUESTIONS

Write the correct answer in the space provided.

1. In the list of styles, what symbol designates a paragraph style?

2. In the list of styles, what symbol designates a character style?

3. How do you print a list of a document's styles?

4. What does the button 🔠 do?

5. How do you display style names in the left margin of a document?

_____

6. How do you open the New Style dialog box?

_____

7. What are the four types of styles?

_____

8. On what style are all standard styles based?

_____

CRITICAL THINKING

Answer these questions on a separate page. There are no right or wrong answers. Support your answers with examples from your own experience, if possible.

1. The <u>B</u>ased On option is often used to create a group of heading styles based on one style and a group of body text styles based on another style. Why aren't heading styles and body text styles typically based on the same style?

2. Create complementary styles for a heading and for body text, using two different fonts. Describe the formatting for each style and provide a sample of the styles used together. Describe the type of document for which this combination would be suited.

Skills Review

EXERCISE 15-12

Apply styles and create new styles.

1. Open the file **Itin2**.
2. Use the Style box on the Formatting toolbar to apply a style to the document title by following these steps:
   a. Position the insertion point in the title (the first line).

   b. Click the down arrow to open the Style box on the Formatting toolbar and choose Heading 1.
3. Use the Styles and Formatting task pane to apply a style by following these steps:

   a. Click the Styles and Formatting button on the Formatting toolbar to display the Styles and Formatting task pane.

   b. Position the insertion point in the next heading, "Travel in the Southwest."

   c. Click Heading 2 in the task pane. Delete the blank paragraph mark below the heading.

4. Apply the Heading 2 style to the heading that begins "Travel to Other Parts" at the bottom of page 1. Delete the blank paragraph mark below the heading.

5. Create a new paragraph style for a heading by following these steps:

   a. Select the heading "Sample Itinerary—The Great Southwest."

   b. Format the text with 6 points of space after paragraphs and change the character formatting to underlined, no bold.

   c. Click in the Style box on the Formatting toolbar to select the style name and key **Itin Head**.

   d. Press Enter.

   e. Apply the Itin Head style to the heading "Sample Itinerary—The California Coast" on page 2.

6. Use the New Style dialog box to create a new style for the itinerary text by following these steps:

   a. Click New Style in the task pane.

   b. If the Underline button U is selected, click it to deselect this formatting.

   c. Key **Itin** in the Name text box.

   d. Click Format and choose Paragraph.

   e. Set the space after to 6 points, set the left indent to 0.45 inch, set the right indent to 0.25 inch, and set a 0.55-inch hanging indent. Click OK.

   f. Click OK in the New Style dialog box.

7. Select all the itinerary text under the heading "Sample Itinerary—The California Coast." Choose the Itin style from the Styles and Formatting task pane.

8. Apply the Itin style to the itinerary text under "Sample Itinerary—The Great Southwest."

9. In place of the space following each day heading ("Day 1", "Day 2", and so on) in both itineraries, insert a tab character.

10. Change the document title to all caps with 72 points of spacing before paragraphs.

11. Save the document as *[your initials]***15-12** in your Lesson 15 folder.

12. Print and close the document.

EXERCISE 15-13

Create, redefine, modify, and rename styles.

1. Open the file **Offices2**. Display the Styles and Formatting task pane.

2. For the document title, create and apply a paragraph style named Office Heading with the formatting 14-point Times New Roman, bold italic.

3. For the text "Golden Gateway Travel," create and apply a paragraph style named Subhead with the formatting 14-point Arial, bold italic, with 12 points of spacing before and 3 points of spacing after paragraphs.

4. Position the insertion point in each line of text that contains a Gateway Travel office name, and apply the Subhead style (press F4 to repeat the style).

5. Redefine the Subhead style by following these steps:

   a. Select the line of text beginning with "Golden Gateway" and remove the italics by clicking I.
   b. Right-click the italic Subhead style in the task pane. Choose Update to Match Selection.

6. Create a new character style for the word "Fax" by following these steps:
   a. Click the New Style button in the task pane.
   b. Key Fax Num in the Name text box.
   c. Choose Character from the Style type drop-down list.

   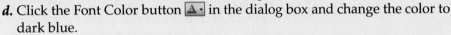

   d. Click the Font Color button A▾ in the dialog box and change the color to dark blue.
   e. Click Format, choose Font, and choose Small caps. Click OK. Click OK again.

7. Select the word "Fax" the first time it occurs and apply the Fax Num style from the task pane. Apply the same style to each occurrence of "Fax" by selecting the word and pressing F4.

8. Rename the Fax Num style by following these steps:
   a. Right-click the Fax Num style in the task pane and choose Modify.
   b. In the Name text box, edit the text to just "Fax" and click OK.

9. Delete the Office Heading style by following these steps:
   a. Right-click the Office Heading style in the task pane.
   b. Choose Delete. Click Yes.

10. At the bottom of the task pane, open the Show drop-down list and choose All styles to display all of Word's styles. Scroll to the bottom of the list and apply the Title style to the document title. Modify the style to include all caps.

11. Change the top margin of the entire document to 2 inches and delete the hard page break before the Ontario office. Adjust the soft page break at the bottom of page 1 if it interrupts an office location.

12. Add a header to number page 2 (include the word "Page"). Center the header and position it 1 inch from the edge of the page.

13. Format all paragraphs below the title with a 2-inch left indent.

14. Save the document as [your initials]15-13 in your Lesson 15 folder.

15. Print and close the document.

Create styles, use style options, and display and print styles.

1. Open the file **NYmemo**. Display the Styles and Formatting task pane. Change the date in the memo heading to today's date.

2. To the text "Peabody Hotel," create and apply a new paragraph style named Hotel Head with the formatting 12-point Arial, bold italic.

3. Apply the new style to "Renaissance Hotel" and "Hotel Lexington."

4. Create a new paragraph style named Subhead that's based on the Hotel Head style by following these steps:

   a. With the insertion point in "Hotel Lexington," click the New Style button on the task pane.
   b. Key **Subhead** in the Name text box.
   c. Check that Style based on is set to the Hotel Head style.
   d. Using the Formatting buttons in the New Style dialog box, change the font formatting to 11 points and turn off bold and italic.
   e. Click OK.

5. Create a new paragraph style named Info that is based on the Normal style. Use the formatting 11-point Times New Roman, italic, with 3 points of spacing after paragraphs.

6. Assign Subhead as the following paragraph style for Hotel Head and assign Info as the following paragraph style for Subhead by following these steps:

   a. Right-click the Hotel Head style in the task pane.
   b. Choose Modify.
   c. Choose Subhead from the Style for following paragraph drop-down list.
   d. Click OK. Repeat the procedure to assign Info as the following paragraph style for Subhead.

7. Position the insertion point at the end of the heading "Peabody Hotel," press Enter to apply the Subhead style, and key **A New York Tradition**.

8. Press Enter to apply the Info style and key **For 50 years, the Peabody Hotel has served visitors to the Big Apple.**

9. At the end of "Renaissance Hotel," press Enter and key the following two lines of text:
   Modern Elegance
   This brand new hotel has the perfect midtown location.

10. Key the following two lines of text under "Hotel Lexington" (using the same styles applied to the other hotel text):
    Best Buy
    This hotel offers the best weekend package.

11. Display the style area for the document by following these steps:

   a. Choose Tools, Options and click the View tab.
   b. Set the Style area width to 0.5 inch and click OK.

12. Format the memo with the correct spacing at the top of the page and a continuation header. Remember to include your reference initials at the end of the document.

13. Restore the style area width to zero.

14. Print the document style sheet by following these steps:

   a. Open the Print dialog box.
   b. Choose Styles from the Print what drop-down list and click OK.

15. Spell-check the document and save it as *[your initials]*15-14 in your Lesson 15 folder.

16. Print and close the document.

EXERCISE 15-15

Use AutoFormat and the Style Gallery

1. Start a new document. Display the Styles and Formatting task pane.

2. Key **Fax Cover Sheet** and press Enter twice. Apply the Heading 1 style to the first line of text.

3. At the second paragraph mark below the heading, key the text shown in Figure 15-13 (on the next page).

4. Use AutoFormat to format the entire document by following these steps:

   a. Choose AutoFormat from the Format menu.
   b. Make sure General document is set as the document type and click OK.

5. Use the Style Gallery to choose the styles from another template by following these steps:

   a. Choose Theme from the Format menu and click Style Gallery.
   b. Choose Professional Fax from the Template list box and click OK.

6. Modify the style Message Header so it has a 0.6-inch left indent, no hanging indent, and a 2-inch left tab only (clear all other tab settings).

7. Apply the style Message Header Last to the "Number of Pages" line. Add a 3.25-inch left tab to this style and clear all other tab settings.

8. Apply the style Heading 2 to the text "Message:".

9. Make sure the paragraph below "Message:" is Body Text style.

10. Format the document with a 2-inch top margin.

11. Spell-check the document.

12. Save the document as *[your initials]*15-15 in your Lesson 15 folder.

13. Print and close the document.

FIGURE 15-13

Date: *[Enter today's date]*

To: Ms. Mary Oliver

Fax Number: (505) 555-4597

From: Frank Youngblood

Fax Number: (505) 555-1244

Number of Pages (including this sheet): 3

Message:

Attached is the itinerary for your upcoming business trip to Orlando, Florida. As you requested, I reserved an early morning flight out of Albuquerque for you. I reserved a return flight for the evening of the 18th, but afternoon flights are also available, if you prefer. Please let me know as soon as possible, because Orlando is a popular destination this time of year. Thank you for using Duke City Gateway Travel and have a wonderful trip.

Insert tab after each colon

Single-space entire document. No blank lines between paragraphs.

Lesson Applications

EXERCISE 15-16

Create, apply, and modify styles; print styles.

1. Start a new document. Key the text shown in Figure 15-14. Create the styles as indicated for the first two lines.

FIGURE 15-14

Family Vacations in the Southwest ← New style: Headline 1
                                         24-pt Arial bold, centered

For More Information: ←

Tom Carey   New style: Headline 2

Duke City Gateway Travel Agency   18-pt Arial Italic, centered

15 Montgomery Boulevard

Albuquerque, NM 87111

(505) 555-1234

Gateway Travel —|— Your Gateway to the World
                  M

2. Apply the Headline 2 style to the last line of text. Modify the style to include 18 points of paragraph spacing before and after.
3. Create a style called Body 1 (based on the Normal style) for the remainder of the text; the style should be 12-point Arial, centered, with 1.5-line spacing.
4. Center the document vertically on the page.
5. Spell-check the document.
6. Save the document as *[your initials]***15-16** in your Lesson 15 folder.
7. Print the document and the style sheet and close the document.

EXERCISE 15-17

Create, modify, and rename styles; print styles.

1. Open the file **NYwkend**.
2. For the first line of text, create and apply a style named HeadCenter using 20-point Times New Roman, bold, all caps, centered, with 72 points of spacing before and 12 points of spacing after paragraphs.
3. For the text containing names and addresses, create and apply a style named Address (based on the Normal style) using 12-point Times New Roman, italic.

4. Apply the Heading 2 style to the text "Group Travel."

5. Modify the Heading 2 style to include center alignment.

6. Apply the style Heading 3 to the text beginning "<u>New York</u>."

7. Rename the style HeadCenter with the name Headline and change the style's font to Arial.

8. Modify the Address style by changing the font from Times New Roman to Arial and adding a 0.75-inch left indent.

9. Center the text that begins "New York."

10. Add the ZIP Code **87223** to the address for Donald Gagliano.

11. Save the document as *[your initials]***15-17** in your Lesson 15 folder.

12. Print the document and the style sheet and close the document.

EXERCISE 15-18

Modify styles, use style options, and print styles.

1. Open the file **Paradise**.

2. Format the document as a letter to the name and address shown below, from Nina Chavez, Travel Counselor. Remember to use the correct top margin and spacing and to include your reference initials.

   Ms. Joan Piazza
   Weathertech Industries
   42 Las Galinas Boulevard
   Albuquerque, NM 87111

3. Edit the opening paragraph of the letter to read:

   These are the Paradise Island hotels we discussed. The dates indicate each hotel's peak season.

4. Delete one blank line after the opening paragraph and the line containing the text "<u>Hotel</u>."

5. Change the left and right margins to 1 inch. Delete the paragraph toward the end of the document that begins "These hotels offer" and the blank line after it.

6. Create a style named Hotel using 11-point Arial Narrow, small caps, bold, and apply the style to each hotel name.

7. Create a style named Description, based on the Hotel style, using italic, no bold, no small caps.

8. Assign Description as the following paragraph style for the Hotel style.

9. Position the insertion point at the end of the text "Crown Beach Resort", press Enter, and key **The largest and oldest hotel on the island**.

10. Use the same method to enter descriptions for the remaining hotels, which are listed as follows:

    Located on a charming and quiet bay
    A snorkeler's delight, located on the reef
    Hear the soothing surf from every room
    Located next to a bird sanctuary and a tropical garden

11. Create a style named Dates based on the Description style but with 6 points of spacing after paragraphs.

12. Apply the Dates style to all the hotel dates. Delete the blank paragraph mark after all the text using the Dates style, except after the last date.

13. Add a closing paragraph to the letter by keying:

    I look forward to speaking with you soon to help you plan your trip.

14. Modify the font size of the Hotel style to 12 points. Modify the Normal style font to Arial.

15. Select the list of hotels and dates and apply a 2-column format. Insert a column break at "Ocean Resort."

16. Spell-check the document.

17. Save the document as *[your initials]***15-18** in your Lesson 15 folder.

18. Print the style sheet and the document and close the document.

EXERCISE 15-19 ✚ *Challenge Yourself*

Create, apply, and modify styles; use AutoFormat; and print styles.

1. Start a new document. Format it as a memo to Tom Carey from Frank Youngblood. Use today's date. The subject of the memo is **Museum Update**.

2. For the opening paragraph, key:

    Please note that the following museums have facilities for handicapped visitors and are accessible to those who use wheelchairs.

3. Continue the memo by keying the text shown in Figure 15-15 (on the next page). Insert a blank line before, but not after, each museum name.

4. Create a style called Museum, using Arial bold. Apply the Museum style to the museum names.

5. AutoFormat the document. Make sure all four lines in the memo heading have the same format.

6. Modify the Message Header style with your choice of font and paragraph changes. Make at least three modifications, such as setting a tab, changing shading, or adding paragraph spacing.

7. If you added a tab to the Message Header style, align text as necessary to ensure that all the header text aligns uniformly.

FIGURE 15-15

```
National Atomic Museum
For more information, contact David Lee at 555-3458.

New Mexico Museum of Natural History
For information about special programs, contact Susan Sheehan at
555-9182.

Albuquerque Art Museum
For more information, contact Angela Rodriguez at 555-8754.
```

8. Modify the style of the museum names to include 12 points of spacing before paragraphs and no spacing after.

9. Create a character style named MemoHead for the text "MEMO TO:" using a font style and at least one font effect (such as embossing or a shadow) of your choice. Apply the style to "FROM:," "DATE:," and "SUBJECT:".

10. Add your reference initials to the bottom of the document.

11. Save the document as *[your initials]***15-19** in your Lesson 15 folder.

12. Print the document and the style sheet and close the document.

On Your Own

In these exercises, you work on your own, as you would in a real-life work environment. Use the skills you've learned to accomplish the task—and be creative.

EXERCISE 15-20
Create your own one-page newsletter. Create and use styles for the different newsletter elements, such as the title, date line, body text, and publisher information. Save the document as *[your initials]***15-20**. Print the document and the styles.

EXERCISE 15-21
Assume you have been on a job interview. Write a simple follow-up letter and use the Style Gallery to apply the Professional, Contemporary, or Elegant Letter template. Apply the appropriate styles to your letter. Save the document as *[your initials]***15-21**. Print the document and the styles.

EXERCISE 15-22

Create a document that includes your five favorite songs. Each song should appear as a three-line description—song title, songwriter, and performer—with each line using a different style. The songwriter and performer styles should be based on the song title style. Specify the songwriter style as the style following the title style, and the performer style as the style following the songwriter style. Create the styles first; then key the text. Modify the styles as desired. Save the document as *[your initials]*15-22. Print the document and the styles.

OBJECTIVES

After completing this lesson, you will be able to:

1. **Use Word templates.**
2. **Create new templates.**
3. **Attach templates to documents.**
4. **Modify templates.**
5. **Use the Organizer.**
6. **Use wizards.**

MICROSOFT OFFICE
SPECIALIST
ACTIVITIES
In this lesson:
WW03S-5-1
WW03E-5-3
See Appendix C.

 Estimated Time: 1½ hours

I f you often create the same types of documents, such as memos or letters, you can save time by using templates. Word provides a variety of templates that contain built-in styles to help you produce professional-looking documents. You can also create your own templates and reuse them as often as you like.

Using Word's Templates

A *template* is a file that contains formatting information, styles, and, sometimes, text for a particular type of document. It provides a reusable model for all documents of the same type. Every Word document is based on a template. You can modify templates to include formatting and text that you use frequently.

The following features can be included in templates:

- Formatting features, such as margins, columns, and page orientation
- Standard text that is repeated in all documents of the same type, such as a company name and address in a letter template
- Character and paragraph formatting that is saved within styles
- Macros (automated procedures) and AutoText entries

Templates also include *placeholder text* containing the correct formatting, which you replace with your own information when you create a new document.

The default template file in Word is called **Normal**. New documents that you create in Word are based on the Normal template and contain all the formatting features assigned to this template, such as the default font, type size, paragraph alignment, margins, and page orientation. The *Normal template* differs from other templates because it stores settings that are available globally. In other words, you can use these settings in every new document even if they are based on a different template. The file extension for template files is .dot.

EXERCISE   **16-1**   **Use a Word Template to Create a New Document**

When you start Word, it opens a new blank document and displays the Getting Started task pane. (If the task pane does not appear, choose New from the File menu.) To display the New Document task pane, click the Other Task Panes button ▼ at the top of the task pane and select New Document. In the New Document task pane, you can open an existing document, create a blank document, create a new document based on an existing document, or create a document based on a template.

1. Choose New from the File menu to open the New Document task pane.

  TIP: Pressing Ctrl+N or clicking the New Blank Document button ▯ opens a new document but does not open the New Document task pane.

2. Under New in the task pane, click Blank Document. Word opens a new document based on the default template Normal and closes the New Document task pane. (See Figure 16-1 on the next page.)

3. Close the document without saving it.

4. Choose New from the File menu to reopen the New Document task pane. Under New in the task pane, click From existing document. Word displays the New from Existing Document dialog box.

5. Locate the file **Memo4** and double-click it. Word opens a copy of the document. Notice that the Title bar displays Document followed by a number. The document can be edited, formatted, and saved.

6. Close the document without saving it.

FIGURE 16-1
New Document
task pane

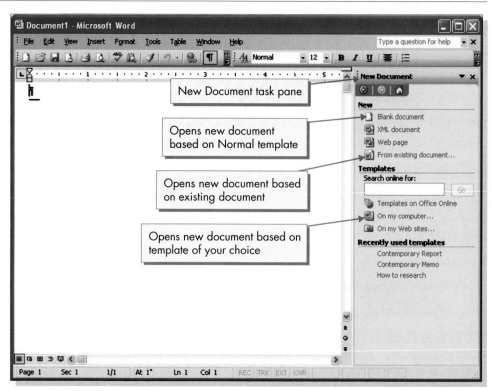

7. Reopen the New Document task pane. Under Templates click On my computer. In the Templates dialog box, double-click the Blank Document icon. A new document opens that is based on the default template Normal. (Clicking Blank Document in the New Document task pane produces the same result.)

8. Close the document without saving it.

9. Reopen the New Document task pane and click On my computer.

10. Click the Letters & Faxes tab. Click the Contemporary Fax icon. Notice the design of the template in the Preview box. (See Figure 16-2 on the next page.)

 NOTE: Some templates might not be installed on your computer. Check with your instructor if you see the message "Template not installed yet" above the Preview box.

11. Use the Preview box to view some other Word templates on other tabs in the dialog box.

12. Return to the Letters & Faxes tab and double-click the Contemporary Letter icon to open a new document based on this template. The document is formatted in a contemporary style. Placeholders prompt you to key the information required to complete the letter. (See Figure 16-3 on the next page.)

FIGURE 16-2
Letters & Faxes tab
of the Templates
dialog box

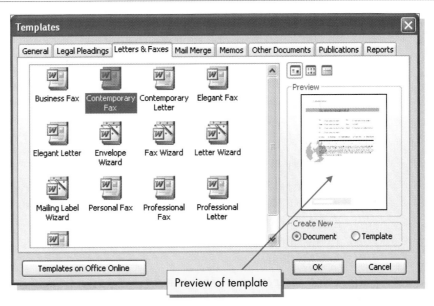

Preview of template

FIGURE 16-3
Creating a
document from the
Contemporary
Letter template

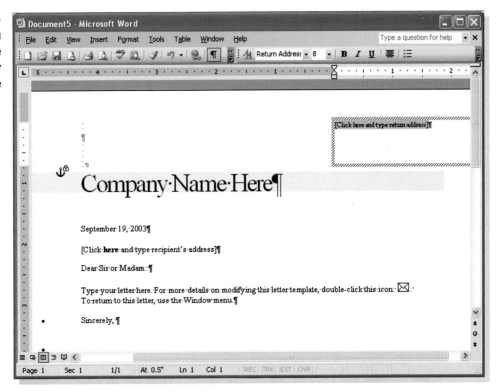

NOTE: The formatting stored in templates is often complex and highly stylized and does not necessarily conform to the traditional formatting for business documents as described in *The Gregg Reference Manual.*

13. In the upper-right corner of the page, click to select the placeholder text that reads "[Click here and type return address]." Key **180 O'Farrell Street**, press (Enter), and key **San Francisco, CA 94102**.

 NOTE: If you have difficulty locating or selecting the placeholder text, increase the zoom to 100%.

14. Use the I-beam to drag and select the placeholder text "Company Name Here." (You cannot click this placeholder to select the text.) Key **Golden Gateway Travel**.

15. Click to select the placeholder text "[Click here and type recipient's address]." Key **Mr. Joseph Fodor**, press (Enter), key **940 Noe Street**, press (Enter), and key **San Francisco, CA 94114**.

16. In the salutation line, use the I-beam to select the text "Sir or Madam" and key **Mr. Fodor**.

17. Select the paragraph that begins "Type your letter here" and key the two paragraphs shown in Figure 16-4. Paragraph formatting for spacing is included in the template, so you don't have to insert a blank line between paragraphs.

FIGURE 16-4

```
Enclosed are the tickets for your business trip to New York City. Your
seats are in business class, as you requested, and we were able to
schedule the special meals you require.

Have a pleasant flight, and thank you for choosing Golden Gateway
Travel. We look forward to helping you again.
```

18. In the letter closing, click the first line of placeholder text and key **Amanda Suarez**. Click the second line of placeholder text and key **Travel Specialist**. Add your reference initials, followed by an enclosures notation for two tickets. Use the Style box on the Formatting toolbar to apply the Reference Initials style and the Enclosure style to the new lines. There should be no blank paragraph marks in the document.

19. At the end of the document, click the slogan placeholder text and key **Your Gateway to the World**.

20. Increase the font size for all text below the company name to 12 points.

21. Increase the paragraph spacing after the date to 36 points.

22. Save the document as *[your initials]***16-1** in a new Lesson 16 folder.

23. Print and close the document.

NOTE: If Word asks if you want to save the changes to the Contemporary Letter template, click No.

Creating New Templates

You can create your own templates for different types of documents by using one of three methods:

- Create a blank template file by using the default template and define the formatting information, styles, and text according to your specifications.
- Open an existing template, modify it, and save it with a new name.
- Open an existing document, modify it, and save it as a new template.

EXERCISE 16-2 **Create a New Template**

1. Open the New Document task pane (File, New) and click On my computer. In the Templates dialog box, click the General tab and click the Blank Document icon.

FIGURE 16-5
General tab
of Templates
dialog box

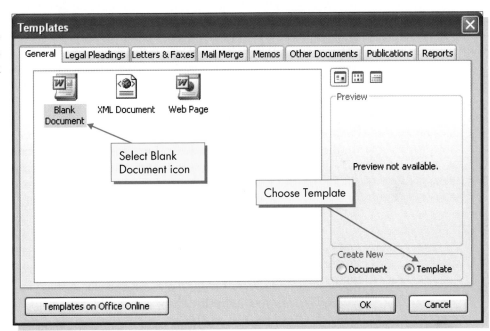

2. In the lower-right corner of the dialog box, under Create New, click Template, and then click OK. A new template file opens with the default name Template1.

NOTE: After Template is selected, it will remain selected the next time the Templates dialog box is opened.

3. Set a 2-inch top margin for a standard business letter.

4. Create a new style named Bodytext that is 12-point Arial.

5. Insert the date as a field at the top of the document by choosing Insert, Date and Time. Use the third date format in the Available formats list in the Date and Time dialog box. Check Update automatically so the date field is updated each time the document is printed.

6. Apply the Bodytext style to the date field.

7. Press Enter four times, and then key the inside address shown in Figure 16-6.

FIGURE 16-6

```
Mr. Roger Forman

Forman & Price Marketing, Inc.

52 Jefferson Boulevard

Albuquerque, NM 87109
```

8. Press Enter twice and key the salutation **Dear Mr. Forman:**.

 NOTE: Because templates are meant to be reused many times, you would create a letter template with a specific inside address and salutation only for someone to whom you write frequently. Otherwise, a letter template would include placeholder text or blank paragraph marks for this information.

9. Press Enter three times and key **Sincerely,**.

10. Press Enter four times and key the name **Nina Chavez**. On the next line, key the title **Senior Travel Counselor**. Press Enter twice and key your reference initials.

11. Choose Save As from the File menu. A folder named "Templates" should appear in the Save in box, and Document Template should appear in the Save as type box. (See Figure 16-7 on the next page.)

 NOTE: By default, Word saves new templates in a User template folder on your hard disk. The specific location is C:\Documents and Settings\Username\ Application Data\Microsoft\Templates (or a similar location in your computer). Before proceeding, ask your instructor where you should save your templates. If you use the default location, you can create new documents from your templates by using the New dialog box. If you use your Lesson 16 folder, you create new documents from your templates by using Windows Explorer or My Computer.

FIGURE 16-7
Save As dialog box

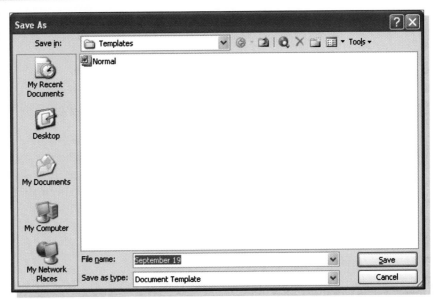

12. Save the template with the filename *[your initials]***letter** in your Lesson 16 folder (unless your instructor advises you to save in the default Templates folder).

13. Close the template.

EXERCISE 16-3 **Create a New Template by Using an Existing Document**

1. Reopen the New Document task pane by choosing New from the File menu. Under New, click From existing document.

2. In the New from Existing Document dialog box, locate the file **Memo4** and click Create New. Word opens a copy of the document.

3. Check that the margins are set for a standard business memo (2-inch top margin, 1.25-inch left and right margins).

4. Delete all text to the right of each tab character in the memo heading.

5. In the date line of the memo heading, insert the date as a field; use the third date format. Check Update automatically so the date field is updated each time the document is printed.

6. Delete the opening paragraph, but include three blank paragraph marks after the subject line.

7. Choose File, Save As. Choose Document Template from the Save as type drop-down list box.

8. Save the file as *[your initials]***memo** in your Lesson 16 folder (unless your instructor advises you to save in the default Templates folder).

9. Close the template.

Attaching Templates to Documents

All existing documents were assigned a template—either Normal or another template that you assigned when you created the document. You can change the template assigned to an existing document by attaching a different template to the document. When you *attach* a template, that template's formatting and elements are applied to the document, and all the template styles and AutoText entries become available in the document.

EXERCISE **16-4** **Attach a Template to a Document**

1. Open the file **Memo2**.

 NOTE: If you do not see a list of Word documents, check that the Files of type drop-down list is set to All Word Documents.

2. Choose Templates and Add-Ins from the Tools menu. The Templates and Add-ins dialog box shows that the document is currently based on the Normal template.

FIGURE 16-8
Templates and
Add-ins dialog box

3. Click <u>At</u>tach. The Attach Template dialog box opens, displaying available templates and folders in the current folder.

4. Change to the folder that contains Word templates. The full default path of this folder is C:\Program Files\Microsoft Office\Templates\1033. (The folder may not be on the C:\ drive on your computer. Check with your instructor.)

FIGURE 16-9
Attach Template
dialog box

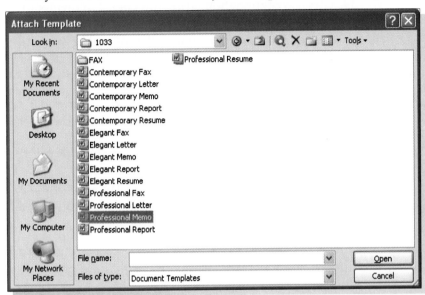

 NOTE: Template icons have a yellow top border. This helps distinguish them from regular document files.

5. Double-click the template called Professional Memo.

6. Click the Automatically <u>u</u>pdate document styles check box and click OK. Formatting from the Professional Memo template is applied to this document, and you can now apply any of the Professional Memo styles.

7. Display the Styles and Formatting task pane.

8. Position the insertion point in the subject line of the memo heading and apply the style Message Header Last. Because this style already includes spacing after paragraphs, delete the two blank lines below the subject line.

9. Position the insertion point immediately to the left of "TO:" in the memo heading and press Enter. Place the insertion point in the first line of text ("MEMO") and apply the style Document Label.

10. Apply the character style Message Header Label to the text "TO:," "FROM:," "DATE:," and "SUBJECT:" in the message header. (Remember, you must select text before applying a character style.) Select all four lines of the entire message header and change the left tab setting to 1.5 inches to improve alignment.

11. Delete the date in the date line and key today's date.

12. Set a 1.75-inch top margin. This margin setting, along with the existing paragraph formatting, places "MEMO" 2 inches from the top of the page.

13. Change the "TO" line so the memo is addressed simply to "Staff." Delete the memo text from "Finalizing the Arrangements" through the end of the document. Add your reference initials.

14. Create a continuation header on page 2 (remember to select the Different first page option on the Layout tab of the Page Setup dialog box, and to set the header 1 inch from the edge). Before keying the heading text, change the Header style to left alignment with no space after paragraphs, and change the Page Number character style to 10 point.

 REVIEW: To modify a style before applying it, right-click the style name in the Styles and Formatting task pane and choose Modify.

15. Check pagination in the document. Make sure a heading is not separated from the text that follows it.

16. Save the document as *[your initials]*16-4 in your Lesson 16 folder.

17. Print and close the document. If you are asked to save changes to the Professional Memo template, click No.

NOTE: Attaching templates is similar to using the Style Gallery, as discussed in the previous lesson. The Style Gallery lets you apply styles from a selected template. Attaching a template replaces the template that is currently attached to the document.

Modifying Templates

After you create a template, you can change its formatting and redefine its styles. You can also create new templates by modifying existing templates and saving them with a new name.

NOTE: Any changes you make to the formatting or text in a template affect future documents based on that template. The changes do not affect documents that were created from the template before you modified it.

EXERCISE **16-5** **Modify Template Formatting**

1. Click the Open button on the Standard toolbar. From the Files of type drop-down list, choose Document Templates.

2. From the Look in drop-down list, locate the folder you used to save your templates (for example, the Templates folder on your hard disk under C:\Documents and Settings\[Username]\Application Data\Microsoft or your Lesson 16 folder).

3. Locate the template file *[your initials]***letter**. Notice the top yellow border that distinguishes a template icon from a document icon. You can also point to an icon to check its file type.

4. Double-click the file *[your initials]***letter** to open it. Display the Styles and Formatting task pane.

TIP: Opening a template through the Open dialog box opens the actual template. Double-clicking a template in Windows Explorer, My Computer, or the Templates dialog box opens a new document based on the template. Changes that you make to the new document do not affect the template.

5. Change the Bodytext style to 12-point Times New Roman.

6. On the first inside address line, replace "Mr. Roger Forman" with **Name**. Replace the three-line address with **Address**.

7. On the salutation line, delete "Mr. Forman" so the line reads "Dear :".

8. In the closing, replace "Nina Chavez" with **Sender** and "Senior Travel Counselor" with **Title**.

9. Replace the reference initials with **Reference initials**.

10. Click the Save button 🔲 to save the changes. The earlier version of the template is overwritten by the new version.

11. Close the template.

NOTE: To create a new template based on an existing template, modify the existing template as desired, and then save the template with a new name.

Using the Organizer

Instead of modifying template styles, you can copy individual styles from another document or template into the current document or template by using the Organizer. The copied styles are added to the style sheet of the current document or template. When you copy styles, remember these rules:

- Copied styles replace styles with the same style names.
- Style names are case-sensitive—if you copy a style named "HEAD" into a template or document that contains a style named "head," the copied style is added to the style sheet and does not replace the existing style.

You can also copy AutoText entries by using the Organizer. In addition to the preceding rules, note that you can copy AutoText entries only among templates, not among documents. To open the organizer, choose Templates and Add-Ins from the Tools menu and click the Organizer button, and then select the appropriate tab.

EXERCISE 16-6 Copy Styles to Another Template

1. Use the Open button 🖾 to open the template *[your initials]*memo created in Exercise 16-3.

2. Choose Templates and Add-Ins from the Tools menu and click the Organizer button.

3. Click the Styles tab in the Organizer dialog box. On the left side of the dialog box, the Organizer lists the template and styles currently in use. You use the right side of the dialog box to copy styles to or from another template.

4. Click the Close File button on the right side of the dialog box. The Normal template closes, and the Close File button changes to Open File.

5. Click the Open File button. In the Open dialog box, make sure Document Templates appears in the Files of type box.

6. Open the Look in drop-down list and change to the folder that contains preinstalled Word templates (the full default path is C:\Program Files\ Microsoft Office\Templates\1033).

7. Double-click the Contemporary Memo template. You can now choose styles from this template to copy into your memo template.

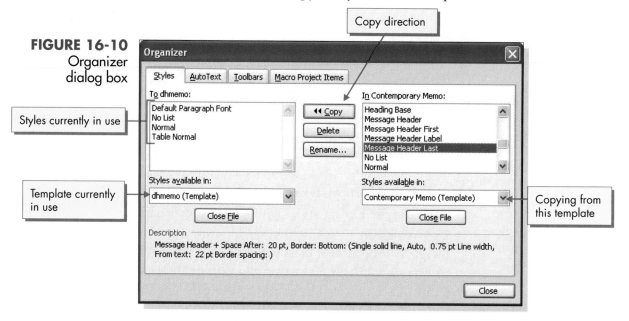

FIGURE 16-10
Organizer
dialog box

Styles currently in use

Template currently in use

Copy direction

Copying from this template

8. Scroll down the list of styles in the Contemporary Memo template. Click Message Header Last and click Copy. Notice that this action also copies the based-on styles, Message Header and Body Text.

9. Choose the Normal style from the Contemporary Memo style list. Notice the style description.

10. Click Copy, and then click Yes to overwrite the existing style, Normal.

11. Close the dialog box. The styles you chose from the Contemporary Memo template are copied to the current template. Notice that the Normal style from the Contemporary Memo template replaced the previous Normal style, so the text is formatted in Times New Roman 10-point with a left indent.

12. Apply the newly copied style Message Header Last to the subject line.

13. Close the template without saving changes.

E X E R C I S E 16-7 Copy AutoText to Another Template

1. Open the template *[your initials]*memo.

2. Click in the third blank line below the subject line and key the following paragraph:

   All materials published by Duke City Gateway Travel are the sole property of Gateway Travel and may not be reproduced without written authorization from an officer of the company.

3. To save the paragraph you just keyed as an AutoText entry, select the paragraph and its paragraph mark. Choose AutoText from the Insert menu, and then click AutoText.

4. Open the Look in drop-down list and choose *[your initials]*memo to save the entry in this template. Key **authorization** in the Enter AutoText entries here text box and click Add. (See Figure 16-11 on the next page).

5. Deselect the text in the document. Choose Templates and Add-Ins from the Tools menu. Click Organizer in the Templates and Add-ins dialog box.

6. In the Organizer dialog box, click the AutoText tab. Make sure the file *[your initials]*memo is open on the left side of the dialog box; you will copy the new AutoText entry from this template to your letter template.

7. Close the Normal.dot file on the right side of the dialog box by clicking Close File. Then click Open File. Locate the folder that contains the templates you created and open *[your initials]*letter.

8. Make sure the AutoText entry "authorization" is selected. Click Copy. The entry is copied from your memo template to your letter template. (See Figure 16-12 on the next page).

FIGURE 16-11
Creating an
AutoText entry

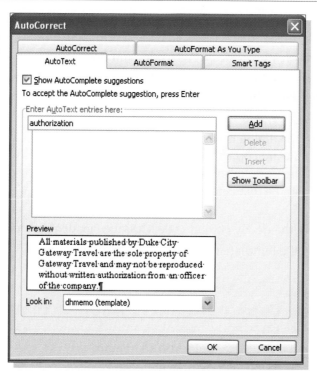

FIGURE 16-12
Copying an
AutoText entry

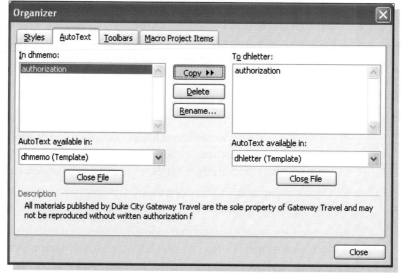

9. Click Close. When prompted, click Yes to save the changes to *[your initials]*letter.

10. Delete the "All materials" paragraph from your memo template. Make sure there are three blank paragraph marks below the subject line. Close the template, saving changes when prompted.

11. Start a new document based on the *[your initials]*letter:

- If you saved your template in the default Templates folder, open the New Document task pane and click On my computer. Click the General tab, select your template, and choose Document under Create New. Click OK.

- If you saved your template in your Lesson 16 folder, right-click the Windows Start button , choose Explore, locate your folder, and double-click your letter template file.

12. In the new document (which should be titled "Document" followed by a number), select the name placeholder text and key **Mr. Tom Kelly**. Select the address placeholder text and key the following three lines:

   Santiago Publishing Company
   300 Calle Del Sol
   Albuquerque, NM 87112

13. Insert the text **Mr. Kelly** in the salutation line. Position the insertion point in the second blank paragraph mark after the salutation line and key the text in Figure 16-13.

FIGURE 16-13

Enclosed are the brochures you requested about our Gateway to the Southwest tours. Feel free to use these materials for background information in your upcoming book, *Undiscovered New Mexico*. We request that you not quote extensively from the articles contained in these brochures.

14. With the insertion point at the end of the paragraph you just keyed, begin keying the AutoText entry **authorization**. When you see the ScreenTip, press Enter to let AutoComplete complete the name for you.

15. Press Enter, and then key the new paragraph in Figure 16-14.

FIGURE 16-14

Please call if I can be of further assistance. Our travel consultants, who wrote these materials, are authorities on New Mexico history and would be happy to answer specific questions about the places and events described.

16. Insert a paragraph mark above "Sincerely." Replace the "Sender" placeholder text with **Tom Carey**. Replace the "Title" placeholder text with **Senior Travel Counselor**. Replace the "Reference initials" placeholder text with your initials. Add an enclosures notation.

17. Save the document as *[your initials]***16-7** in your Lesson 16 folder.

18. Print and close the document.

Using Word's Wizards

A *wizard* is an automated and interactive template. Wizards can format letters, memos, and reports quickly. After you create a document by using a wizard, you can make additional formatting changes. Like other Word templates, wizards are included in the Templates dialog box and organized in categories under the dialog box tabs.

EXERCISE **16-8** **Use a Wizard to Create a New Document**

A wizard presents a series of dialog boxes, each containing options to choose and elements to include or exclude in the final document.

1. Open the New Document task pane. Click On my computer to open the Templates dialog box.

2. Click the Memos tab. Make sure Document is selected under Create New. Double-click the Memo Wizard icon.

3. In the Memo Wizard dialog box, a list at the left shows the steps you follow to complete the wizard. Click Next.

4. Choose the style Professional and click Next.

FIGURE 16-15
Memo Wizard
dialog box

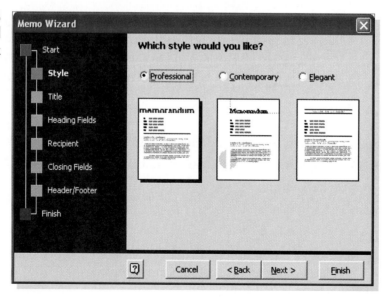

5. Use the default title text ("Interoffice Memo") and click Next.

6. In the dialog box for Heading Fields, check the Date box, key your name in the From box, and key **Sample California Itinerary** in the Subject box. Click Next.

7. In the To box, key **Frank Youngblood**. Clear the Cc check box and click No to create a distribution list. Click Next.

8. For closing items, clear all the check boxes and click Next.

9. For header and footer items, clear all the check boxes and click Next. Click Finish. Cancel the Office Assistant.

10. For the opening paragraph of the memo, click the placeholder text and key **This is the sample itinerary you wanted to review.**

11. At the next paragraph mark, insert the file **Itin1**. To insert a file, choose Insert, File, locate and select the filename, and then click Insert.

12. Make any formatting changes to improve the memo's appearance, such as applying and modifying styles.

13. Save the document as *[your initials]***16-8** in your Lesson 16 folder.

14. Print and close the document.

USING ONLINE HELP

How you access the templates you create depends on where you store them and also on the location settings for User Templates on the computer you are using. If you work on a computer that is shared by others, such as an office workstation or a computer lab, you might be able to change your template file location settings to access your templates. You can also create a new tab for your templates in the Templates dialog box to better organize your template files.

Use Help to learn how to change template location settings:

1. Key **templates** in the Ask a Question box and press Enter.

2. Click the topic About template locations.

3. Locate and read the information on how to check your template file location settings.

4. Click the See Also link to read about global templates. Close Help when you finish.

LESSON 16 Summary

➤ A template is a reusable model for a particular type of document. Templates can contain formatting, text, AutoText entries, and other elements. By default, all new documents are based on the Normal template.

➤ Word provides a variety of templates upon which you can base a new document or a new template. You can modify any existing template and save it with a new name. You can also modify any existing document and save it as a new template.

➤ Every document is based on a template. You can change the template assigned to an existing document by attaching a different template to the document, thereby making the new template's styles and AutoText entries available in the document.

➤ To modify a template you created, open it from the Open dialog box, choosing Document Templates from the Files of type drop-down list.

➤ Instead of modifying template styles, use the Organizer to copy individual styles from one document or template to another. The Organizer can also copy AutoText from one template to another.

➤ A wizard is an automated and interactive template. A wizard walks you through the process of creating a letter, memo, or other type of document. Wizards are included in the Templates dialog box and organized in categories under the dialog box tabs.

LESSON 16   Command Summary

FEATURE	BUTTON	MENU	KEYBOARD
Use a template or wizard		File, New, On my computer	
Attach template		Tools, Templates and Add-Ins, Attach	
Copy styles or AutoText		Tools, Templates and Add-Ins, Organizer	

Concepts Review

Each of the following statements is either true or false. Indicate your choice by circling T or F.

T F **1.** A wizard is a sophisticated template that automates document creation.

T F **2.** You can open the New Document task pane by pressing Ctrl+N.

T F **3.** After a template is assigned to a document, it cannot be changed.

T F **4.** A Word template can contain placeholder text that you replace with your own information.

T F **5.** When copying styles to a template, if the style names do not match the existing styles, they are added to the style sheet.

T F **6.** You can copy AutoText entries among templates.

T F **7.** If you do not specify a template when you create a new document, the document is created without one.

T F **8.** You can use the Organizer to copy styles among templates or documents.

Write the correct answer in the space provided.

1. What is the file extension assigned to a template filename?

_____

2. When you choose New from the File menu, what happens?

_____

3. Which commands would you use to open the dialog box to attach a different template to a document?

_____

4. How can you create a new template by using an existing document?

_____

5. How can you change the styles in a template to match the styles in another template or document?

_____

6. When adding an AutoText entry to a template, how do you specify the template you want to use to store the entry?

_____

7. Which button do you click in a wizard dialog box to move from one dialog box to another?

_____

8. How do you open the Templates dialog box?

_____

CRITICAL THINKING

Answer these questions on a separate page. There are no right or wrong answers. Support your answers with examples from your own experience, if possible.

1. Review Word's templates for letters, memos, and reports. How do they compare with the standard business format for these documents as described in *The Gregg Reference Manual* (or a similar handbook)?
2. Many businesses create templates that are used by all employees for internal and external correspondence. Why would a business take this approach? What advantages does it offer to a business?

Skills Review

EXERCISE 16-9

Use an existing Word template to create a letter.

1. Create a letter based on Word's Elegant Letter template by following these steps:
   a. Choose New from the File menu.
   b. In the New Document task pane, click On my computer.
   c. In the Templates dialog box, click the Letters & Faxes tab. Make sure Document is selected under Create New.
   d. Double-click the Elegant Letter template.
2. At the top of the document, click the placeholder text for the company name and key **Duke City Gateway Travel Agency**.

3. Modify the Company Name style to 14-point Garamond bold with 24 points of spacing before the paragraph.

4. Replace the placeholder text in the document with the text shown in Figure 16-16. Click or select each selection before entering the appropriate text. You might want to increase the zoom when keying the last two lines.

FIGURE 16-16

```
Ms. Ann Foster

11 Pueblos Terrace

Albuquerque, NM 87501

Dear Ms. Foster:

Enclosed are the tickets and itinerary for your trip to Los Angeles.
Please note that we requested special meals.

We also requested bulkhead seating; however, the airline is holding
those seats for assignment at the gate. When you arrive at the
departure gate, airline personnel will assist you with your seat
assignment.

We wish you a very pleasant trip and hope you call on Duke City
Gateway Travel for your future travel needs.

Nina Chavez
Senior Travel Counselor

15 Montgomery Boulevard, Albuquerque, NM 87111
(505) 555-1234 (505) 555-1244
```

5. At the blank paragraph marks below the closing, add your reference initials and an enclosures notation. Apply the Reference Initials and Enclosure styles to the appropriate text.

6. Modify the Normal style to 12 points. (This changes all the styles used below the company name, which are based on the Normal style.)

7. Change the paragraph spacing for the Date paragraph to 24 points before and 36 points after.

8. Save the document as *[your initials]*16-9 in your Lesson 16 folder.

9. Print and close the document.

EXERCISE 16-10

Create a new template, attach it to another document, and modify a template.

1. Create a new template by following these steps:

   a. Choose New from the File menu and click On my computer in the New Document task pane.

   b. In the Templates dialog box, click the General tab, and then click the Blank Document icon.

   c. Under Create New, click Template and click OK.

2. Modify the Normal style to be 12-point Arial.

3. Modify the Heading 1 style so the paragraph formatting is center-aligned and the font size is 18-point.

4. Modify the Heading 2 style so the paragraph formatting is center-aligned (do not remove the bold and italic formatting).

5. Modify the Heading 3 style so the font size is 12-point.

6. Save the template as *[your initials]*agenda in your Lesson 16 folder or in the default Templates folder on the hard disk, whichever your instructor told you to use. Close the template.

7. Start a new document based on the Normal template by clicking the New button.

8. Key the text shown in Figure 16-17. Use single spacing. Apply the Heading styles to the paragraphs indicated.

FIGURE 16-17

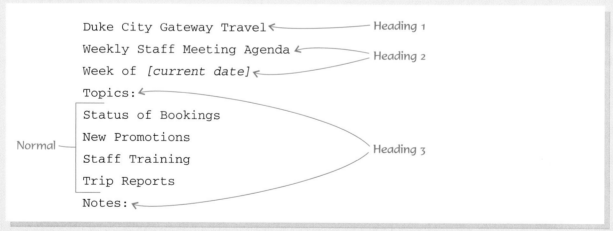

Duke City Gateway Travel ← Heading 1
Weekly Staff Meeting Agenda ← Heading 2
Week of [current date] ←
Topics: ←
Status of Bookings
New Promotions
Normal — Staff Training   Heading 3
Trip Reports
Notes: ←

9. Attach the template you created and automatically update the styles by following these steps:

    a. Choose Templates and Add-Ins from the Tools menu.

    b. Click Attach.

    c. Locate *[your initials]*agenda and click Open.

    d. Check the Automatically update document styles check box and click OK.

10. Save the document as *[your initials]*16-10 in your Lesson 16 folder.

11. Print and close the document.

EXERCISE 16-11

Use the Organizer to copy styles from another template.

1. Open the file **DukeHistory**.

2. Apply Heading 1 style to the first line (the agency name). Apply Heading 2 style to the bold text and Heading 3 style to the italic text.

 TIP: Click the bold text, click Select All in the Styles and Formatting task pane, and then apply the style. Do the same for the italic text.

3. Apply the Body Text style to all paragraphs currently formatted with the Normal style. (First change the Show setting in the Styles and Formatting task pane to All styles to display the Body Text style.)

4. Delete the blank paragraph marks in the document.

5. Use the Organizer to copy styles from another template by following these steps:

    a. Choose Templates and Add-Ins from the Tools menu, and then click the Organizer button.

    b. In the Organizer dialog box, click the Styles tab. Click the Close File button on the right side of the dialog box (under Normal.dot).

    c. Click the Open File button on the right side of the dialog box.

    d. In the Open dialog box, locate the folder that contains Word templates (C:\Program Files\Microsoft Office\Templates\1033) and choose the Contemporary Report template. Click Open.

    e. Click Body Text in the list on the right side of the dialog box (under Contemporary Report), and then click Copy. Click Yes to overwrite the existing style entry.

    f. Repeat the previous step to copy the Footer, Heading 1, Heading 2, and Heading 3 styles from the Contemporary Report template on the right to the DukeHistory document on the left. To copy the styles simultaneously, hold down Ctrl while selecting each of the styles to copy. When prompted, overwrite the existing style entries in the DukeHistory document.

    g. Click the Close button.

6. Insert page numbers at the bottom right of the document, starting with 2 on page 2.

7. Modify the paragraph spacing for the Heading 1 style to 72 points before and 24 points after.

8. Save the document as *[your initials]*16-11 in your Lesson 16 folder.

9. Print and close the document.

EXERCISE 16-12

Use a wizard and create a new template.

1. Choose New from the File menu. Click On my computer in the task pane.

2. In the Templates dialog box, click the Memos tab.

3. Click the Memo Wizard icon. Under Create New, click Template. Click OK to start the Memo Wizard.

4. In the opening Memo Wizard dialog box, click Next.

5. Choose Contemporary and click Next.

6. Change the title text to **confidential memo** (in lowercase) and click Next.

7. Change the date format and delete the text in the From and Subject text boxes. Click Next.

8. Key **Staff** in the To text box. Clear the Cc check box. Click Next.

9. Click the Typist's initials check box and key your reference initials in lowercase. Clear all other check boxes. Click Next.

10. Clear all check boxes for header and footer items. Click Finish. Cancel the Office Assistant.

11. Save the template as *[your initials]*confmemo in your Lesson 16 folder or in the default Templates folder on the hard disk, whichever your instructor told you to use.

12. Print and close the template.

Lesson Applications

Use a template to create a document.

1. Use the Professional Letter template to create a new document.

2. Replace the company name placeholder text with **Duke City Gateway Travel**. Click the return address placeholder text and key the following four lines:

   15 Montgomery Boulevard
   Albuquerque, NM 87111
   Phone (505) 555-1234
   Fax (505) 555-1244

3. Click the placeholder text for the recipient's address and key the following:

   Ms. Barbara Almy
   Almy-Price, Inc.
   4200 Constance Place NE
   Albuquerque, NM 87109

4. Replace the text "Sir or Madam" with **Ms. Almy**.

5. Open the file **SantaFe**. Copy all the text below the title and paste it into the letter, replacing the sample body text of the letter.

6. Delete the headings and paragraph descriptions of The Plaza, Canyon Road, and Sena Plaza.

7. Apply the Heading 2 style to the remaining underlined headings. Apply the Body Text style to the paragraph descriptions you pasted and to the first paragraph after the salutation.

8. Delete the blank paragraph marks above each Heading 2 heading and above "Sincerely" (if there is one).

9. For the first paragraph under the salutation, add the opening sentence **We reserved a place for you on our June 9 walking tour of Santa Fe.**

10. Add as a closing paragraph **We hope you enjoy the tour.**

11. Replace the signature name and job title placeholder text with the following:

    Alexis Johnson
    Travel Counselor

12. On the next line, key your reference initials and apply the Reference Initials style.

13. Modify the Company Name style to 14 points. Modify the Date style to include 33 points spacing after it.

14. Spell-check the document and save it as *[your initials]*16-13 in your Lesson 16 folder.

15. Print the document. Close both open documents.

EXERCISE 16-14

Use a template and copy styles by using the Organizer.

1. Start a new document based on the template you created, *[your initials]*letter.

2. Replace the name and address placeholder text with **Mr. Henry Suarez, 142 North Terrace, Albuquerque, NM 87051** and add **Mr. Suarez** to the salutation.

3. Press Enter twice after the colon and key the text shown in Figure 16-18, pressing Enter twice between paragraphs.

FIGURE 16-18

```
Thank you for calling Duke City Gateway Travel. Enclosed are the
travel brochures you requested, as well as your itinerary and tickets.
Your boarding passes for both flights are also enclosed.

Have a wonderful trip, and please feel free to call us with any
questions you may have.

We look forward to assisting you again.
```

4. Replace the placeholder text in the closing with the name **Nina Chavez** and the title **Senior Travel Counselor**. Change the placeholder text for initials to your reference initials. On the next line, key **Enclosures**.

5. Open the Organizer dialog box by choosing Templates and Add-Ins from the Tools menu. On the right side of the dialog box, close the Normal template file and open the Elegant Letter template file (located in C:\ Program Files\Microsoft Office\Templates\1033).

6. Copy the styles Company Name, Normal, and Return Address to the current document, replacing the Normal style entry when prompted. Close the Organizer dialog box.

7. Insert a new paragraph mark above the date. At the new paragraph mark, key **Duke City Gateway Travel** and apply the Company Name style to it.

8. Insert a new paragraph mark after the enclosure notation and apply the Return Address style. Increase the zoom if needed. Key **15 Montgomery Boulevard** and press `Spacebar` twice; then key **Albuquerque, NM 87111**. On the next line, key **Phone (505) 555-1234** and press `Spacebar` twice; then key **Fax (505) 555-1244**.

9. Modify the Normal style by increasing the font size to 12 points. Delete the extra paragraph mark above "Sincerely" (if there is one).

10. Save the document as *[your initials]***16-14** in your Lesson 16 folder.

11. Print and close the document.

EXERCISE 16-15

Modify a template and attach a template.

1. Open the template you created, *[your initials]***letter**. (Use the Open dialog box—do not create a document based on this template.)

2. Modify the Normal style to a 12-point font of your choice.

3. Save the template as *[your initials]***letter2** in the folder where you saved the other templates. Close the template.

4. Open the file **Caliente**.

5. Attach the template *[your initials]***letter2**, updating document styles automatically.

6. Copy the Date style from the Professional Letter template to the current template. Modify the Date style so that its font and point size match the Normal style. Add 36 points of spacing after paragraphs to the Date style.

7. Key the date at the top of the document and apply the Date style.

8. Add your reference initials and an enclosure notation.

9. Set a 2-inch top margin. If the document does not fit on one page, change the left and right margins to 1 inch. You might also need to switch to Print Preview and shrink to fit the document on one page.

10. Save the document as *[your initials]***16-15** in your Lesson 16 folder.

11. Print and close the document.

EXERCISE 16-16   *Challenge Yourself*

Use a wizard to create a document.

1. Display the New Document task pane. Under Templates, click Templates on Office Online to open the Microsoft Office Template Gallery.

 NOTE: You need to be connected to the Internet to use the Microsoft Office Template Gallery.

2. Click the category Calendars and Planners.

3. Click the link for the calendar templates for the current year (for example, 2004 Calendars).

4. Notice that each template in the list of calendar templates is preceded by a program icon such as PowerPoint, Visio, or Word.

5. Click the Word template for 2004 Calendar (Mon-Sun, 12pp.).

6. Click Download now. The template will open in a Microsoft Word window.

 NOTE: If the Security Warning dialog box appears, click Yes.

7. On the appropriate month in the calendar that you just created, key the event text shown in Figure 16-19. To key the text, click to the right of the date in the calendar, press Enter, and key the text.

FIGURE 16-19

9/2	Arrive SFO noon, check-in Fairmont Hotel 3pm, dinner 7pm
9/3	Depart hotel 9am for full-day tour of San Francisco, Muir Woods, Sausalito
9/4	Depart hotel 8:30am for all-day Mt. Tamalpais hike, dinner 7pm at Muir Beach
9/5	Depart hotel 8am for all-day trip to Carmel and Monterey, dinner on Cannery Row
9/6	Depart SFO 9am, arrive LAX 10am

8. Preview the month of September.

9. Save the document as *[your initials]*16-16 in your Lesson 16 folder.

10. Log off, if necessary. Print the page for September, and then close the document.

On Your Own

In these exercises, you work on your own, as you would in a real-life work environment. Use the skills you've learned to accomplish the task—and be creative.

EXERCISE 16-17

Use the Resume Wizard or a resume template (located on the Other Documents tab in the Templates dialog box) to create your resume. Include as much detail about yourself as possible. Modify the formatting as needed. Save the document as *[your initials]*16-17 and print it.

EXERCISE 16-18

Create a cover letter for your resume, using a matching template style. Address the cover letter to a prospective employer. Print an envelope for the letter. Save the document as *[your initials]*16-18 and print it.

EXERCISE 16-19

Using the New Document task pane, go to Templates on Office Online on the Web and choose a template from any category. Preview the template on the Web site, and then edit it in Word, using your own information. Save the document as *[your initials]*16-19 and print it.

17

Mail Merge

OBJECTIVES

MICROSOFT OFFICE
SPECIALIST
ACTIVITIES
In this lesson:
WW03E-2-6
WW03E-2-7
See Appendix C.

After completing this lesson, you will be able to:

1. **Create a main document.**
2. **Create a data source.**
3. **Insert merge fields into a main document.**
4. **Perform a mail merge.**
5. **Use data from other applications.**
6. **Edit an existing main document.**
7. **Sort and filter a data source.**
8. **Create mailing labels.**

 Estimated Time: 1½ hours

Businesses and organizations often want to send the same letter to several people. *Mail merging* combines a document such as a form letter with a list of names and addresses to produce individualized documents. Using this process, you can create hundreds of personalized letters with just two documents:

- The *main document,* which contains special merge fields that act as placeholders for the recipient's name and address
- The *data source,* which lists the specific recipient information (including the name, address, and any additional data such as the phone number) to be inserted in the merge fields

You can also create mailing labels or envelopes by using Word's mail merge feature.

Creating a Main Document

The Mail Merge task pane guides you through the three steps for completing a mail merge:

1. Create or identify the main document.
2. Create or identify the data source.
3. Merge the data source with the main document.

You can create the main document and data source after displaying the Mail Merge task pane, or you can use existing files. The first step is to identify the main document.

EXERCISE **17-1** **Select a Starting Document**

You can mail-merge different types of documents, including letters, e-mail messages, envelopes, and labels. In this exercise, your main document will be a business letter.

1. Start a new document.
2. Open the Tools menu and choose Letters and Mailings, Mail Merge. The Mail Merge task pane opens. The first step in mail merge is to select the document type. Choose Letters if it is not already selected.

> **NOTE:** A document must be open before you start a Mail Merge. If you have a document open and want to use it in the mail merge, you will start at Step 3 in the mail merge task pane. The Mail Merge task pane guides you through a mail merge in six steps.

3. At the bottom of the Mail Merge task pane, click the Next: Starting document link. (See Figure 17-1 on the next page.)

> **NOTE:** You can also perform a mail merge by using the Mail Merge toolbar. However, the Mail Merge task pane is easier and faster for simple mail merge projects.

4. Under Select starting document in the Mail Merge task pane, you have three choices: you can use the current document, a template, or an existing document. Choose the first option, Use the current document (if it is not already selected).
5. Format the current document as a business letter. Set a 2-inch top margin, and insert a date field at the top of the document, using the third format in the Date and Time dialog box. Make sure Update automatically is checked.

FIGURE 17-1
Mail Merge
task pane

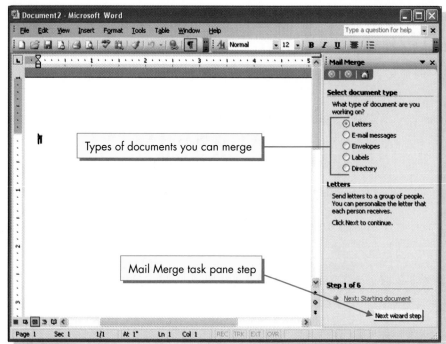

Types of documents you can merge

Mail Merge task pane step

6. Open the document header pane. Create a letterhead for Duke City Gateway Travel by keying the text shown in Figure 17-2. Use 11-point Arial, centered. To separate the phone and fax numbers, insert a symbol, such as the bullet character shown.

FIGURE 17-2
Creating letter from
current document

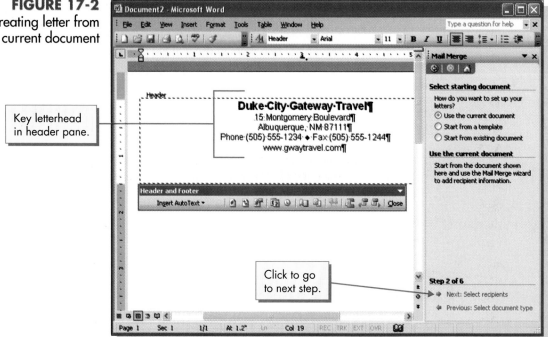

Key letterhead
in header pane.

Click to go
to next step.

7. Close the header pane, and then save the document as *[your initials]***17-1main** in a new folder for Lesson 17.

8. At the bottom of the task pane, click the next step, Next: Select recipients. Now you'll create your data source.

Creating a Data Source

A *data source* is a file that contains information such as names and addresses. The information is organized in a table. Each column of the table represents a category of information, such as last names. The column heading of each category is called a *field name*. Each row of the table represents a *record*, which is usually a person's name and his/her contact information. Each piece of information in a record is a *field*.

To perform a mail merge, you can use a preexisting data source in the form of a Word table, an Excel worksheet, an Access table, or an Outlook contact list. Or, you can create your own data source during the mail merge process, which creates an Access database file. To do this, you first define the data (fields) you need for each record (person), such as title, name, and address, and then you enter each item of information.

FIGURE 17-3
Sample data
source table

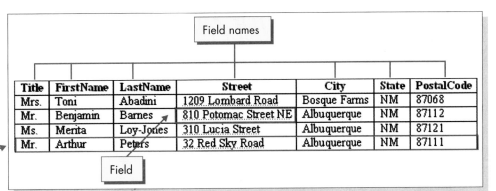

Title	FirstName	LastName	Street	City	State	PostalCode
Mrs.	Toni	Abadini	1209 Lombard Road	Bosque Farms	NM	87068
Mr.	Benjamin	Barnes	810 Potomac Street NE	Albuquerque	NM	87112
Ms.	Merita	Loy-Jones	310 Lucia Street	Albuquerque	NM	87121
Mr.	Arthur	Peters	32 Red Sky Road	Albuquerque	NM	87111

EXERCISE **17-2** **Create a Data Source**

Under Select recipients in the Mail Merge task pane, you have three choices: you can use an existing list, such as an Excel or Access data file; you can select from contacts you might have entered in Outlook; or you can type a new list.

1. Choose Type a new list in the Mail Merge task pane. You will create your own data source by entering the data needed for this mail merge, which creates an Access database file.

2. Under Type a new list, click the Create link. Word opens the New Address List dialog box. (See Figure 17-4 on the next page.)

FIGURE 17-4
Preparing to
create your own
data source

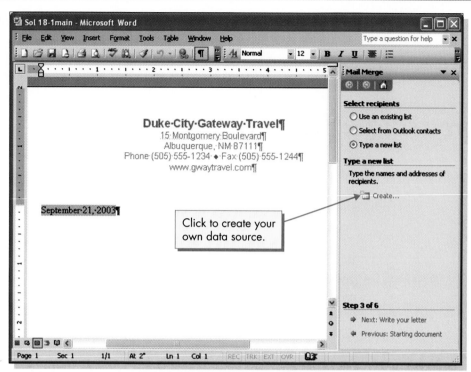

3. Use the scroll bar in the dialog box to see a list of commonly used field
   names. Because you won't be using many of these fields, you will
   customize this address list to include only the fields you need.

FIGURE 17-5
New Address List
dialog box

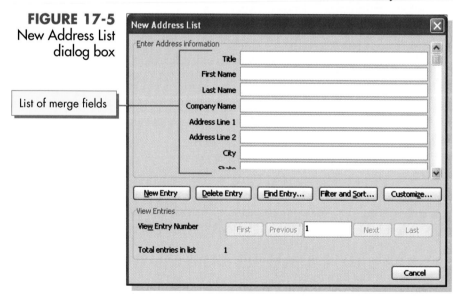

4. Click Customize. Word opens the Customize Address List dialog box,
   where you can delete, add, or rename field names and change their order.
   (See Figure 17-6 on the next page.)

5. In the list of field names, click Company Name. Click <u>D</u>elete. Click Yes when Word asks if you're sure you want it deleted. The field name is removed from the list.

6. Using the same technique, remove the following fields from the list: Address Line 2, Country, Home Phone, Work Phone, E-mail Address.

FIGURE 17-6
Customizing the list of address information

Field names you defined

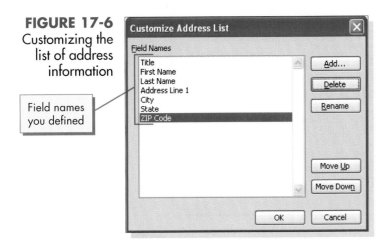

NOTE: Check that you have only the field names shown in Figure 17-6. If you don't, click Cancel and start again.

7. Click OK to close the Customize Address List dialog box. The New Address List dialog box lists only the fields you'll need for your mail merge letters. You are now ready to enter data.

NOTE: When preparing to enter data, it's best to customize the address list so you can quickly key each line of data and move from record to record without worrying about a missed or blank field.

EXERCISE **Enter Records in a Data Source**

After defining field names for your data source, you can create records for mail merging.

1. In the New Address List dialog box, key **Mr.** in the Title field and press (Tab) or (Enter). Notice that the Vie<u>w</u> Entry Number box is set to 1, which means you're entering record number 1.

TIP: To move from field to field within a record, press (Tab), (Enter), or (↓) to move to the next field and press (Shift)+(Tab) or (↑) to move to the previous field. You can also click the mouse.

2. Key the information shown in Figure 17-7 in the appropriate text boxes, pressing ⟨Tab⟩ or ⟨Enter⟩ after each field entry.

FIGURE 17-7
Entering data for record number 1

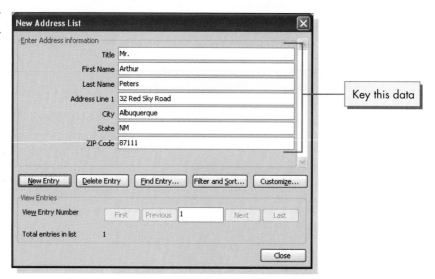

3. Click the <u>N</u>ew Entry button. The record is added to the data source and the blank fields appear for record 2.

TIP: If you press ⟨Enter⟩ after the last field, and then press ⟨Enter⟩ again, you'll display the next blank record. This is helpful when you're keying a lot of data quickly—you can keep your hands on the keyboard.

4. Key the data shown in Figure 17-8 in the appropriate field text boxes. Click <u>N</u>ew Entry (or ⟨Enter⟩) after you complete each record, including record 4.

FIGURE 17-8

Field names	Record 2	Record 3	Record 4
Title:	Ms.	Mr.	Mrs.
First Name:	Merita	Benjamin	Toni
Last Name:	Loy-Jones	Barnes	Abadini
Street:	310 Lucia Street	9854 Reigor Lane	1209 Lombard Road
City:	Albuquerque	Albuquerque	Bosque Farms
State:	NM	NM	NM
ZIP Code:	87121	87109	87068

5. Click the Fir<u>s</u>t button to view the first record you keyed.
6. Click the Ne<u>x</u>t button to view the second record you keyed.

7. In the Vie**w** Entry Number text box, key **3** and press (Enter) to display record 3 for editing. Change the address to **810 Potomac Street NE**. Change the ZIP Code to **87113**.

8. Click the **L**ast button to display the last record. If blank record 5 appears, click the **D**elete Entry button. Click Yes when asked if you want to delete it.

 TIP: You can use the **F**ind Entry button to find a record by name or other field.

9. Click Close to close the New Address List dialog box.

10. In the Save Address List dialog box, open your Lesson 17 folder, key *[your initials]***17-3data** as the filename, and click **S**ave. The Mail Merge Recipients dialog box appears, showing the mail merge database you created.

FIGURE 17-9
Mail Merge
Recipients
dialog box

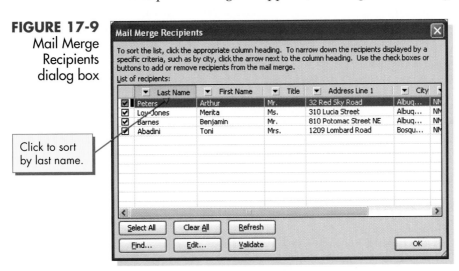

Click to sort
by last name.

11. Click the Last Name column heading (not the down arrow to the left of the column heading) to sort the list alphabetically by last name.

 NOTE: You can add or remove a recipient to include in your mail merge by selecting or clearing the check box next to the recipient's name.

12. Click OK to close the Mail Merge Recipients dialog box. Notice the addition of the Mail Merge toolbar at the top of your screen. You will use the toolbar later in the lesson.

Inserting Merge Fields into a Main Document

Now that you have created a data source, you can complete the main document by keying text and inserting placeholders for data called *merge fields*. Merge fields appear in the main document as *field codes,* which show the field name, such as

«Title». Mail merging replaces these fields with information from your data source, changing «Title» to Mr., for example.

When a field name has a space in its title (such as "Last Name"), Word displays the merge field in the document as «Last_Name», with an underscore for the space.

EXERCISE **17-4** **Insert Merge Fields into the Main Document**

1. In the Mail Merge task pane, click Next: Write your letter. Now you'll begin to assemble your letter.

2. Position the insertion point four lines below the date. In the Mail Merge task pane, click Address block. The Insert Address Block dialog box opens.

3. Clear the Insert company name check box, because you aren't using that field. The Preview box shows the field elements for the address block: title, first and last names, street address, city, state, ZIP Code, and country. Click OK. The address block is inserted in the document and contains the recipient's name and address.

FIGURE 17-10
Inserting the
address block

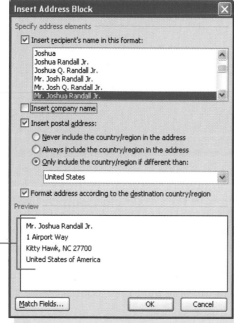

Address format
for merge letters

4. Position the insertion point two lines below the address block, and then click Greeting line in the Mail Merge task pane. The Greeting Line dialog box opens. (See Figure 17-11 on the next page.)

5. Change the comma in the greeting line format to a colon. The Preview box shows how the greeting will appear in your letters. The greeting includes "Dear," followed by the title field, the last name, and a colon. Click OK.

FIGURE 17-11
Inserting the
greeting line

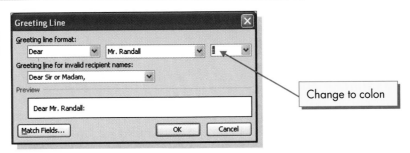

FIGURE 17-11
Inserting the
greeting line

6. Two lines below the greeting line, key the following opening paragraph:

   Enclosed are your airline tickets and the itinerary for your trip to Israel and Egypt. Please call me if you have any questions or concerns.

7. Press Enter twice to start a new paragraph.

8. In the Mail Merge task pane, click More items. The Insert Merge Field dialog box appears. You'll use this to insert individual fields into your letter.

9. With the Title field already selected, click Insert. Choose Last Name from the list of fields and click Insert. Click Close.

10. Click between «Title» and «Last_Name» and press Spacebar.

TIP: When you insert fields in a main document, make sure to include the correct spacing and punctuation.

11. Click after «Last_Name». Key a comma, insert a space, and then continue keying the paragraph with the text **I am sure you and your family will have a great time. Bon voyage!**

12. Press Enter twice and key **Sincerely,**.

13. Press Enter four times and key **Alexis Johnson**. Key **Travel Counselor** on the next line. Add your reference initials, followed by an enclosure notation.

14. Preview the letter in Print Preview. Because this is a very short letter, improve its appearance by changing the left and right margins to 2 inches, making the header 1 inch from the edge, and adding three blank lines before the date field.

15. Click the Save button 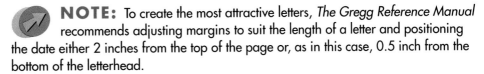 to save these changes to the main document.

NOTE: To create the most attractive letters, *The Gregg Reference Manual* recommends adjusting margins to suit the length of a letter and positioning the date either 2 inches from the top of the page or, as in this case, 0.5 inch from the bottom of the letterhead.

Performing a Mail Merge

Now that you've created both the main document and the data source, you can begin the mail merge. The mail merge will create one copy of the main document customized for each record. In each copy, the merge fields will be replaced by data from one record in the data source.

The simplest way to perform the mail merge is to:

● Preview the merged letters onscreen to see how they look with the merged data.

● Complete the merge by merging directly to the printer, or merge to a new document that you can save and print later.

EXERCISE **17-5** **Preview and Complete the Merge**

After you preview the merged letters and are satisfied with the results, complete the merge. You can merge all the records or a certain range of records, such as records 2 through 4.

1. In the Mail Merge task pane, click Next: Preview your letters.

FIGURE 17-12
Previewing
merged letters

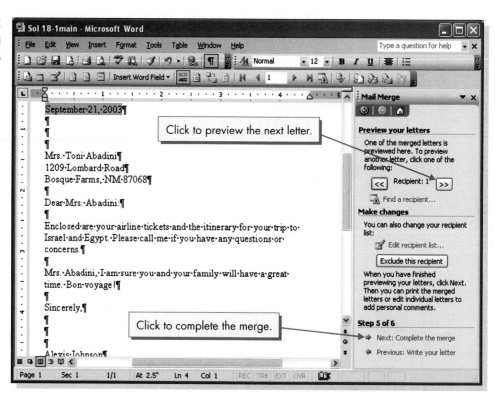

2. In the Mail Merge task pane, click the double-arrow button to the right of Recipient 1 to see the next letter. Continue to click the button to review each merged letter.

3. At the bottom of the Mail Merge task pane, click Next: Complete the merge. The task pane now offers two merge options: Print will send the merged letters directly to the printer; Edit individual letters will merge the letters to a new document that you can edit or print later.

4. Click the merge option Print.

5. In the Merge to Printer dialog box, you can choose to print all merged letters or specific records. Key **1** in the From box and key **2** in the To box to print just the first two merged letters. Click OK.

FIGURE 17-13
Preparing to print
merged letters

Specify records
to print.

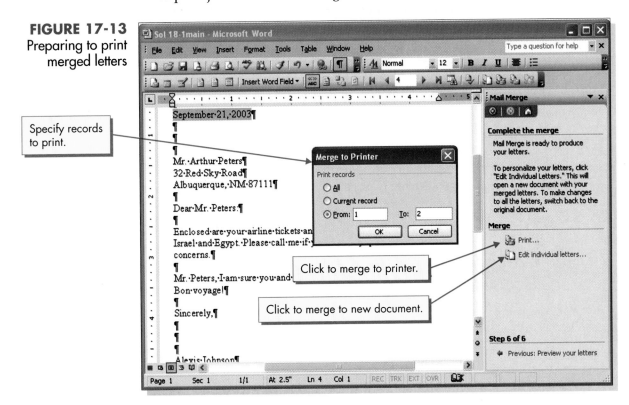

6. Click OK in the Print dialog box. Word prints records 1 and 2 as merged letters.

7. Click the merge option Edit individual letters in the Mail Merge task pane.

8. In the Merge to New Document dialog box, choose All and click OK. Word creates a new document, temporarily called Letters1, containing four merged letters. Scroll through the document to check the letters. Each letter appears as a separate section.

9. Save the merged document as *[your initials]***17-5merged** in your Lesson 17 folder.

10. Print pages 3–4 so you have a complete printout of all four letters.

REVIEW: Because each letter is a separate section, key **s3-s4** in the Pages box of the Print dialog box.

11. Close the merged document and the main document, saving changes.

Using Data from Other Applications

A data source can be a different file type, such as a Word table, an Excel worksheet, or an Outlook contact list. Using the Mail Merge task pane to create a data source creates an Access database file automatically. You can also use the Mail Merge toolbar and the Database toolbar to help you edit your mail merge files and perform all mail merge functions.

TIP: Generally, the Mail Merge task pane is best for creating mail merge documents, and the Mail Merge and Database toolbars are useful for managing existing files.

EXERCISE 17-6 Use Data from a Word Document

1. Open the file **Names2**. This is a Word table that will be used as a database in a mail merge.

2. Use the View menu to display the Database toolbar. The Database toolbar displays buttons specifically for managing database fields and records.

NOTE: Because Names2 was created as a Word document, you must open it in Word to change the fields. You can edit the table as you would any table, or you can use the Database toolbar buttons.

3. Insert a column on the far left of the table. In the top cell of the new column, key **Title**. This is a new field name you are adding to the data source file **Names2**.

4. Key **Mr.** as the title for the first two records and key **Dr.** as the title for the last record. (Remember to press ↓, not Enter, when going to the next cell in the column.)

FIGURE 17-14
Using and editing a Word data source file

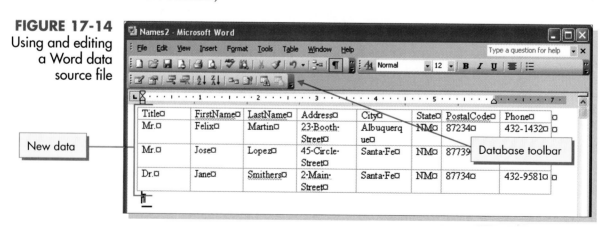

5. Click the Data Form button 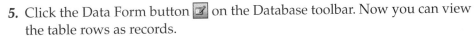 on the Database toolbar. Now you can view the table rows as records.

6. In the Data Form dialog box, click the Next Record button ▶ to display the record for Jose Lopez and change the address to **7534 Main Avenue**.

7. Display the next record and change the city for Jane Smithers to **Albuquerque**. Click Close. Notice the changes in the Word table.

TABLE 17-1   Database Toolbar Buttons

BUTTON	DESCRIPTION	FUNCTION
	Data Form	Opens the Data Form dialog box, which displays one record.
	Manage Fields	Deletes, renames, or inserts field names in the data source.
	Add New Record	Inserts a blank row at the bottom of the table for a new record.
	Delete Record	Deletes the record that contains the insertion point.
	Sort Ascending	Sorts data in a column in order from the beginning of the alphabet, the lowest number, or the earliest date.
	Sort Descending	Sorts data in a column in order from the end of the alphabet, the highest number, or the latest date.
	Insert Database	Inserts information from a data source as a table in your document.
	Update Field	Updates a selected field or record in the database.
	Find Record	Searches for specific data records within a selected field.
	Mail Merge Main Document	Switches to the main document associated with this data source.

8. Save the revised database as *[your initials]***17-6data** in your Lesson 17 folder. Close the document. Close the Database toolbar.

9. Open the file **Trip1**. This is a previously created main document with field codes.

10. Replace "xx" with your reference initials.

11. Display the Mail Merge toolbar, if it does not already appear.

12. Click the Open Data Source button ▦ on the Mail Merge toolbar.

13. In the Select Data Source dialog box, locate and open the file *[your initials]*17-6data. The file is now open for use with Trip1, but it is not visible.

14. Click the Mail Merge Recipients button 🖉 on the Mail Merge toolbar. The Mail Merge Recipients dialog box displays the data from the edited Word file (with the columns slightly reordered).

15. Click OK to close the Mail Merge Recipients dialog box.

16. Click the Merge to New Document button 🗐 on the Mail Merge toolbar. Click OK to merge all records. View the merged documents in Print Preview. Close Print Preview.

17. Save the merged document as *[your initials]*17-6merged in your Lesson 17 folder, print only the first letter, and close the document.

TABLE 17-2 Mail Merge Toolbar Buttons

BUTTON	DESCRIPTION	FUNCTION
	Main document setup	Formats a main document for a specific type of mail merge, such as letters or labels, or restores the document to a normal Word document.
	Open Data Source	Attaches an existing data source to the active document, making the active document a mail merge main document if it is not one already.
	Mail Merge Recipients	Displays a dialog box where you can sort, search, filter, add, and validate mail merge recipients.
	Insert Address Block	Provides different address formats (including title, name, and address) to insert into a main document for a mail merge.
	Insert Greeting Line	Provides different greeting line formats (including salutation, name, and punctuation) to insert into a main document for a mail merge.
	Insert Merge Field	Inserts merge fields from your data source or from a list of standard address fields into the main document.
Insert Word Field ▾	Insert Word Field	Inserts into the main document Word fields, such as Ask and Fill, to control how Word merges data.
«»ABC	View Merged Data	Displays the main document merged with information from the associated data source.
	Highlight Merge Fields	Applies background shading to mail merge fields in a main document to make them more visible.
	Match Fields	Displays a dialog box where you can match fields from a data source to fields in a main document that have different field names.

continues

TABLE 17-2 Mail Merge Toolbar Buttons *continued*

BUTTON	DESCRIPTION	FUNCTION
	Propagate Labels	Copies the field codes for one label into the rest of the labels in a mail merge main document.
	First Record	Displays the main document merged with information from the first record in the data source, if the View Merged Data button is clicked.
	Previous Record	Displays the main document merged with information from the previous record in the data source, if the View Merged Data button is clicked.
1	Go to Record	Shows the currently merged record number. To display the main document merged with a specific record, key the record number and press Enter.
	Next Record	Displays the main document merged with information from the next record in the data source, if the View Merged Data button is clicked.
	Last Record	Displays the main document merged with information from the last record in the data source, if the View Merged Data button is clicked.
	Find Entry	Searches the database for text contained in a specified mail merge field.
	Check for Errors	Reports errors in the main document or data source that prevent merging.
	Merge to New Document	Opens a new document with the mail merge results.
	Merge to Printer	Sends merged documents directly to the printer.
	Merge to E-mail	Sends merged documents to e-mail addresses.
	Merge to Fax	Sends merged documents to fax numbers.

EXERCISE 17-7 **Use an Excel Worksheet as a Data Source**

1. With **Trip1** still open, display the Mail Merge toolbar if it does not already appear.

 2. Click the Open Data Source button ▦ on the Mail Merge toolbar.

3. In the Select Data Source dialog box, locate and open the Excel file **TripData**. This is an Excel workbook that contains fields necessary for use with Trip1.

 NOTE: In the Files of type drop-down list, choose Excel Files to locate the file more quickly.

4. In the Select Table dialog box, click OK to use data from the first worksheet in the Excel workbook. The file is now open for use with Trip1, but it is not visible.

 5. Click the Mail Merge Recipients button on the Mail Merge toolbar. The Mail Merge Recipients dialog box displays the data from the Excel worksheet.

6. Click OK to close the Mail Merge Recipients dialog box.

7. Click the Merge to New Document button on the Mail Merge toolbar. Click OK to merge all records.

 NOTE: The field names in this letter match the field names in the data source. If you tried to merge this main document with a data source that had different field names, you would get errors. You could correct these errors through a matching fields process, but it's best to match a main document with the correct data source. Naming fields consistently is also a good idea.

8. Preview the merged document.

9. Save the merged document as *[your initials]***17-7merged** in your Lesson 17 folder. Print only the first letter, and then close the document.

EXERCISE **17-8** **Use Outlook Data as a Data Source**

1. If you already have at least two Outlook contacts created that contain a title, first name, last name, street address, city, state, and ZIP Code, you can move on to step 8. Otherwise, follow steps 2 through 7 to create two Outlook contacts.

 2. Open Microsoft Outlook. In the Outlook Bar (on the left of the Outlook window), click the Contacts shortcut.

3. Click New on the Standard toolbar. In the Untitled-Contact dialog box, display the General tab.

4. Click the Full Name button. In the Check Full Name dialog box, choose a Title, key a first name in the First field and a last name in the Last field, and click OK.

5. Click the Business button in the Addresses section. In the Check Address dialog box, key a street name, city, state, and ZIP Code. Click OK.

6. Save and Close the contact.

7. Repeat steps 3 through 6 to add another contact; then close Outlook.

8. With **Trip1** still open, display the Mail Merge toolbar, if it does not already appear.

9. Open the Tools menu and choose Letters and Mailings, Mail Merge. The Mail Merge task pane opens the Mail Merge task pane at Step 3 of 6.

10. Under Select recipients, click Select from Outlook contacts.

11. Under Select from Outlook contacts, click Choose Contacts Folder.

12. If the Choose Profile dialog box appears, click OK to accept Outlook as the Profile Name.

13. In the Select Contact List folder dialog box, click OK to choose Contacts Personal Folders.

 NOTE: Your contacts might be located in a different folder. If you cannot locate your contacts, ask your instructor for help.

14. In the Mail Merge Recipients dialog box, click OK to choose the recipients from Outlook.

15. Click the Merge to New Document button on the Mail Merge toolbar and click Current record. Click OK to merge the current record.

 NOTE: The field names in this letter do not match the field names in the Outlook data source. You must correct these errors through a matching fields process.

16. In the Invalid Merge Field dialog box, replace the FirstName field with First from the drop-down list. Click OK. Replace LastName with Last from the drop-down list. Click OK. Replace PostalCode with ZipPostal_Code. Click OK. Click OK.

17. Save the merged document as *[your initials]***17-8merged** in your Lesson 17 folder. Print and close the document.

18. Close Trip1 without saving changes.

Editing an Existing Main Document

After you have created a main document, you can edit and reuse it. In the main document, you might want to change the text or add a new field from the data source. In a previous exercise, you edited the Word data source table that you will use now.

EXERCISE 17-9 **Edit an Existing Main Document**

1. Reopen **Trip1** and click to the left of «FirstName» in the address block. Click the Open Data Source button on the Mail Merge toolbar.

2. Locate **17-6data** and open it. Click the Insert Merge Fields button . Notice that Title is now the first field because you added it to the database file earlier in this lesson when you edited the Word table.

3. Click Insert. Click Close, and then insert a space after the title field.

FIGURE 17-15
Adding the Title
merge field to the
main document

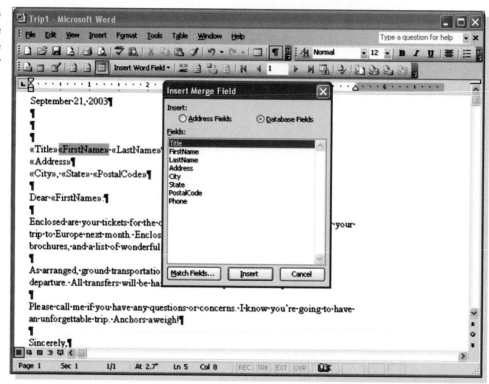

4. Edit the second paragraph by replacing "pick you up" with **arrive**.

5. Click the View Merged Data button on the Mail Merge toolbar. Use the Next Record button ▶ to preview each letter. Notice that the titles are included.

6. Click the View Merged Data button again to return the display to field codes.

7. Save the main document as *[your initials]***17-9main** in your Lesson 17 folder and print it.

8. Click the Merge to New Document button 🗐 on the Mail Merge toolbar. Click OK to merge all records.

9. Preview the three-page document. Print the merged documents using the pages per sheet print option to print four pages per sheet.

10. Close the merged document without saving it. Close the main document saving changes.

Sorting and Filtering a Data Source

At times, you might want to sort your data source before merging with your main document. You can also filter the data so that only records with certain characteristics are merged.

EXERCISE 17-10 Sort a Data Source

1. Open **Trip2**, change the top margin to 2 inches, and add a date line to the letter containing the current date.
2. Click the Open Data Source button 🖹 on the Mail Merge toolbar. Locate and open the file **Names3**.
3. Click the Mail Merge Recipients button 🗹.
4. Click the LastName column heading (not the down arrow to the left of the column heading) to sort the list alphabetically by last name.
5. Click the down arrow next to the LastName column heading. Choose (Advanced...) to open the Query Options dialog box.
6. Click the Sort Records tab. In the Then by drop-down list, choose FirstName. Notice that the default order is Ascending.
7. Click OK. The records are sorted first by last name, and then by first name.

EXERCISE 17-11 Filter a Data Source

1. Click the down arrow next to the LastName column heading.
2. Choose (Advanced...) to open the Query Options dialog box.

 NOTE: You can click the down arrow next to any of the field names to access the Query Options dialog box. Also, you can click the down arrow next to a field name and choose one of the fields listed. This filters the data based on the field you choose.

3. Display the Filter Records tab if it is not active. In the first Field text box, choose LastName from the drop-down list.
4. Make sure the Comparison text box is set to Equal to.
5. In the Compare to text box, key **Albert**.
6. In the text box below and to the left of LastName, choose And from the drop-down list if it is not selected. (See Figure 17-16 on the next page.)

7. In the Field text box to the right of And, choose PostalCode from the drop-down list.

8. In the Comparison text box, choose Greater than.

9. In the Compare to text box, key **87200**.

FIGURE 17-16
Filtering data by using comparisons

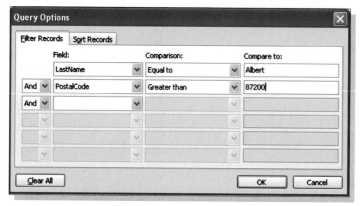

10. Click OK to filter the data—Word looks for records with the last name "Albert" and a postal code greater than "87200" and finds one record. Click OK again to accept the filtered data.

11. Add your reference initials to the letter and make "Enclosure" plural (more than one item will accompany the letter).

12. Save the document as *[your initials]***17-11main** in your Lesson 17 folder.

13. Click the Merge to New Document button. Click OK. The record for "Donna Albert" is merged with the main document to create a new document.

14. Save the document as *[your initials]***17-11merged** in your Lesson 17 folder.

15. Print the letter and close all open documents, saving changes if prompted.

Creating Lists and Mailing Labels

You can merge data from a data source to create a list or a directory. You can also create mailing labels from a document to address envelopes and packages. Word enables you to designate the style of the label and insert the merge fields for the addresses.

EXERCISE 17-12 **Create a List from a Data Source**

To create a list from a data source, you can use the Mail Merge toolbar or the Mail Merge task pane.

1. Start a blank document and display the Mail Merge toolbar.

2. Click the Main document setup button ▣ on the Mail Merge toolbar.
3. Choose Directory from the list and click OK.
4. Locate and open **Names3** as a data source.
5. Open the Mail Merge Recipients dialog box and sort the data source alphabetically by LastName. Click OK.
6. Click the Insert Merge Fields button ▣, choose LastName from the list of fields in the Insert Merge Field dialog box, and click Insert.
7. Choose PostalCode from the list, click Insert, and then click Close. Both fields are inserted in the document.
8. Insert a 1.5-inch left tab between the two fields and insert a new paragraph mark after the PostalCode field.

> **NOTE:** You must insert a paragraph mark after the PostalCode field or the last field in the paragraph. Otherwise, each record will be listed one after the other in one paragraph.

9. Click the Merge to New Document button ▣ and allow all the records to be merged.
10. Save the merged list as *[your initials]***17-12merged** in your Lesson 17 folder and print it.
11. Close only the merged document.

EXERCISE 17-13 Create a Catalog-type Main Document

Not only can you merge data to create a list, but you can also create more extensive catalog-type documents or directories.

1. With your insertion point directly after the LastName field, click the Insert Merge Fields button ▣.
2. Choose FirstName from the list and click Insert.
3. Choose State from the list and click Insert.
4. Click Close to close the Insert Merge Fields dialog box.

EXERCISE 17-14 Add Formatting to a Catalog-type Main Document

1. Insert tabs before each of the fields except the first field.
2. Select the paragraph containing the fields, clear the 1.5-inch tab, and set 1.75-inch, 3-inch, 3.75-inch, and 4.75-inch left tabs.
3. Select the «LastName» field and make it bold.
4. Apply numbering to the paragraph containing the fields.

5. Apply double-line spacing to the paragraph containing the fields.
6. Set a 2-inch top margin and apply a single line page border to the document.

EXERCISE 17-15 Use an If...Then...Else Field to Refine a Merge

The If...Then...Else field compares two values, and then inserts the text appropriate to the result of the comparison as the field result in your document. For example, the statement "If the weather is good, we'll go to the zoo; if not, we'll go to the movies" specifies a condition that must be met (good weather) for a certain action to take place (going to the zoo). If the condition is not met, an alternative action occurs (going to the movies).

1. Place the insertion point immediately to the right of the «PostalCode» field and press Tab.
2. Click Insert Word Field on the Mail Merge toolbar.
3. Choose If...Then...Else from the menu list. The Insert Word Field: IF dialog box opens.
4. Under Field name, choose City from the drop-down list.
5. Under Comparison, choose Equal to.
6. In the Compare to text box, key **Albuquerque**.
7. In the Insert this text text box, key **Local**.
8. In the Otherwise insert this text text box, key **Not local**.
9. Click OK and the field is inserted in the document.
10. Save the document as *[your initials]*17-15main in your Lesson 17 folder.
11. Click the Merge to New Document button and allow all the records to be merged. All records with "Albuquerque" as the city are labeled "Local." All others are labeled "Not local."
12. Create a title at the top of the catalog directory: **Catalog Directory of Frequent Traveler Members in New Mexico**. Format it as 14-point Arial bold, centered. (It will also be double-spaced, and you will need to remove line numbering.)
13. Change the paragraph spacing of the title to 12 points after.
14. Save the merged catalog directory as *[your initials]*17-15merged in your Lesson 17 folder and print it.
15. Close all open documents, saving changes if prompted.

EXERCISE 17-16 Create Mailing Labels

You can use the Mail Merge task pane or the Mail Merge toolbar to create labels. In this exercise, you will use the Mail Merge task pane.

The process of creating mailing labels requires that you:

- Create a mailing label main document.
- Choose a data source.
- Specify label size and type.
- Insert merge fields.
- Merge the main document and the data source.

1. Start a blank document.
2. Choose Tools, Letters and Mailings, Mail Merge.
3. In the task pane, choose Labels as the document type.

 TIP: To create envelopes instead of labels, choose Envelopes as the document type, and then follow the Mail Merge steps.

4. Click Next: Starting document.
5. Under Change document layout, click Label options.
6. In the Label Options dialog box, choose Avery standard from the Label products drop-down list.
7. Scroll the Product number list and choose 5160 – Address. Click OK. The document is now formatted for labels.
8. In the task pane, click Next: Select recipients, and then click Browse.
9. In the Select Data Source dialog box, locate and open **Names1**.
10. In the Mail Merge Recipients dialog box, display the last column and click the column heading, PostalCode. The records are now sorted by postal code (ZIP Code).
11. Click OK. Word adds the merge field «Next Record» to your labels.
12. In the task pane, click Next: Arrange your labels. Using the document's horizontal scroll bar, scroll to the left so that you can see the first label.
13. Click Address block in the task pane and click OK in the Insert Address Block dialog box. Word adds the «Address Block» field to the first label. (See Figure 17-17 on the next page.)
14. Scroll to the bottom of the task pane by pointing to the down arrow at the bottom of the task pane. Click Update all labels. Word automatically adds the «Address Block» field to each label.
15. Click Next: Preview your labels. Word merges the data from **Names1** with your label document. Show table gridlines (Table, Show Gridlines), if they do not already appear.
16. Click the final Mail Merge step, Next: Complete the merge. (See Figure 17-18 on the next page.)
17. Save the merged labels as *[your initials]*17-16labels in your Lesson 17 folder.

FIGURE 17-17
Adding the
Address block field

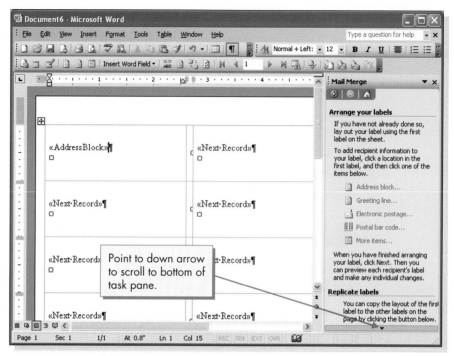

18. Click Print under Merge in the task pane. Prepare your printer with the correct label sheet or use a blank sheet of paper.

19. In the Merge to Printer dialog box, click OK. In the Print dialog box, click OK. Close and save the document.

FIGURE 17-18
Merged labels

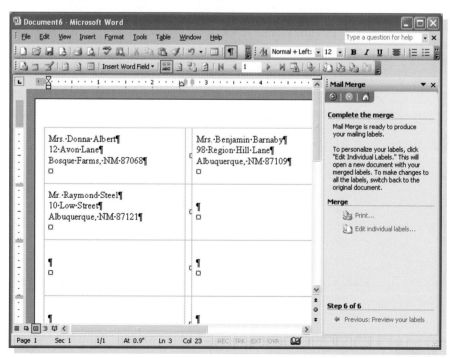

TIP: You can also use the Mail Merge toolbar to merge labels. Open a blank document, click the Main document setup button , and choose Labels. Set the label size. Click the Open Data Source button and choose the data file. Use the Insert Address Block button to insert the address block in the first label. Click the Propagate Labels button, and then click the View Merged Data button.

USING ONLINE HELP

FIGURE 17-19
Help about troubleshooting mail merges

Mail merge is an extremely valuable, yet complex feature. Help is available to troubleshoot any problems you encounter.

Troubleshoot mail-merge problems in Help:

1. In the Ask a Question box, key **mail merge** and press Enter.
2. Click the topic Troubleshoot mail merge.
3. Review the information in the Help window.
4. Close Help when you finish.

LESSON 17 Summary

➤ Mail merging combines a main document (such as a form letter) with a data source (such as a list of names and addresses) to produce individualized documents. The main document contains merge fields. The data source is a table—column headings are field names and rows are records. Each cell of a record is a field that appears in the matching merge field of the main document.

➤ The main document is a Word document; the data source can be a table in Word, an Excel worksheet, an Access database, or an Outlook contact list.

➤ Use the Mail Merge task pane to step you through the mail merge process and create the mail merge documents. Use the Mail Merge toolbar and Database toolbar to manage mail merge documents.

➤ You can create a new data source in one of the Mail Merge task pane steps by keying records into a data form called the New Address List dialog box. You can customize the fields in this dialog box to match the fields you'll use in your main document. A data source created by using the task pane produces an Access database file.

➤ You insert merge fields in a main document individually or as field blocks, such as Word's built-in address block.

➤ You can preview the data merged into your main document before performing the merge. Complete the merge by merging directly to a printer (or merging to e-mail addresses or fax numbers) or to a new document.

➤ At any point before or after merging, you can edit mail merge documents. Edit the main document or the data source file (if it is a Word table) as you would any Word document.

➤ Use the Database toolbar when editing a data source file that is a Word table. See Table 17-1.

➤ To create a main document by using the Mail Merge toolbar, click the Main document setup button 🔲. To open a data source, click the Open Data Source button 🔲. To view or edit the contents of a data source, click the Mail Merge Recipients button 🔲. To insert a merge field in the main document, click the Insert Merge Field button 🔲. To view merged data, click the View Merged Data button 🔲. See Table 17-2.

➤ You can sort and filter data source files to produce only the merged files you want. Use more than one field for advanced sorting capability.

➤ You can merge data from a data source to create a list or a catalog-type document or directory. To do this, you can use the Mail Merge task pane.

➤ You can create mail merge envelopes or mailing labels. Choose envelopes or labels as the main document type in the Mail Merge task pane, choose or create a data source, choose envelope or label options, insert merge fields into the main document, and then perform the merge.

LESSON 17 Command Summary

FEATURE	BUTTON	MENU	KEYBOARD
Mail Merge		Tools, Letters and Mailings, Mail Merge	

Concepts Review

Each of the following statements is either true or false. Indicate your choice by circling T or F.

T   F   **1.** Field names are the column headings in a data source table.

T   F   **2.** A field contains several pieces of unrelated information.

T   F   **3.** Mail merging inserts the information from the merged document into copies of the data source.

T   F   **4.** Each record in a data source table must contain the same number of fields.

T   F   **5.** A row in a data source table is called a record.

T   F   **6.** You cannot edit an existing main document.

T   F   **7.** When you create a data source by using the Mail Merge task pane, you create an Access database.

T   F   **8.** Data source files can only be Word documents.

Write the correct answer in the space provided.

1. What serves as a placeholder in the main document for information found in the data source?

_____

2. List an example of a document type for a mail merge main document.

_____

3. What are the two major documents used in mail merging?

_____

4. What is a record composed of?

_____

5. Which toolbar can you use to add and delete records in an active data source?

_____

6. What is the procedure for sorting a list of names in the Mail Merge recipients dialog box?

_____

7. What is the mail merge field called that contains a salutation, a name, and punctuation for letters?

_____

8. What do you click in the New Address List dialog box to modify the list of field names?

_____

CRITICAL THINKING

Answer these questions on a separate page. There are no right or wrong answers. Support your answers with examples from your own experience, if possible.

1. Businesses use a mail merge process to send personalized letters for fund-raising or sales promotions. For a business, what are the advantages and disadvantages of sending personalized mail of this type?

2. Have you or a member of your family ever received a personalized mailing from a business? Describe the reaction you had to receiving it.

Skills Review

EXERCISE 17-17

Create a main document and a data source, insert merge fields, and perform a mail merge.

1. Create a memo main document by following these steps:
   a. Create a new blank document and change the top margin to 2 inches. This document will be a memo.
   b. Open the Tools menu and choose Letters and Mailings, Mail Merge.
   c. In the Mail Merge task pane, choose Letters, if it is not already selected.
   d. Click Next: Starting document.
   e. Make sure Use the current document is selected.

2. Create a data source by following these steps:
   a. Click Next: Select recipients.
   b. Choose Type a new list and click Create.
   c. Click the Customize button.

  d. With the first field name selected (Title), click the Rename button and key **DateSent**. Click OK.

  e. Select the second field name (First Name), click Rename, and key **FullName**. Click OK.

  f. Rename the third field name, Last Name, as **Leaves**.

  g. Rename Company Name as **Returns**.

  h. Delete the remaining fields by choosing each one and clicking Delete. (Hint: Start by deleting the last field so you won't accidentally delete the Returns field.)

  i. Click OK.

  j. Key the data shown in Figure 17-20 (entering the current year where "20--" is shown). Click New Entry after completing each record.

FIGURE 17-20

DateSent	FullName	Leaves	Returns
March 6, 20--	Frank Murillo	March 29	April 6
March 30, 20--	Nina Chavez	April 16	April 27
April 28, 20--	Susan Allen	May 9	May 14
May 4, 20--	Tom Carey	May 18	May 27

  k. Click Close to return to the main document.

3. Save the data source as *[your initials]***17-17data** in your Lesson 17 folder.

4. Delete blank record 5 by following these steps:

  a. In the Mail Merge Recipients dialog box, click in blank row 5 (which should contain a check mark box).

  b. Click the Edit button.

  c. Click the Delete Entry button. Click Yes to confirm the deletion.

  d. Click Close. Click OK to close the Mail Merge Recipients dialog box.

5. In the Mail Merge task pane, click Next: Write your letter.

6. With the insertion point at the top of the document, set a 1-inch left tab. Key the memo heading shown in Figure 17-21. Use double-spacing.

FIGURE 17-21

```
MEMO TO:      All Branch Offices
FROM:         Frank Youngblood
DATE:
SUBJECT:      Vacation Schedule of Duke City Gateway Travel
```

7. Insert a merge field into the document by following these steps:

  a. Insert a tab character after "DATE:".

  b. In the Mail Merge task pane, click More items.

  c. In the Insert Merge Field dialog box, choose the field DateSent and click Insert. Click Close.

  NOTE: You can also open the Insert Merge Field dialog box by clicking the Insert Merge Field button 🗐 on the Mail Merge toolbar.

8. Key the body of the memo as shown in Figure 17-22. Use correct spacing after the subject line and between paragraphs. Insert merge fields in place of field names by using the Insert Merge Fields dialog box. Include spaces around merge fields where needed.

FIGURE 17-22

```
Please be advised that «FullName» will be out of the office from
«Leaves» through «Returns». All calls to «FullName» should be directed
to me.
If you are unable to reach me by telephone, please send me an e-mail
message or a fax. Thank you.
```

9. Add your reference initials.

10. Save the main document as *[your initials]*17-17main in your Lesson 17 folder.

11. Preview the merged data by following these steps:

  a. In the Mail Merge task pane, click Next: Preview your letters.

  b. Click the double-arrow button to the right of Recipient: 1 to display each merged record.

  c. Make any corrections to the main document that might be needed and resave the document.

12. Complete the merge by following these steps:

  a. In the Mail Merge task pane, click Next: Complete the merge.

  b. Click Edit individual letters to merge the data to a new document.

  c. In the Merge to New Document dialog box, choose All and click OK.

13. Save the merged document as *[your initials]*17-17merged in your Lesson 17 folder and print four pages per sheet.

14. Close all open documents, saving changes if prompted.

EXERCISE 17-18

Use and edit a data source from Word, edit an existing main document, sort and filter data, and merge the documents.

1. Edit an existing Word data source and add a new record by following these steps:

   a. Open the file **Travelers1**. This is a Word table used as a data source.

   b. Display the Database toolbar.

   c. Change the abbreviation "St." to **Street** wherever it occurs.

   d. Change the "City" of Mrs. Toni Abadini to **Albuquerque**.

    e. Insert a new row to the bottom of the table or click the Add New Record button on the Database toolbar.

   f. Key the data shown in Figure 17-23.

FIGURE 17-23

Title:	Ms.
FirstName:	Linda
LastName:	Del-Reo
Address:	12 East Main Street
City:	Santa Fe
State:	NM
PostalCode:	87345

2. Add a new field by following these steps:

   a. Insert a new column before the "Address" column. Key the column heading **Company** at the top of the new column.

   b. Key the company names shown in Figure 17-24 directly into the table.

FIGURE 17-24

LastName	Company
Peters	Franklin Cable Works
Loy-Jones	Millcreek Consultants
Barnes	Oakwood Planning Association
Abadini	B & G Enterprises
Del-Reo	Old Town Café

3. Center the table horizontally on the page. Save the file as *[your initials]*17-18data in your Lesson 17 folder. Print and close the document.

4. Open the file **Trip2**.

5. Insert the date field at the top of the document, with three blank lines below it. Set a 2-inch top margin and 1.5-inch left and right margins.

6. Use the Mail Merge toolbar to choose a data source by following these steps:

   a. Click the Open Data Source button [image].
   b. Locate and choose *[your initials]*17-18data as the data source file.

7. Modify the main document by following these steps:

   a. Insert a new line above «Address».
   b. Click the Insert Merge Fields button [image] on the Mail Merge toolbar, choose the field Company, and click Insert. Close the dialog box.
   c. Add your reference initials to the main document and make "Enclosure" plural (more than one item will accompany the letter).

8. Save the document as *[your initials]*17-18main in your Lesson 17 folder.

9. Use the Mail Merge toolbar to sort and filter data by following these steps:

   a. Click the Mail Merge Recipients button [image].
   b. Click the LastName column heading to sort by last name.
   c. Click the down arrow to the left of LastName and choose (Advanced…).
   d. Click the Sort Records tab and click Descending next to the Sort by text box. Make sure the Then by text box is empty. Click OK.
   e. Click the down arrow next to the LastName column heading again.
   f. Choose (Advanced…) to reopen the Query Options dialog box.
   g. Click the Filter Records tab if it is not active.
   h. In the first Field text box, choose City from the drop-down list.
   i. In the Comparison text box, choose Equal to if it is not selected.
   j. In the Compare to text box, key **Albuquerque**.
   k. In the text box below and to the left of the City, choose And from the drop-down list.
   l. In the Field text box next to And, choose PostalCode from the drop-down list.
   m. In the Comparison text box, choose Less than.
   n. In the Compare to text box, key **87100**.
   o. Click OK twice.

10. Use the Mail Merge toolbar to preview and complete the merge by following these steps:

    a. Click the View Merged Data button [image] to see the data merged with the main document.
    b. Click the button again to see the fields.

    c. Click the Merge to New Document button [image].
    d. With All selected, click OK to merge the record.

11. Save the merged document as *[your initials]***17-18merged** in your Lesson 17 folder.

12. Print the letter and close all open documents, saving changes if prompted.

EXERCISE 17-19

Create a catalog-type main document, format the document, and refine a merge by using If...Then...Else.

1. Create a catalog-type main document by following these steps:

   a. Start a blank document and display the Mail Merge toolbar.

   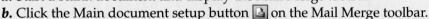*b.* Click the Main document setup button 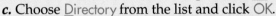 on the Mail Merge toolbar.

   c. Choose Directory from the list and click OK.

   d. Locate and open **Clients** as a data source. Use the Contacts table.

   e. Open the Mail Merge Recipients dialog box and sort the LastName field alphabetically. Click OK.

   f. Click the Insert Merge Fields button 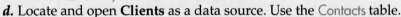, choose LastName, and click Insert.

   g. Choose FirstName from the list and click Insert.

   h. Choose Phone from the list, click Insert, and then click Close.

   i. Press (Enter) to insert a blank paragraph mark after the last field.

2. Format the document by following these steps:

   a. Set 2-inch, 3.5-inch, and 5.5-inch left tabs in the fields line.

   b. Insert a tab before the «FirstName» and «Phone» fields.

   c. Apply numbering to the field paragraph.

   d. Apply 1.5-line spacing to the field paragraph.

   e. Make the «Phone» field italic.

3. Insert an If...Then...Else field by following these steps:

   a. With the insertion point to the right of «Phone», turn off italic formatting and press (Tab).

   b. Click Insert Word Field on the Mail Merge toolbar.

   c. Choose If...Then...Else.

   d. Under Field name, choose State from the drop-down list.

   e. Under Comparison, choose Equal to if it is not active.

   f. In the Compare to text box, key **NM**.

   g. In the Insert this text text box, key **No**.

   h. In the Otherwise insert this text text box, key **Yes**.

   i. Click OK.

4. Save the document as *[your initials]***17-19main** in your Lesson 17 folder.

5. Click the Merge to New Document button  and allow all the records to be merged.

6. Create a title at the top of the catalog directory: **Tour Participants Needing New Mexico Hotel Accommodations**. Format it as 14-point Arial bold, small caps, centered. (Remove the line numbering.)

7. Insert one blank paragraph mark after the title. Set a 1.5-inch top margin.

8. Save the merged catalog directory as *[your initials]***17-19merged** in your Lesson 17 folder and print it.

9. Close all open documents, saving changes if prompted.

EXERCISE 17-20

Create labels.

1. Create a main document for labels that will be used as name badges by following these steps:

   a. Start a new document and open the Mail Merge task pane.
   b. Choose Labels. Click Next: Starting document.
   c. Click Label options.
   d. In the Label Options dialog box, choose Avery standard from the Label products drop-down box.
   e. Choose 5383 - Name Badge from the Product number drop-down list and click OK.

2. Go to the next Mail Merge step. Click Browse to locate and open the data source **Names1**. Click OK to close the Mail Merge Recipients dialog box.

3. Insert merge fields for a name badge by following these steps:

   a. Scroll in the main document to display the top left label (where the insertion point is).
   b. Go to the next Mail Merge step (Arrange your labels).
   c. Use the More items option to open the Insert Merge Field dialog box and insert the fields «FirstName» and «LastName». Close the dialog box, and then insert a space between the fields.
   d. Scroll to the bottom of the task pane and click Update all labels.

4. Go to the next Mail Merge step to preview your labels.

5. Go to the next Mail Merge step to complete the merge.

6. In the merged document, select the text in the table and format it as 24-point, bold, centered. Apply a 1/2-point Grid border to the four name badges that contain names.

7. Save the merged document as *[your initials]***17-20merged** in your Lesson 17 folder.

8. Print the document. Close all open documents without saving.

Lesson Applications

Create a main document and a data source, and merge the documents.

1. Create a new form-letter main document with a 2-inch top margin.
2. Create a data source with the following field names: **SalesRep**, **Date**, **Contact**, **Company**, and **Commit**.
3. Key the two records shown in Figure 17-25. In the "Date" row, enter today's date for both records.

FIGURE 17-25

Field name	Record 1	Record 2
SalesRep	Doris Simms	Doris Simms
Date	[today's date]	[today's date]
Contact	Ethel Lewis, President	Mark Hunter, V.P. of Marketing
Company	Congregational Sisterhood	Taos Hotels, Inc.
Commit	No commitment was made by Lewis at this time.	Hunter will speak with MIS and accounting divisions next week.

4. Save the data source as *[your initials]***17-21data** in your Lesson 17 folder.

 NOTE: Be sure to check for, and delete, any blank records that might exist in the data source.

5. In the main document, key the following heading followed by three blank lines:

   Duke City Gateway Travel
   15 Montgomery Boulevard
   Albuquerque, NM 87111

6. Format the heading as 14-point bold, centered, small caps with a 1-point shadow paragraph border and 10% gray shading.
7. On the third blank line below the heading, key **CALL REPORT** as uppercase bold, centered.
8. Press (Enter) three times, left-align the paragraph mark, if necessary, and set a 2.25-inch hanging indent.
9. Key the text and insert the merge field codes shown in Figure 17-26. Insert one blank line after each line. The merge field codes should align at the 2.25-inch indent.

FIGURE 17-26

```
Sales Representative:        «SalesRep»
Date of Contact:             «Date»
Name and Title of Contact:   «Contact»
Company/Organization:        «Company»
Commitment:                  «Commit»
```

10. Save the main document as *[your initials]***17-21main** in your Lesson 17 folder.

11. Preview the merged data. Correct the main document and data source, if necessary, and save any changes.

12. Complete the merge by merging both records to a new document.

13. Save the new document as *[your initials]***17-21merged** in your Lesson 17 folder.

14. Print two pages per sheet. Close all open documents, saving changes if prompted.

EXERCISE 17-22

Create a main document and use your Outlook contacts as a data source.

1. Open Microsoft Outlook and create two new contacts if you do not have any contacts available to use as a data source. In your contacts, include the following:
   - First and last name
   - Street address
   - City, state, ZIP Code

2. Return to Word and open the file **Trip3**.

3. Open the document header pane. Create a simple letterhead for Duke City Gateway Travel and make sure it is 1 inch from the top of the page. You can use the letterhead you created earlier in this lesson, making sure you include the phone number and Web address.

4. Because this will be a short letter, insert blank lines above the date to position it at about 3 inches from the top of the page (check the Status bar) and change the left and right margins to 2 inches.

5. Replace "xx" with your reference initials at the bottom of the page.

6. Save the document as *[your initials]***17-22main** in your Lesson 17 folder and print it.

7. With the Mail Merge task pane, use your Outlook contacts as your data source.

8. Using the Mail Merge toolbar, sort the records by ZIP Code and merge them all to a new document.

9. Save the document as *[your initials]***17-22merged** in your Lesson 17 folder. Print the first two letters (if you have two or more merged letters) two pages per sheet.

10. Close all documents, saving any changes.

EXERCISE 17-23

Edit existing mail merge documents, filter data, and perform a mail merge.

1. Open the file **Invoice1**. Change the top margin to 1 inch.

2. Format the heading (the first six lines) as bold, centered. Under the heading, change the word "Invoice" to 14-point uppercase, centered.

3. Edit the rest of the document as follows:

    - Replace each colon (:) with a tab character.
    - Convert the paragraphs from "Date" through "Total Due" (including the tabs) into a two-column table by selecting the text and clicking the Insert Table button on the Standard toolbar.
    - Make the first column 1.75 inches wide and the second column 2.5 inches wide.
    - Make the first-column text bold.
    - Change the row height for all rows to at least 0.5 inch.
    - Center the table horizontally on the page and center the cell contents vertically.
    - Apply 1-point gridlines inside the table and a 3-point double-line (one thick, one thin) outside border to the table.
    - Move the "Ticket Number" row to just under the "Invoice Number" row.
    - In the first row, second column, insert the date as an automatically updating field, using the December 25, 2004, format.
    - Delete the "City/State/ZIP Code" row and insert three new rows labeled **City**, **State**, and **ZIP Code**.

4. Save the document as *[your initials]***17-23main** in your Lesson 17 folder.

5. Open the file **Accounts1**. This file will be the data source for the main document. Edit the table as shown in Figure 17-27 (on the next page). Change the page orientation to landscape to make space for the extra columns created by reorganizing the data in the Address2 field. (Hint: To avoid rekeying data, create the new columns, and drag and drop data to the new cells).

6. Use the AutoFit to Contents option to set the width of the columns.

7. Save the data source as *[your initials]***17-23data** in your Lesson 17 folder and print it. Close the document.

8. Switch to your main document and choose *[your initials]***17-23data** as its data source.

9. Edit the main document to insert the appropriate merge fields from the data document in the second column of the table.

10. Save changes to the main document and print it.

FIGURE 17-27

Break data in Address2 into
three fields: City, State, and ZIP.

AcctNo	InvNo	Name	Street ~~Address1~~	Address2	TktNo	Total
2037L	797411	Nicole R. Sanchez	798 Armden Drive	Chama, NM 87520	33679085	$598.~~00~~
2943L	1856429	Emery Ellis	2720 San Pedro ~~Dr.~~ Drive	Albuquerque, NM 87110	8731108	$875.~~00~~
1243S	617831	Sara O'Neill	83 Montgomery Boulevard	Albuquerque, NM 87111	2289408	$458.~~00~~

11. Filter the source data to show the one record containing "Chama" as the city.

 REVIEW: You do not have to use the Query Options dialog box to filter the data. You can simply choose the city name from the list of fields under City.

12. Merge to create a new document. Save the new document as *[your initials]***17-23merged** in your Lesson 17 folder and print it.

13. Close all open documents, saving changes.

EXERCISE 17-24 ➕ *Challenge Yourself*

Use an Excel worksheet as a data source, sort the data, create and edit the main document, merge the documents, and create mailing labels.

1. Open the file **Europe2**. This file will be your main document.

2. Use the Excel file **TourEurope** (Sheet1$) as the data source.

3. Create a two-level sort of the recipients by last name, and then by first name.

4. Make the following changes to the main document to format it as a letter:
   - Set a 2-inch top margin.
   - Insert the date as an updating field, using the correct date format.
   - Insert the address block.
   - Insert the greeting line. Make sure to change the greeting line punctuation to a colon.
   - Insert your reference initials.

5. Make these additional changes in the body of the main document:
   - In the last sentence of the first paragraph, delete "France" and insert the merge field «Tour».
   - In the third paragraph, delete the sentence "We will give you directions." In its place, key **Fall is a great time of year to start planning your next trip.**

6. Save the main document as *[your initials]***17-24main** in your Lesson 17 folder.

7. Preview the merged data; then merge to a new document.

8. Save the new document as *[your initials]***17-24merged** in your Lesson 17 folder and print four pages per sheet.

9. Close the merged document and main document, saving changes.

10. Create mailing labels. Use the 5160 product number and the data source you used earlier for this exercise. Use the Address Block field in the labels.

11. Merge the labels to a new document. Format the label text as 14-point Arial Narrow bold.

12. Save the labels as *[your initials]***17-24labels** in your Lesson 17 folder.

13. Print the labels and close the file. Close the main label document without saving.

On Your Own

In these exercises, you work on your own, as you would in a real-life work environment. Use the skills you've learned to accomplish the task—and be creative.

EXERCISE 17-25

Create a main document that is a form letter you will send to five friends, family members, or business contacts. Create a data source for the contacts. Include at least one merge field in the body of the form letter. Save the main document and the data source with appropriate filenames. Merge the documents to a new file called *[your initials]***17-25** and print it.

EXERCISE 17-26

Increase the size of the data source created in the previous exercise, or create a new data source as a Word table that contains ten records. Create name tag labels that contain just the first and last name, formatted attractively. Save the merged labels as *[your initials]***17-26** and print.

EXERCISE 17-27

Launch Internet Explorer and research five or more companies where you would like to work. Write a form letter that you will send to these companies, and attach your resume to the form letter. Your database should include a contact name, title, and department. Your form letter should include a few merge fields in the body of the letter to further customize the letter for each company. Save the main document and the data source with appropriate filenames. Merge the documents to a new file called *[your initials]***17-27** and print it.

18

Graphics and Charts

After completing this lesson, you will be able to:

1. **Insert clip art.**
2. **Move and format clip art.**
3. **Create WordArt.**
4. **Create text boxes.**
5. **Create shapes.**
6. **Create diagrams.**
7. **Create charts.**
8. **Edit and modify charts.**

 Estimated Time: 1¼ hours

A picture is worth a thousand words, even in word processing. Word provides ready-to-use pictures, called *clip art,* to add impact to your documents. Word also provides additional graphics tools: WordArt, used to make a dramatic statement with text, and charts and diagrams, used to display information visually. You can also draw various shapes and text boxes in a word document. These exciting tools enable you to add attention-grabbing images to a text document.

Inserting Clip Art

Microsoft Office provides access to a wide variety of drawings, photographs, sound effects, music, videos, and other media files, called *clips.* This lesson focuses mainly on clip art, which are drawings.

When you want to insert clips in a document, you can use one of two search methods:

- Open the Clip Art task pane and search for pictures by keyword.
- Open the Microsoft Clip Organizer window and view collections of clips organized by category (such as Business, People, and Transportation).

EXERCISE 18-1 Find Clips by Using Keywords

Each clip in the Microsoft Office collection has keywords associated with it. Using a keyword is an easy way to narrow your search for an appropriate clip.

1. Open the file **Balloon2**.

    NOTE: The documents you create in this course relate to the Case Study about Duke City Gateway Travel, a fictional travel agency (see pages 1 through 4).

2. Position the insertion point at the second paragraph mark below the title.

3. Open the Insert menu and choose Picture; then choose Clip Art from the submenu. The Clip Art task pane opens.

     TIP: You can also click the Insert Clip Art button 🖼 on the Drawing toolbar to insert clip art. To display the Drawing toolbar, choose Toolbars from the View menu and click Drawing.

4. In the Search box in the task pane, key **balloon**. The task pane has two search option boxes, one for searching specific collections of clips and the other for searching for specific types of clip files. You would use these boxes to narrow your search.

5. Make sure the Search in text box is set for Selected collections. If it is not, click the down arrow and clear the Everywhere box. A check should appear in My Collections and Office Collections. Make sure the Results should be text box is set for All media file types. If it is not, click the down arrow and check the All media types box. (See Figure 18-1 on the next page.)

6. Click the Go button. The task pane displays clips containing a balloon.

    NOTE: If the Search in text box is set for Everywhere, the search results will include clips from the Microsoft Web collection, which requires an Internet connection. These clips are identified by a tiny globe at the bottom-left corner of the picture.

7. Position the mouse pointer over the clip of the balloons and light bulb. A ScreenTip displays the image size, file format (WMF), and some of the keywords associated with the clip.

8. Click the clip to insert it into the document. (See Figure 18-2 on the next page.)

FIGURE 18-1
Using a keyword
to find clips

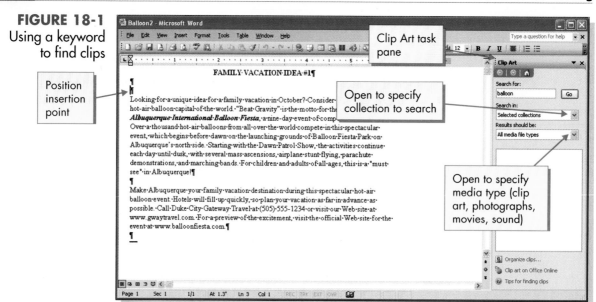

Clip Art task pane

Position insertion point

Open to specify collection to search

Open to specify media type (clip art, photographs, movies, sound)

NOTE: To search for additional clips, enter another keyword in the Search for text box.

FIGURE 18-2
Inserting a clip
from the task pane

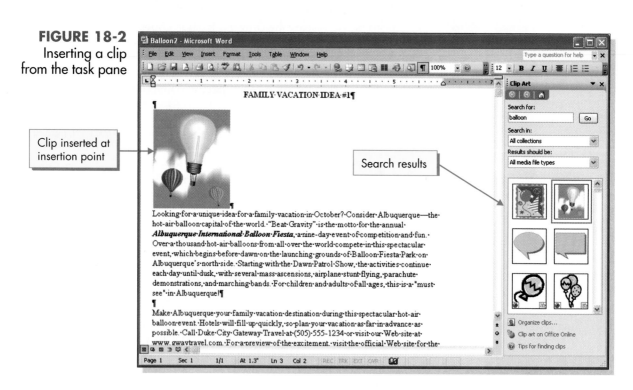

Clip inserted at insertion point

Search results

EXERCISE **18-2** **Find Clips by Browsing Categories**

All clips in the Microsoft Office collection are organized by category. You can browse clips by category in the Microsoft Clip Organizer window.

1. Click Organize clips at the bottom of the Clip Art task pane. The Microsoft Clip Organizer window appears. Click Now if the Add Clips to Organizer dialog box displays.

2. Under Collection List, click the plus sign next to Office Collections to expand the list of categories.

3. Click various category names to display the available clips. Expand a category, if necessary, to display subcategories.

4. To locate a computer picture, scroll to the Technology category. Click the plus sign to expand the category, and then click Computing.

5. Point to the computer picture selected in Figure 18-3. Click the down arrow that appears to the right of the picture. A pop-up menu appears with several options. For example, you can copy the clip, find clips of similar style, or preview a larger version of the clip.

FIGURE 18-3
Inserting a clip from the Microsoft Clip Organizer window

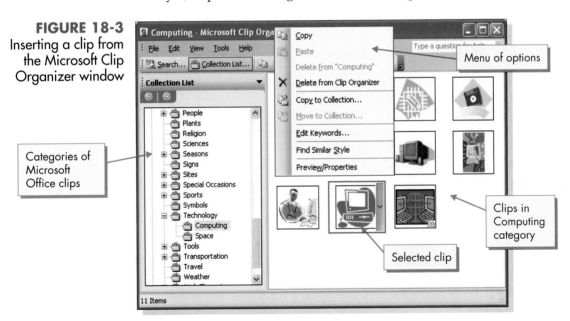

6. Click Copy to copy the clip to the Clipboard.

7. Click within the document and position the insertion point at the last paragraph mark. Paste the clip.

 REVIEW: Right-click where you want to paste the clip, and then choose Paste from the shortcut menu, or use the Paste button 🔳 or Ctrl+V.

8. Save the document as *[your initials]***18-2** in a new folder for Lesson 18.

9. Close the Insert Clip Art task pane. Right-click the Microsoft Clip Organizer on the Windows taskbar and close it. Click No when asked if you want the clip to remain on the Clipboard.

Moving and Formatting Clip Art

After you insert a clip in a Word document, there are many ways to manipulate it. You can change its size, trim it, change its position, and apply formatting options. To manipulate any graphic, you must select it first.

EXERCISE   18-3   **Select Clip Art**

To select clip art, you click it. A selected graphic has a *selection rectangle* around it, formed by eight small squares at each side and corner of the object. These squares are *sizing handles,* which you drag to resize the graphic.

1. Scroll to the balloon picture near the top of the document.

2. Click the picture to select it. Notice the selection rectangle and the sizing handles. Also notice that the Picture toolbar appears when a picture is selected. (See Figure 18-5 on the following page.)

   NOTE: If the Picture toolbar does not appear, open the View menu and choose Toolbars, Picture.

FIGURE 18-4
Clicking a picture
to select it

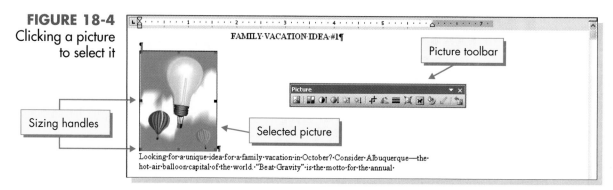

3. Click outside the borders of the picture to deselect it. The Picture toolbar disappears.

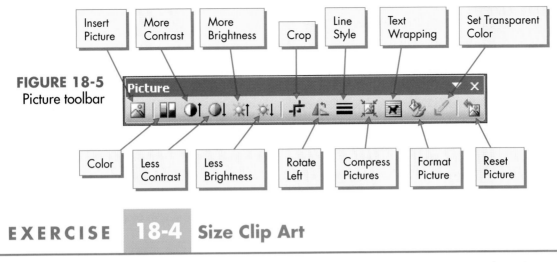

FIGURE 18-5
Picture toolbar

EXERCISE **18-4** **Size Clip Art**

When you *size* a picture, you reduce or enlarge it. You can *scale* a picture to be a percentage of its original size. *Proportional sizing* resizes a picture while maintaining its relative height and width.

1. Click the balloon picture to select it.
2. Move the pointer to the sizing handle in the top-right corner until the pointer looks like this: ↖
3. Drag the sizing handle toward the center of the picture and notice that a dotted-line box appears. The box represents the new size.

FIGURE 18-6
Sizing a picture

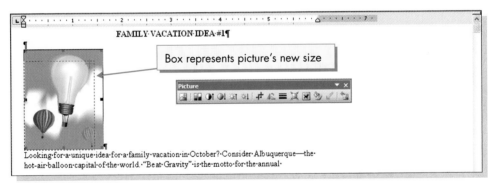

4. Release the mouse button and notice that the picture size changes proportionately.
5. Drag the top-center sizing handle down and release the mouse button. Notice that the picture size changes disproportionately.

NOTE: The four corner sizing handles resize the image proportionately. The top center, bottom center, and side middle handles distort the image's original proportions as the size changes.

FIGURE 18-7
Picture sized
proportionately and
disproportionately

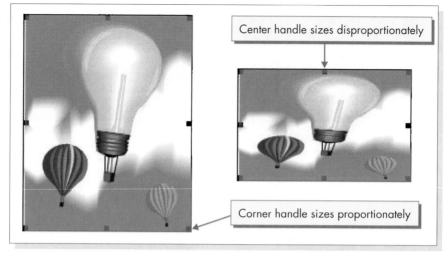

6. Click the Undo button to undo the disproportionate resizing.

7. Select the computer picture. Instead of using the sizing handles to size it, click the Format Picture button on the Picture toolbar.

8. In the Format Picture dialog box, click the Size tab. Under Scale, change the Height to 50%. Press Tab. The Width automatically changes to 50% because you are scaling proportionately.

FIGURE 18-8
Format Picture
dialog box, Size tab

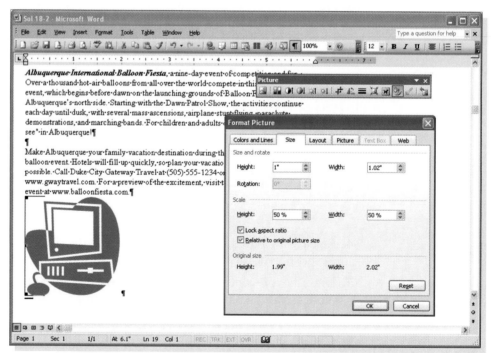

NOTE: In the Format Picture dialog box, when the Lock aspect ratio box is checked, Word scales a picture proportionately, and when the Relative to original picture size box is checked, Word scales the picture from its original size.

9. Click OK. The computer picture is half its original size.

TIP: You can use the Size tab of the Format Picture dialog box to size a picture. Simply enter the desired measurements in the appropriate text boxes.

EXERCISE 18-5 Crop Clip Art

When you trim or *crop* a picture, you hide part of the picture. For example, if you have a picture of a person standing next to a computer, you can crop either the person or the computer out of the picture. What you crop is neither displayed on screen nor printed, but remains part of the original image.

1. Select the computer picture.

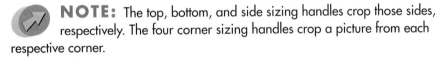

2. Click the Crop button ⊹ on the Picture toolbar. Word switches to Print Layout view.

3. Place the cropping tool over the bottom middle sizing handle and drag toward the top of the picture until you have only the monitor showing.

4. Release the mouse button and notice that the graphic is cropped from the bottom.

5. Click the Undo button ↺ ▾ to undo the cropping.

NOTE: The top, bottom, and side sizing handles crop those sides, respectively. The four corner sizing handles crop a picture from each respective corner.

EXERCISE 18-6 Restore Clip Art to Its Original Size

1. Select the balloon picture.

2. Click the Reset Picture button 🖼 on the Picture toolbar. The picture changes back to its original size, as it appeared when you first inserted it.

3. Select the computer picture. Reset the picture to its original size.

4. Change the computer picture back to the 50 percent size.

5. Save the document as *[your initials]*18-6 in your Lesson 18 folder.

EXERCISE **18-7** Move Clip Art

You can move a picture by cutting and pasting or by using drag-and-drop.

1. Select the computer picture and cut it, placing it on the Clipboard.

2. Paste the picture at the paragraph mark above the paragraph beginning "Make Albuquerque."

3. Select the picture, point to it with the arrow pointer, hold down the mouse, and drag the picture back to the last paragraph mark. The pointer changes to the drag-and-drop shape , the pointer you use for dragging-and-dropping text.

FIGURE 18-9
Dragging a picture

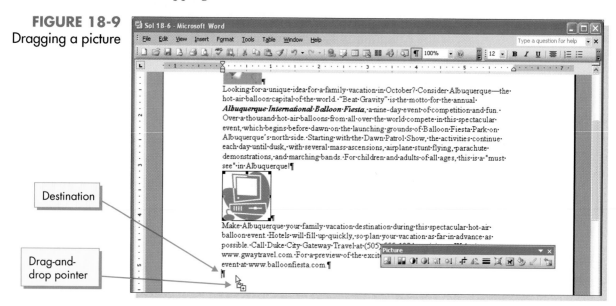

4. Release the mouse button. The picture is positioned at the last paragraph mark. If it is not, click the Undo button and drag it again.

NOTE: To delete a picture, select it and press Delete or Backspace.

EXERCISE **18-8** Change Wrapping Style

When you insert a graphic, it appears in the document as an *inline graphic,* by default. An inline graphic is treated like a character and aligns with the current paragraph. You can change an inline graphic to a *floating graphic.* A floating graphic is placed on the drawing layer of the Word document. You can move the floating graphic freely and layer it behind or in front of text or other objects, changing how text wraps around the graphic. This feature is called *text wrapping.*

1. Select the balloon picture. The picture is an inline graphic, treated by Word like any text object. It is left-aligned with its own paragraph mark.

2. Click the Text Wrapping button on the Picture toolbar. Notice the wrapping options available.

3. Choose Square. Text now wraps around the picture. Notice the anchor symbol and the different selection handles around the picture (now circles instead of solid squares). These are characteristics of a selected floating picture.

FIGURE 18-10
Changing the wrapping style

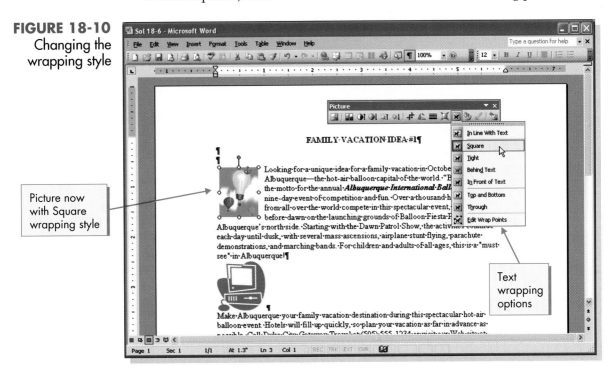

Picture now with Square wrapping style

Text wrapping options

NOTE: The green circle above the picture is the rotation handle. You can drag this handle to rotate the picture either left or right.

4. Move the pointer over the picture. The pointer is now a four-headed arrow. You use this pointer to drag the floating picture freely on the page.

5. Using the four-headed arrow, drag the picture to different locations in the document. Notice how the text wraps squarely around the picture.

TIP: As you drag a floating picture, Word displays a dotted-line outline of the picture. You can use the top edge of the outline as a positioning guide.

6. Change the picture back to an inline graphic by choosing the In Line With Text wrapping style. The picture returns to the area where it was first inserted. If it appears on the same line as the first paragraph, click before "Looking" and press Enter twice, leaving a blank line below the picture.

7. Select the picture and click the Center button 📄 on the Formatting toolbar to center-align the picture.

> **TIP:** To align an inline picture, use the Alignment buttons on the Formatting toolbar. To align a floating picture, open the Format Picture dialog box and select the Layout tab.

8. Select the computer picture and change the wrapping style to Top and Bottom.

9. Move the picture around. Notice how text wraps only around the top and bottom, and not around the sides of the picture.

10. Change the wrapping style to Tight. This style is like Square, but the text wraps more closely around the contours of the picture.

11. Position the picture to the left of the last paragraph.

> **TIP:** To fine-tune the position of a floating graphic, use the Arrow keys. To move a graphic by very small increments, hold down [Ctrl] as you move the Arrow keys.

12. Increase the balloon picture size by 25 percent (key **125%** in the Height text box in the Scale section).

13. Format the document title as 16 points with 72 points of spacing before the paragraph.

14. Save the file as *[your initials]***18-8** in your Lesson 18 folder and print it. Leave it open for the next exercise.

FIGURE 18-11
Preview of
final document

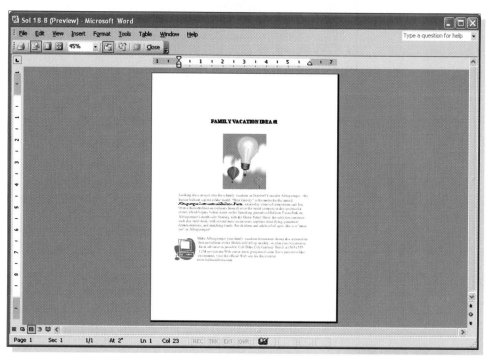

Creating WordArt

WordArt is a drawing tool you can use to create special effects with text. You can choose from a variety of WordArt styles, and then modify the object by editing the text or changing the shape, wrapping style, size, color, position, and so on.

E X E R C I S E **18-9** **Create WordArt**

1. Select the title text "FAMILY VACATION IDEA #1" up to, but not including, the paragraph mark. Click the Cut button ✂ to place the text on the Clipboard. (Because you'll be creating WordArt from this text, not including the paragraph mark will retain the spacing before the title.)

2. Open the Insert menu and choose Picture, WordArt. The WordArt Gallery dialog box appears, displaying a number of WordArt styles.

  TIP: You can also click the Insert WordArt button 🅰 on the Drawing toolbar to open the WordArt Gallery dialog box.

FIGURE 18-12
WordArt Gallery
dialog box

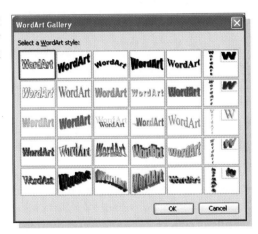

3. Choose one of the styles and click OK. The Edit WordArt Text dialog box appears. You can key or paste text in the Text box when creating WordArt. (See Figure 18-13 on the next page.)

4. Use Ctrl+V to paste the title text in the Text box. Click OK. The WordArt object is inserted in the document as an inline graphic.

FIGURE 18-13
Edit WordArt Text
dialog box

5. Click the WordArt object to select it. The WordArt toolbar appears.

FIGURE 18-14
WordArt toolbar

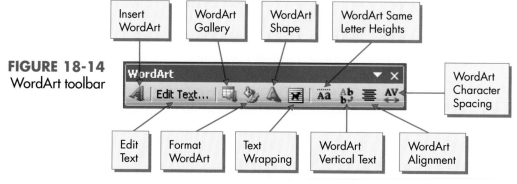

6. Click the WordArt Gallery button 🔲 on the WordArt toolbar and choose another style. Some styles are more readable than others, particularly for all uppercase text. Experiment until you find the style you like best; then close the WordArt Gallery dialog box.

7. Click the WordArt Shape button 🔺 and choose another shape. If you don't like the results, click the Undo button. (See Figure 18-15 on the next page.)

NOTE: As with any graphic object, there are many options for modifying a WordArt object. Several options appear on the WordArt toolbar, and you can use the WordArt toolbar to open the Format WordArt dialog box, where you can change the color, size, and position.

8. Save the document as *[your initials]*18-9 in your Lesson 18 folder.

9. Print and close the document.

FIGURE 18-15
Title text using
a simple
WordArt style

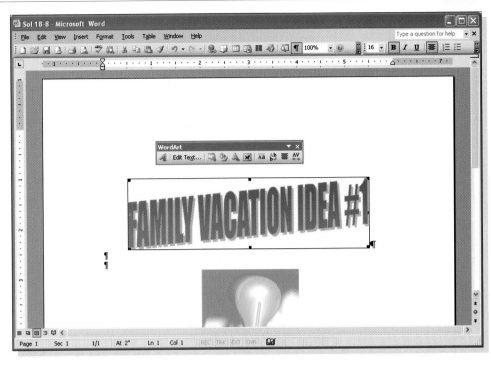

Inserting Text Boxes

In this exercise, you insert a photograph in a document and add text to the layout. The easiest way to do this is by inserting a *text box*—a free-floating rectangular object. You can position a text box anywhere on a page and apply formatting to it.

EXERCISE 18-10 Insert a Text Box

You can create a text box from existing text or insert a blank text box and key text inside it (similar to drawing a shape and adding text to the shape).

1. Start a new document. Use the Page Setup dialog box to change the left and right margins to 1 inch and the page orientation to Landscape.

2. Display the Clip Art task pane. Key the search text **Golden Gate Bridge**.

3. Change the Search in text box to All collections. (Click the down arrow and check the Everywhere box.) Change the Results should be text box to Selected media file types. (Open the drop-down list, check Photographs, and clear all other check boxes. Click Go.

4. Click to insert one of the available pictures. Be sure to choose one that is landscape in orientation (wider than it is tall) rather than portrait. Close the Clip Art task pane.

5. Click the picture to select it and click the Text Wrapping button on the Picture toolbar. Choose Square.

6. Use the Zoom drop-down list on the Standard toolbar to change the view to Whole Page.

7. Drag the picture to the upper-left corner, about 1 inch from the top and left margins.

8. Double-click the picture to open the Format Picture dialog box.

9. Click the Size tab. Make sure Lock aspect ratio and Relative to original picture size are checked. Under Size and rotate, set the Width box to 8 inches.

10. Click the Layout tab. Under Horizontal alignment, choose Center and click OK. Change the zoom to Page Width to see the resized and center-aligned photograph. Deselect the picture.

> **NOTE:** The Alignment buttons on the Formatting toolbar can be used only for text and in-line pictures (which have no text-wrapping style applied). For pictures that you can move freely on the page, align by using the Format Picture dialog box, Layout tab.

11. Display the Drawing toolbar, and scroll to display the lower-left corner of the photograph.

> **NOTE:** There are two ways to display the Drawing toolbar: Choose Toolbars from the View menu and choose Drawing or click the Drawing button on the Standard toolbar.

12. Click the Text Box button on the Drawing toolbar, and press Esc to turn off the Drawing Canvas, which appears automatically when you select a drawing tool. The pointer changes to a crosshair +.

13. Click the crosshair pointer immediately below the bottom-left corner of the picture. A one-inch square text box is inserted. The text box has a black border and contains an insertion point for keying text. The Text Box toolbar appears.

14. Key **Explore San Francisco** in the text box.

15. Point to the text box border. When you see the four-headed arrow pointer, drag to move the text box over the lower-right corner of the picture. Notice that the text box has a white fill color.

16. To draw another text box to a specified size, click the Text Box button . Position the crosshair pointer below the bottom-left corner of the picture, and drag down and across to create a rectangle about ½ inch high and 5 inches wide.

> **NOTE:** When you draw a text box, an anchor symbol appears on-screen, indicating that the text box is anchored to the nearest paragraph. This gives the text box a relationship to the surrounding document text.

FIGURE 18-16
Inserting text boxes

EXERCISE 18-11 Select and Size a Text Box

When you click within a text box, you activate it. The text box is then in Text Edit mode—you can add, edit, or format the text.

When you click the text box border, you select the text box. You can then move or format the text box. There is a subtle difference between the shaded border of a selected text box and an activated text box.

1. Click inside the text box containing text to activate it. Notice the border, made of slanted lines. You can add, edit, or format text in an activated text box.

 NOTE: An activated text box has its own ruler, which you can use to change margins, indents, and tab settings for text within the text box.

2. Select the text "Explore San Francisco." Apply an interesting handwriting-type font to the selected text and make it 48-point bold.

 NOTE: A text box does not expand automatically when you add or enlarge text. You must resize the text box.

3. Point to the top-left resize handle of the corner text box. When you see the two-headed arrow pointer, drag the handle diagonally up and to the left until the text fits on two lines.

4. Delete the space character between "Explore" and "San," and start "San" on a new line. Make the text box wider as needed to accommodate the text so it fits on two lines.

5. Using the middle-bottom handle, drag the bottom border up to fit the text. Notice that clicking on a text-box border to resize it also selects the text box. The selected text-box border has a dotted pattern, rather than slanted lines.

6. Drag the selected text box until it is inside the bottom-right corner of the picture.

7. Click the border of the empty text box to select it. Resize the box to the width of the picture.

8. Click within the empty text box to activate it and key the text shown in Figure 18-17. Use the same handwriting font you applied earlier at a 26-point font size (or a font size appropriate to the font). Format "Golden Gateway Travel" as small caps (if readable in your chosen font). Before the address and phone number, insert the sun-shaped symbol (Wingdings font, fourth row, third character) shown in the figure and change the font size of the symbols to 18 points.

 REVIEW: To insert a symbol, open the Insert menu, choose Symbol, click the Symbols tab, choose a font, and then choose a symbol.

FIGURE 18-17
Resized text boxes with formatted text

9. Resize and position the bottom text box as needed to resemble the one shown in the figure.

EXERCISE 18-12 Format a Text Box

As with other drawing objects, you can change the border, fill, and alignment of a text box and size the box to exact measurements. You can also change the internal text box margins. For simple formatting, you can use the Formatting and Drawing toolbars; for more complex settings, you use the Format Text Box dialog box.

1. Select the "Explore San Francisco" text box (click the text box border). Click the Align Right button ▣ on the Formatting toolbar to right-align the text.

2. With the text box still selected, click the Line Color button arrow ◢▾ on the Drawing toolbar and choose No Line.

3. Using the Font Color button A▾, change the text color to white. The text disappears against the white fill color.

4. Using the Fill Color button ◈▾, choose No Fill. The text appears against the picture background.

5. Select the bottom text box and use the Center button ▤ on the Formatting toolbar to center-align the text.

6. Right-click the border of the bottom text box to display the shortcut menu. Choose Format Text Box.

   TIP: You can also open the Format Text Box dialog box by double-clicking the text box border.

7. On the Colors and Lines tab, under Line, click the Style drop-down arrow and choose the 4½-point solid line. Click the Color drop-down arrow and choose Patterned Lines.

8. In the Patterned Lines dialog box, choose the second pattern in the third row and click OK. (See Figure 18-18 on the next page.)

9. Click the Size tab and set the Width of the text box to 8 inches. Click the Layout tab and change the horizontal alignment to Center. Click OK. (Remember that these are the same settings you applied to the picture.)

10. View the document in Print Preview.

11. Close Print Preview. Save the document as *[your initials]*18-12 in your Lesson 18 folder, print the document, and leave it open for the next exercise.

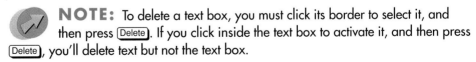

   NOTE: To delete a text box, you must click its border to select it, and then press Delete . If you click inside the text box to activate it, and then press Delete , you'll delete text but not the text box.

FIGURE 18-18
Choosing options
in the Format
Text Box dialog box

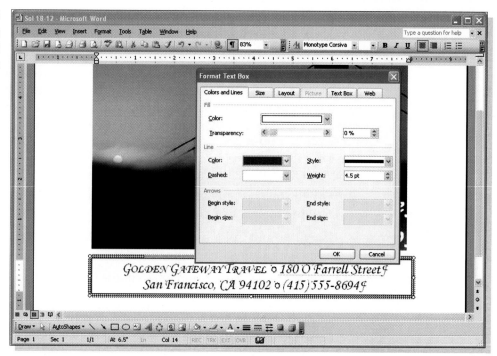

EXERCISE 18-13 Change Text Direction

Changing the orientation of text is a special effect used in desktop publishing. For example, a newsletter might have text running down the side of the page instead of across the top.

You change text direction by rotating text 90 degrees to the left or right. Use the Change Text Direction button on the Text Box toolbar.

1. In *[your initials]*18-12, select the "Explore San Francisco" text box.

2. Click the Change Text Direction button on the Text Box toolbar. The text rotates so it reads from top to bottom.

TIP: You can also rotate text by choosing Text Direction from the Format menu.

3. Click the Change Text Direction button again to make the text read from bottom to top.

NOTE: The arrows and letter in the Change Text Direction button rotate to indicate the direction in which the next click will turn the text.

4. Resize the text box to the height of the picture. Make it wide enough to fit the text on two lines.

5. Move the text box to the inside left of the picture (aligning the left edge of the text box with the left edge of the picture).

6. Change "Explore" to Arial Black, all caps. Change "San Francisco" to Arial bold. You might experiment with Font effects in the Font dialog box, such as applying a shadow and an outline effect.

7. Change the alignment of the text by clicking the Align Bottom button on the Formatting toolbar. Notice the other rotated buttons on the Formatting toolbar.

FIGURE 18-19
Changing
text direction

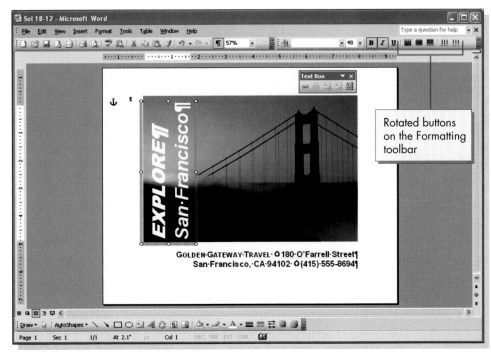

8. Format the text in the bottom text box as 22-point Arial bold, right-aligned. Remove the line from the text box.

9. Change the zoom to Whole Page to see the entire document.

10. Save the document as *[your initials]*18-13 in your Lesson 18 folder, print and close the document.

Working with Shapes and the Drawing Canvas

An easy way to add a graphic element to a document is to use the Drawing toolbar to create a shape. You can draw simple geometric shapes or more complex predesigned shapes. As the lesson progresses, you'll use tools to customize your shapes for dramatic results.

When you draw shapes, Word automatically places them in a *drawing canvas*, a bordered area in your document where you can size, move, and change the objects as a group.

To display the Drawing toolbar:

- Click the Drawing button 🔲 on the Standard toolbar.
- Choose Toolbars from the View menu and choose Drawing.

TABLE 18-1   Buttons on the Drawing Toolbar

BUTTON	PURPOSE
Draw ▾ Draw	Choose commands for changing shapes.
▷ Select objects	Select an object or a group of objects.
AutoShapes ▾ AutoShapes	Choose categories of shapes.
＼ Line	Draw straight lines.
↘ Arrow	Draw lines with arrowheads.
▢ Rectangle	Draw rectangles or squares.
◯ Oval	Draw ovals or circles.
▣ Text Box	Draw a text box.
◢ Insert WordArt	Create text with special effects.
♺ Insert Diagram or Organization Chart	Insert diagram or organization chart.
▣ Insert Clip Art	Insert a clip from the Microsoft Clip Gallery.
▨ Insert Picture	Insert a picture.
▨ ▾ Fill Color	Fill shapes with color.

continues

TABLE 18-1   Buttons on the Drawing Toolbar   *continued*

BUTTON	PURPOSE
Line Color	Change line color.
Font Color	Change color of text.
Line Style	Change style of lines.
Dash Style	Change style of dashes.
Arrow Style	Change arrow style.
Shadow Style	Add a shadow.
3-D Style	Add a 3-D effect.

EXERCISE **18-14** **Create Shapes by Using the Drawing Toolbar**

1. Start a new document. Choose Toolbars from the View menu, and then click Drawing to display the Drawing toolbar.

    NOTE: Another way to display a toolbar is to right-click any displayed toolbar button, and then choose a toolbar from the shortcut menu.

2. Close the task pane if it is open and display the rulers (View, Ruler).

3. Click the Rectangle button 🔲 on the Drawing toolbar. The mouse pointer changes to a crosshair ╋. A drawing canvas, along with the Drawing Canvas toolbar, now appears in your document. (See Figure 18-20 on the next page.)

    NOTE: If the drawing canvas does not appear, choose Tools, Options, General tab, and click Automatically create drawing canvas when inserting AutoShapes. Click OK. Reclick the Rectangle button 🔲. Right-click the drawing canvas to show the Drawing Canvas toolbar.

4. Position the crosshair pointer in the upper-left corner of the drawing canvas. Drag to draw a rectangle that extends across the top of the drawing canvas and is about ½ inch high.

FIGURE 18-20
Preparing to
draw an object

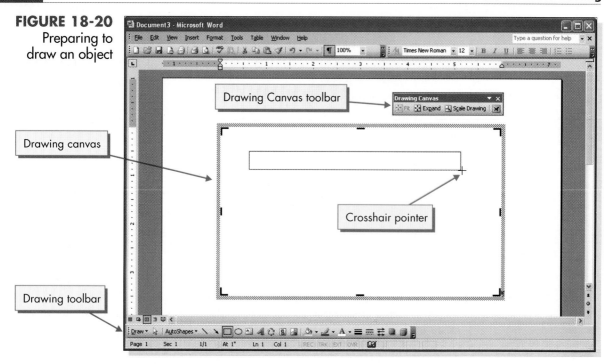

5. Release the mouse button. The rectangle is displayed with eight round sizing handles around its sides. An extra handle with a green circle extends from the top of the rectangle. You can use this handle to rotate an object. When you see an object's sizing handles, the object is selected.

6. Click the Line button ◥. Draw a horizontal line ½ inch below the rectangle and the same length as the rectangle.

 NOTE: To draw a straight line, press (Shift) as you draw the line.

7. Click the Oval button ◯ and draw a ½-inch-high oval below the line on the left side of the Drawing Canvas. Press (Delete) to delete the selected oval.

8. Click the Oval button again ◯. Press (Shift) and draw a ½-inch-high oval below the line. Notice that the oval is a circle. Press (Delete).

 **NOTE:** Press (Shift) to maintain an object's width-to-height ratio as you draw. Press (Ctrl) to draw from the center point instead of from an edge.

9. Click the Oval button ◯, and then click anywhere on the drawing canvas. A 1-inch circle appears on the drawing canvas.

 NOTE: To insert a shape with a predefined size without dragging to draw it, click the Drawing tool and click once in the document.

10. Double-click the Rectangle button ▢. Press (Shift) and draw a small square on the canvas. Draw another square that slightly overlaps this square. Double-clicking a drawing tool keeps the tool active for repeated drawing of the object.

11. Click the Rectangle button ▢ to deactivate it.

12. Click the first square to select it and press (Delete). Select and delete each of the remaining objects, leaving just the empty drawing canvas.

EXERCISE 18-15 Draw AutoShapes

AutoShapes are ready-made shapes you can use for special effects in a document. The seven AutoShape categories are lines, connectors, basic shapes, block arrows, flowchart symbols, stars and banners, and callouts.

When you draw these shapes, the drawing canvas keeps them all together and enables you to format or modify the objects as a group. You can use the drawing canvas to create a drawing you can easily copy to another document. The drawing canvas is also an object you can modify. You can change its location and size and how text wraps around it.

TABLE 18-2   Buttons on the Drawing Canvas Toolbar

BUTTON	PURPOSE
⊞ Fit | Fit Drawing to Contents	Fits drawing canvas snugly around objects.
⊞ Expand | Expand Drawing	Expands drawing canvas around objects.
⊞ Scale Drawing | Scale Drawing	Scales all objects on drawing canvas simultaneously.
⊞ Text Wrapping	Enables you to move the drawing canvas and change its shape around text.

1. Click on the drawing canvas if it is not active.

NOTE: If the drawing canvas is not selected (you would not be able to see it) when you start to draw a shape, Word opens a new drawing canvas above or below the current drawing canvas. You can drag a shape from one drawing canvas to another.

2. Double-click the Line button ╲ on the Drawing toolbar. Press (Shift) and draw a line across the top of the drawing canvas. Draw another line about 1 inch below and the same length as the first.

3. Click the Line button to deactivate it.

4. Click Au<u>t</u>oShapes on the Drawing toolbar.

5. Choose <u>B</u>asic Shapes, and then click the Sun shape in the sixth row.

6. Draw a sun on the left side of the drawing canvas, between the two lines. Don't worry if the sun is not positioned exactly between the two lines; you will format the shape in the next exercise.

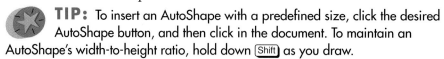

TIP: To insert an AutoShape with a predefined size, click the desired AutoShape button, and then click in the document. To maintain an AutoShape's width-to-height ratio, hold down Shift as you draw.

7. Click Fit on the Drawing Canvas toolbar. The height of the drawing canvas now fits the height of the objects you've drawn.

NOTE: If the Drawing Canvas toolbar does not appear, right-click the drawing canvas and choose Show Drawing Canvas Toolb<u>a</u>r.

8. Click Ex<u>p</u>and on the Drawing Canvas toolbar. The drawing canvas expands slightly around the objects.

FIGURE 18-21
Drawing an
AutoShape

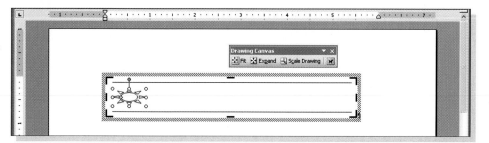

9. Click S<u>c</u>ale Drawing on the Drawing Canvas toolbar. Move the mouse pointer to the bottom-right sizing handle of the drawing canvas. The mouse pointer turns into a diagonal two-headed arrow ↖.

10. Drag the handle diagonally up and to the left corner of the drawing canvas. The objects are scaled smaller.

11. Click the Undo button .

12. Click within the drawing canvas. On the Drawing Canvas toolbar, click the Text Wrapping button and choose <u>S</u>quare. Now you can move the drawing canvas and its contents freely on the page.

13. Drag the drawing canvas to the center of the page.

14. Click the Undo button twice.

TIP: To draw an object independent of the drawing canvas, click the drawing button for the shape you want to draw, press Esc to delete the drawing canvas, and then draw the object. You can also delete a drawing canvas by selecting it and pressing Delete. (If you want to keep any objects, drag them off the canvas before deleting the canvas.)

EXERCISE **18-16** **Resize and Move Shapes**

You can resize, rotate, flip, color, combine, and add text to shapes. To modify a shape, you must first select it by clicking it. To select multiple objects, press (Shift) and click each object or click the Select Objects button [⬚] on the Drawing toolbar and draw a box around the objects you want to select. A selected drawing object has sizing handles for reshaping.

To size a shape, place the mouse pointer on a sizing handle (the mouse pointer becomes a two-headed arrow) and drag the handle. To move a shape, click anywhere inside the shape and drag with the four-headed arrow. If a shape has a yellow diamond, you can change the contour of the shape by dragging the diamond.

1. Move the pointer to the center of the sun shape. Use the four-headed arrow [⬚] to drag the shape to the center of the drawing canvas, positioned evenly between the two horizontal lines.

> **NOTE:** You might notice that when you drag an object to position it, the object moves in fixed increments. By default, objects align to an invisible grid that helps you position them evenly. If you want to precisely position an individual object, you can turn off the Snap to Grid feature or hold down (Alt) as you position the object.

2. Point to the yellow diamond on the sun. When the pointer turns into a small arrowhead, drag the yellow diamond into the center of the shape. The sun now resembles a star shape.

FIGURE 18-22
Changing the
shape of an object

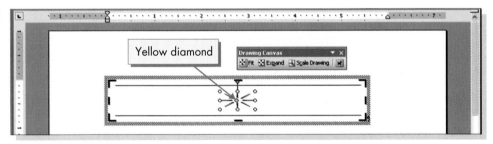

3. Move the mouse pointer to the left-pointing ray of the sun. The mouse pointer changes to a two-headed arrow ↔. Drag the ray to the left side of the drawing canvas. When you start to drag the ray, the mouse pointer becomes a crosshair ╋. (See Figure 18-23 on the next page.)

4. Drag the right-pointing ray to the right side of the drawing canvas. The shape should be centered and should extend from the left side to the right side of the drawing canvas.

FIGURE 18-23
Resizing an object

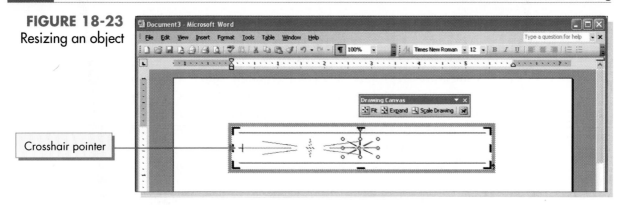

Crosshair pointer

EXERCISE **18-17** **Rotate and Flip Shapes**

You can rotate an object left or right 90 degrees, flip it horizontally or vertically, or rotate it freely by choosing Rotate or Flip from the Draw menu (on the Drawing toolbar). You can also turn the rotate handle on an object. The rotate handle extends from an object and has a green circle.

1. With the shape selected, click Draw on the Drawing toolbar, choose Rotate or Flip, and then choose Rotate Right. The sun shape rotates 90 degrees to the right.

2. Click outside the drawing canvas to deselect the shape and the canvas.

3. Preview the shape in Print Preview; then return to the document in Print Layout view.

4. Select the sun shape by clicking its center or one of its rays.

5. Place the mouse pointer over the green circle on the rotate handle of the selected shape. Rotate the shape to the left, back to its original position between the two lines. The shape and the lines are now in the center instead of at the top of the drawing canvas.

6. Click Fit on the Drawing Canvas toolbar. The drawing canvas resizes snugly to the objects and moves them back to the top of the document. (If the drawing canvas is still in the middle of the document, drag it by its border to the paragraph mark at the top of the document.)

EXERCISE **18-18** **Format Shapes**

You can format shapes by using either the Drawing toolbar or the Format AutoShape dialog box. Examples of formatting include changing fill color, line color, or font color; changing line width and style; and applying a shadow or 3-D effects.

NOTE: To format the drawing canvas (for example, to add a border or background color), select the drawing canvas and use buttons on the Drawing toolbar. Or, right-click the drawing canvas, open the Format Drawing Canvas dialog box, and choose options.

1. Select the sun shape. Click the Fill Color button arrow on the Drawing toolbar and choose Fill Effects.

2. On the Gradient tab, click Two colors. Open the Color 1 list and choose Light Orange. Open the Color 2 list and choose Yellow. Click OK. The sun's rays are colored with a gradient from orange to yellow. The gradient will be more visible later in the exercise.

3. Click the Line Color button arrow on the Drawing toolbar and choose Tan.

4. Click the Shadow Style button 🔲 on the Drawing toolbar. Experiment with other shadow styles to see the effect on the sun shape; then apply Shadow Style 4. (The shadow style number appears a second or two after you place the mouse pointer on the style.)

FIGURE 18-24
Choosing a
shadow style

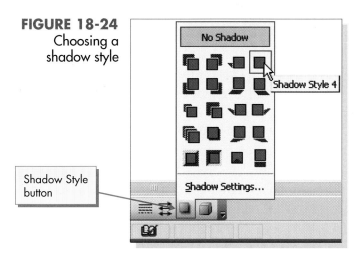

Shadow Style
button

5. Click the Shadow Style button 🔲 and choose Shadow Settings. The Shadow Settings toolbar appears. (See Figure 18-25 on the next page.)

6. Experiment with the Nudge buttons on the toolbar—click each button once to nudge the shadow up, down, left, and right.

7. Click the Shadow Color button arrow on the Shadow Settings toolbar and choose Light Yellow.

8. Change the shadow to Shadow Style 17 and change the shadow color to Gold. Close the Shadow Settings toolbar.

9. Click the 3-D Style button 🔲 on the Drawing toolbar and choose 3-D Style 7. Change the style to 3-D Style 15. (The 3-D style number appears a second or two after you place the mouse pointer on the style.)

FIGURE 18-25
Shadow Settings
toolbar

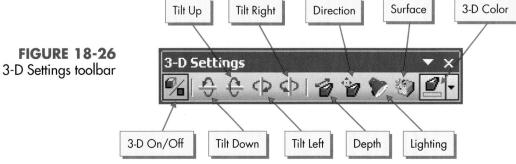

10. Click the 3-D Style button 🔲 again and choose 3-D Settings to display the 3-D Settings toolbar.

FIGURE 18-26
3-D Settings toolbar

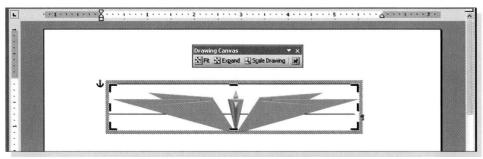

11. Click the Tilt Down button 🔲 once to shorten the object.

12. Click the Lighting button 🔲 and click the top light that points directly down.

13. With the modified sun object now extended out of the drawing canvas, click Fit on the Drawing Canvas toolbar to fit the canvas around the object.

FIGURE 18-27
Modified shape

14. Save the document as *[your initials]*18-18 in your Lesson 18 folder. Print and close the document.

Creating Diagrams

Using Word's diagram tools, you can insert six types of diagrams:

- *Organizational,* sometimes called *org chart,* illustrates the top-down relationship of members of an organization
- *Cycle,* illustrates a process that has a continuous cycle
- *Target,* illustrates steps towards a goal
- *Radial,* illustrates relationships of objects to a main object
- *Venn,* illustrates where areas overlap between objects
- *Pyramid,* illustrates foundation-based relationships

EXERCISE 18-19 Create a Diagram

 To insert a diagram, choose Diagram from the Insert menu or click the Insert Diagram or Organization Chart button on the Drawing toolbar. Choose a diagram type from the Diagram Gallery dialog box.

1. Start a new document and key the title **Gateway Travel Offices in the United States.** Center the text and change it to 14-point Arial Black. Press Enter twice.

2. Choose Diagram from the Insert menu to open the Diagram Gallery dialog box.

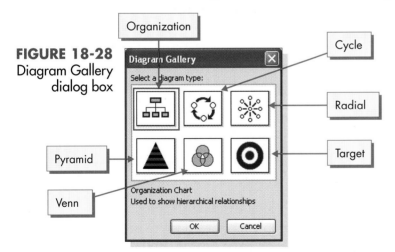

FIGURE 18-28
Diagram Gallery
dialog box

3. Click the Radial diagram (the third diagram in the first row) and click OK. The radial diagram is inserted in the document. The diagram displays with a drawing space, a nonprinting border, and sizing handles. The Radial Diagram toolbar also appears.

4. Select the innermost circle in the Radial diagram, if it is not already selected. To select a circle, point to the circle's edge and click when you see the four-headed arrow pointer. The selected circle has round selection handles.

FIGURE 18-29
Inserting a
Radial diagram

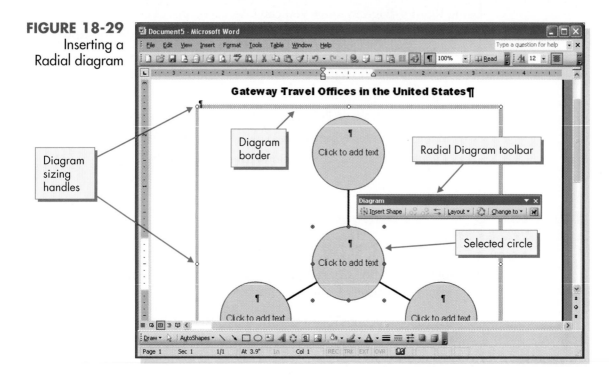

5. With the innermost circle selected, click the Insert Shape button on the Diagram toolbar four times. The Radial diagram should now have seven outer circles.

EXERCISE 18-20 Add Text to a Diagram

1. Click inside the center circle and key **Albuquerque**. Don't worry if the text doesn't fit on one line—you'll resize the diagram later in this exercise.

2. Click in the topmost circle and key **San Francisco**.

3. Moving clockwise, key the following cities, one per circle, in each of the remaining circles: **Chicago**, **New York**, **Boston**, **Atlanta**, **Washington, D.C.**, **Honolulu**.

4. Click a blank area inside the diagram border. The border and sizing handles should display.

5. Point to the middle-right sizing handle on the border. The two-headed arrow pointer appears.

FIGURE 18-30
Sizing the
radial diagram

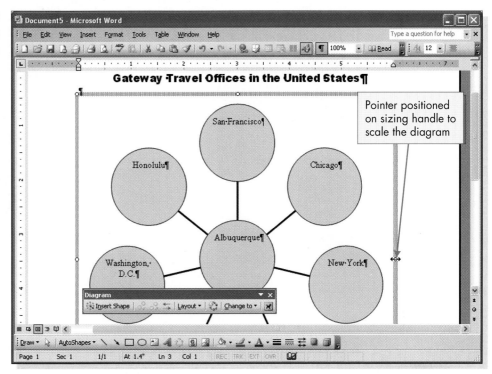

6. Use the pointer to drag the sizing handle slightly to the right. Release the mouse button. "Albuquerque" and "Washington D.C." should fit on one line. If necessary, drag the sizing handle a little more to the right.

7. Click the AutoFormat button 🖸 on the Radial Diagram toolbar. In the Diagram Style Gallery dialog box, click to preview some of the styles. Then choose 3-D Color and click OK.

8. Select the diagram and click the Text Wrapping button 🖼 on the Diagram toolbar. Choose Square and drag the diagram approximately one-half inch below the title. Choose Diagram from the Format menu, and select the Layout tab. Click Center and click OK.

9. Select the Albuquerque circle and apply 24 points spacing before the text. Repeat for each of the office circles. See Figure 18-31 on the next page.

10. Center the document vertically on the page (Page Setup, Layout tab).

11. Save the file as *[your initials[***18-20** in your Lesson 18 folder.

12. Print and close the document.

FIGURE 18-31
Completed
radial diagram

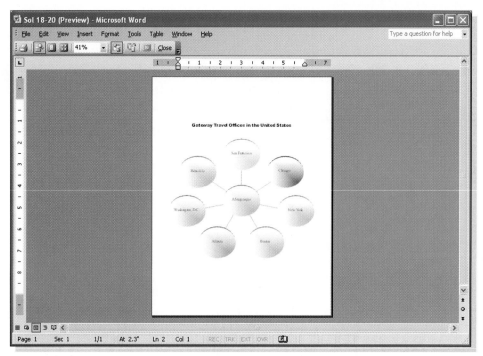

Creating Charts

A *chart* is a visual representation of numeric data. Charts often make values in a table easier to understand. You can usually look at a chart and quickly grasp the meaning of the numbers.

You create a chart in a Word document by inserting a generic chart as an object. Word uses Microsoft Graph to create and edit the chart. You enter the chart data in a table called a *datasheet*. The datasheet is similar to an Excel worksheet. When you first create the chart, the datasheet contains sample data, and the chart is displayed in a 3-D column graph format. You replace the sample data with your own numbers and column labels.

EXERCISE 18-21 Insert a Chart in a Document

1. Open the file **Growth**. Add the date and make the letter from Doris Simms, Travel Counselor. Include your reference initials and an enclosure notation.

2. Position the insertion point in the blank line below the paragraph that begins "As you requested."

3. Open the Insert menu and choose Picture, Chart. A sample chart is inserted in the document between the second and third paragraphs. Word displays the graph editing screen, which includes the Microsoft Graph menu and toolbars

and the datasheet. The chart is based on sample data in the datasheet. You'll replace this data with your own labels and numbers in the next exercise.

FIGURE 18-32
Inserting a chart

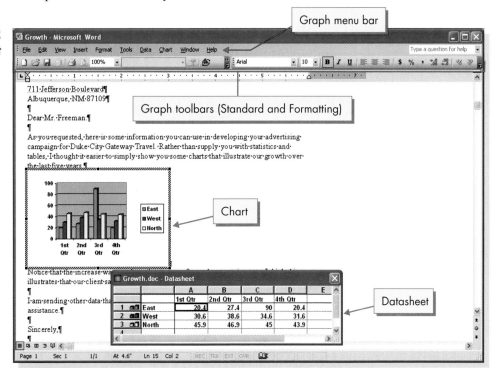

4. Click the Toolbar options button on the Graph toolbar to see the available Standard and Formatting chart toolbar buttons.

5. Click anywhere in the document. The datasheet disappears and the Word menu and previously displayed toolbars reappear.

6. Click the chart once to select it. Notice the selection handles. Word treats the chart as an inline graphic object. You can move or resize it. To edit the chart, you must redisplay the graph editing screen and the datasheet.

EXERCISE **18-22** **Key Data in the Graph Datasheet**

1. Double-click the chart to open the graph editing screen.

2. If necessary, drag the datasheet by its Title bar to another location so you can see more of the underlying graph.

3. Click the upper-left box of the datasheet to select the entire datasheet (or press Ctrl+A).

4. Press Delete. This deletes the sample data. The entire datasheet is now blank and ready for you to key new data. Notice that the chart disappears when you erase the sample data.

NOTE: Although you can perform multiple Undo actions elsewhere in Word, in Microsoft Graph, you can only undo your last action. Be careful when making changes!

5. Key the data and headings shown in Figure 18-33. The chart grows as you key the data. Notice that you can't see the entire row 1 label "Total Revenues" after you enter the value in the adjacent column—this doesn't affect how the label is displayed on the chart.

6. If necessary, press Enter after keying the last figure to make sure it is entered in the chart. Data is not entered until you move to another cell.

FIGURE 18-33
Datasheet with new data

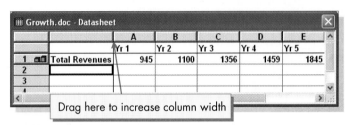

TIP: You can change the width of datasheet columns to display text that might be hidden. Move the pointer to the divider line between the columns until the two-headed arrow appears; then hold down the mouse button and drag to the desired width (or double-click the divider line). If the datasheet is too small to show all the data, point to the side or bottom border of the datasheet window, and then use the two-headed arrow pointer to drag to a larger size.

7. While still in the graph editing screen, choose Datasheet from the View menu or click the View Datasheet button 📇 on the Graph toolbar. The datasheet disappears. Notice that you are still in the graph editing screen, but you can see more of your document.

8. Click the View Datasheet button 📇 to display the datasheet again.

9. Click anywhere in the document to redisplay the Word menu and previously displayed toolbars.

10. Save the document as *[your initials]***18-22** in your Lesson 18 folder. Leave it open for the next exercise.

TABLE 18-3   Microsoft Graph Standard and Formatting Toolbar Buttons

BUTTON	FUNCTION
👆 Import File	Imports files or data from another application
📇 View Datasheet	Shows or hides the datasheet window

continues

TABLE 18-3　Microsoft Graph Standard and Formatting Toolbar Buttons　*continued*

BUTTON	FUNCTION
By Row	Plots data series by rows
By Column	Plots data series by columns
Data Table	Displays the values for each data series in a grid below the chart
Chart Type	Changes the chart type for an individual data series, chart group, or entire chart
Category Axis Gridlines	Shows or hides category axis gridlines
Value Axis Gridlines	Shows or hides value axis gridlines
Legend	Adds or deletes a chart legend
Drawing	Displays or hides the Drawing toolbar
Fill Color	Applies a color to the selected chart element
Microsoft Graph Help	Starts Microsoft Graph Help
Currency Style	Applies the Currency style ($100.00) to selected cells
Percent Style	Applies the Percent style (100 %) to selected cells
Comma Style	Applies the Comma style (1,100.00) to selected cells
Increase Decimal	Increases the number of digits displayed after the insertion point for selected cells
Decrease Decimal	Decreases the number of digits displayed after the insertion point for selected cells
Angle Clockwise	Rotates selected text down at a 45-degree angle
Angle Counterclockwise	Rotates selected text up at a 45-degree angle

EXERCISE 18-23 Create a Chart from a Word Table

You can create a chart by selecting data in a Word table, and then choosing Insert, Picture, Chart. The selected data will appear automatically in the chart's datasheet.

1. In the document *[your initials]*18-22, select the chart and press Delete. (Remember, click a chart once to select it or double-click a chart to activate the chart editing window and display the datasheet.)

2. Open the file **Revenue1**. The table in this file has more up-to-date revenue figures for the travel agency and displays the data differently (showing the data in decimal figures representing millions).

3. Select the table. Open the Insert menu and choose Picture, Chart. Word inserts a chart, using the table data, and displays the chart's datasheet.

 NOTE: You can also copy the data from a Word table, and then paste it into an empty datasheet.

FIGURE 18-34
Charting data
from a Word table

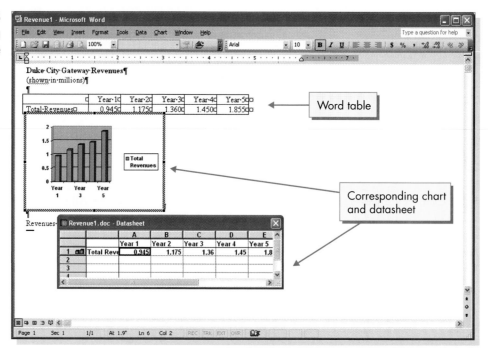

4. Click outside the chart to deactivate the chart editing window.

5. Click the chart to select it. Drag the chart's middle-right sizing handle to the right until you can see all five of the year labels, with the text all on one line.

6. Copy the chart to the Clipboard.

7. Paste the chart into the document *[your initials]*18-22, at the same paragraph mark as the previous chart (above "Notice that").

8. Close the file **Revenue1** without saving.

EXERCISE 18-24 Edit Chart Data

After you create a chart, you might need to edit the data. Numbers in a chart can be changed only on the datasheet.

1. In the document *[your initials]*18-22, double-click the chart to open the graph editing screen.

2. Click the View Datasheet button ▦ if the datasheet is not displayed.

3. Resize the datasheet so you can see all columns at once (drag the right edge).

4. Change the Year 5 total revenue to **2.155** (press [Enter] after keying the text to enter the value). Notice that the chart changes with the new value.

> **NOTE:** If you inadvertently key a value in the wrong cell of the datasheet, Microsoft Graph might add another column or row to your chart. To delete the additional column or row, click the gray box at the beginning of the column or row that contains the erroneous data.

5. Add two new rows of data to the datasheet for New Business and Repeat Business, as shown in Figure 18-35. (You don't have to type the zero before each decimal point; they'll be entered automatically.) Adjust the width of the first column to accommodate "Repeat Business."

FIGURE 18-35
Datasheet with
two new rows

C:\2003 Word\Core\Solution... - Datasheet

			A	B	C	D	E
			Year 1	Year 2	Year 3	Year 4	Year 5
1	◰	Total Revenues	0.945	1.175	1.36	1.45	2.155
2	◰	New Business	0.661	0.762	0.816	0.799	0.742
3	◰	Repeat Business	0.284	0.413	0.544	0.651	1.413
4							
5							
6							

EXERCISE 18-25 Switch Data Series from Rows to Columns

Data is plotted in a chart in groups of *data series*, which are collections of related data points. These values are usually found within the same column or row in the datasheet. Each data series is distinguished by a unique color. Sometimes it's hard to know in advance if it's better to arrange your data in rows or in columns on the datasheet. Microsoft Graph lets you enter it either way, and switch back and forth.

1. Double-click the chart if it is not activated.

2. Click the By Column button ▦ on the Graph toolbar. The graph columns are grouped by column data instead of row data. Notice that the data is now grouped into three categories, showing the yearly changes within each of the three categories.

TIP: Depending on the type of data in your chart, switching data series between row and column orientations is often a good idea. Sometimes a different view can emphasize and improve chart data in a way you didn't anticipate.

3. Click the By Row button 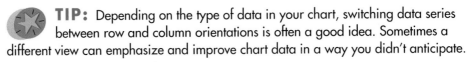 to change the data series groupings back to a row orientation.

TIP: You can also change data series by choosing Series in Rows or Series in Columns from the Data menu.

4. Click anywhere in the document to redisplay the Word document.

5. Click the chart once to select it. You might need to resize the chart again to display the year labels on one line.

6. Click the Center button 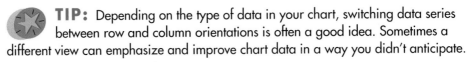 on the Formatting toolbar to center the chart horizontally.

7. Create a letterhead for the document by keying the following information in the header pane, formatted as 10-point Arial, right-aligned:
   Duke City Gateway Travel
   15 Montgomery Boulevard
   Albuquerque, NM 87111
   (505) 555-1234
   www.gwaytravel.com

8. Change the top margin to 2 inches.

9. Preview the document; then save it as *[your initials]***18-25** in your Lesson 18 folder.

FIGURE 18-36
Document with
modified chart
and letterhead

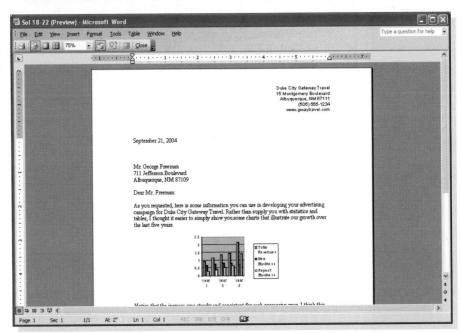

10. Print the document. Leave it open for the next exercise.

Changing Chart Types and Adding Options

When selecting a chart type, the data you are charting generally determines the choices you can use. Although you might like the look of a certain chart type, it might not be appropriate for the numbers or the message you are trying to communicate.

Microsoft Graph offers you a variety of chart types, including column, bar, and line charts. For charts with a single data series, the pie chart is a popular choice. Many charts can be displayed using a 3-D visual effect. You can also combine different types within a single chart, such as column and line.

EXERCISE **18-26** **Change Chart Types**

There are 14 different chart types, and each type has 2 or more sub-types.

1. Double-click the chart to open the graph editing screen. Hide the datasheet.

2. Click the arrow next to the Chart Type button on the Graph toolbar. A pop-up menu displays some basic chart types.

3. Choose Bar Chart , the first option on the second row. The chart changes to a one-dimensional bar chart.

4. Click the Chart Type button arrow again, and choose Line Chart , the first option on the fourth row.

5. Choose some of the other chart options, to see how they display your data.

6. Choose Chart Type from the Chart menu. In the Chart Type dialog box, look at the Standard Types tab. This dialog box provides many more chart types and chart sub-types than the Chart Type button offers.

7. Under Chart type, choose Column. Seven sub-types are available for column charts.

8. Under Chart sub-type, click the first sub-type on the second line, which is the clustered column with a 3-D visual effect. This is the default chart type you started with.

FIGURE 18-37
Chart Type dialog box

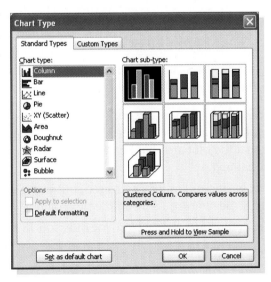

9. Use the Press and Hold to View Sample button to see a preview of the chart. The Chart Sub-type window is replaced by a Sample window, displaying the selected sub-type as it will appear. Release the button.

10. Select the second sub-type on the first row—the Stacked Column—and use the Press and Hold to View Sample button to preview the sub-type.

11. Select the first sub-type—Clustered Column—and click OK. The chart type is changed to a simple one-dimensional column chart.

12. Click anywhere in the document to return to Word. Notice that removing the 3-D effect makes the chart easier to read.

EXERCISE 18-27 Add Chart Options

Charts are very versatile. You can add, remove, or modify chart elements (such as a chart title or gridlines), change colors, and format chart data.

1. Double-click the chart to open the graph editing screen. Hide the datasheet.

2. Position the pointer over various parts and areas of the chart. A ScreenTip will identify the parts by name.

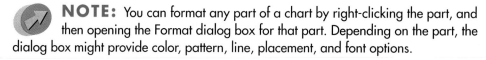

NOTE: You can format any part of a chart by right-clicking the part, and then opening the Format dialog box for that part. Depending on the part, the dialog box might provide color, pattern, line, placement, and font options.

3. Choose Chart Options from the Chart menu. The Chart Options dialog box appears.

4. Click the Titles tab.

5. For the Chart title, key **Growth at Duke City Gateway Travel**.

6. For the Value (Y) axis, key **(Millions)**.

NOTE: In this chart, the Value (Y) axis is the vertical axis, listing the revenue amounts. The Category (X) axis is the horizontal axis, showing the years.

7. Click the Gridlines tab. Under Category (X) axis, click Major gridlines. Click OK. Chart titles and a set of vertical gridlines are added to the chart. Notice that the plot area has been compressed to compensate for the additional text.

8. Make the chart wider and taller, as desired, but make sure the letter fits on one page. (See Figure 18-38 on the next page.)

9. Save the file as *[your initials]***18-27** in your Lesson 18 folder.

10. Print and close the file.

FIGURE 18-38
Chart with title and
vertical gridlines

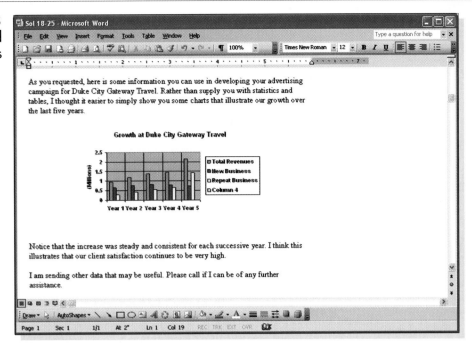

USING ONLINE HELP

FIGURE 18-39
Getting help
about org charts

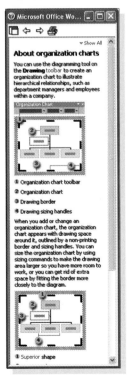

An organization chart (org chart) is a type of diagram that shows the hierarchy of a company or organization, from superiors down to subordinates, coworkers, and assistants. You can create a simple or complex org chart.

Use Help to learn how to create an org chart:

1. In the Ask a Question box, key **org chart** and press Enter.
2. Click the topic About organization charts.
3. Review the information in the Help window.
4. Click the topic Add an organization chart.
5. Review the information.
6. Click each of the related topics about org charts.
7. Close Help. Start a new document and try creating a simple 3-level org chart for your job, family, or school, adding subordinates, coworkers, or assistants, as needed.

LESSON 18 Summary

➤ Add ready-to-use pictures to your document in the form of drawings (clip art) or photographs. You can also add sound effect, music, and video clips.

➤ Search for clips by keyword, using the Clip Art task pane, or view collections of clips by category from the Organize clips window (click the link at the bottom of the Clip Art task pane).

➤ Click a clip to select it. You can copy, paste, resize, crop, and format a selected clip.

➤ A selected clip is surrounded by a selection rectangle with sizing handles at each side and corner. Drag a corner sizing handle to make a clip larger or smaller, while keeping the original proportions of the clip. Click the Format Picture button on the Picture toolbar, and then choose the Size tab to resize a clip to exact measurements or a percentage of the original size.

➤ Crop a clip to hide part of it from view. Click the Crop button ⊞ on the Picture toolbar, position the crop tool on a sizing handle, and drag until you have hidden the part of the picture you want cropped.

➤ You can undo any action applied to a clip. You can also restore a clip to its original size by clicking the Reset Picture button 🖼 on the Picture toolbar.

➤ By default, a clip is inserted in a document as an inline graphic and treated like a character or word. To move a clip that is an inline graphic, cut-and-paste the clip or drag-and-drop it. You can use the alignment buttons on the Formatting toolbar to change the horizontal alignment of a clip.

➤ To move a clip freely on the page, change it from an inline graphic to a floating graphic by changing its wrapping style. Use the Text Wrapping button 🖼 on the Picture toolbar. You can drag a floating graphic to position it anywhere on the page, and have the document text wrap around the graphic. The Square wrapping style wraps text around all sides of a graphic.

➤ Use the WordArt feature to create special effects with text. Choose from a variety of WordArt styles and modify the WordArt object by using the WordArt toolbar buttons (to change shape, size, color, wrapping style, alignment, and so on).

➤ A text box is a free-floating rectangular object that you can position anywhere on a page and apply formatting to it.

➤ Size a selected text box by selecting it and dragging one of its sizing handles. Format a text box by changing its border, fill, alignment, size, and internal margins.

➤ Change text direction in a text box by rotating the text 90 degrees to the left or right.

➤ The drawing canvas is a bordered area in your document that contains objects you draw. You can size, move, and change the objects on the drawing canvas as a group.

➤ You can draw shapes such as rectangles, ovals, and lines by using buttons on the Drawing toolbar. Use the AutoShapes button on the Drawing toolbar to draw ready-made AutoShapes such as stars and arrows.

➤ Click a shape to select it. To select multiple objects, hold down (Shift) and click each object or use the Select Objects button ▨ on the Drawing toolbar to draw a box around the objects.

➤ A selected shape has eight sizing handles. Drag a sizing handle to make a shape larger or smaller. Move a shape by dragging it with the four-headed arrow. Drag the yellow diamond in a shape to change its contour.

➤ Drag a shape's green rotate handle to manually rotate the object. You can also flip an object horizontally or vertically or rotate it 90 degrees left or right.

➤ Use the Format AutoShape dialog box or Drawing toolbar to apply formatting to shapes, including fill color, line color, shadows, and 3-D effects.

➤ Use Word's diagram feature to insert six different types of diagrams, including an organization chart or a target or radial diagram. Add text to a diagram by clicking in each box or circle in the diagram and keying the text.

➤ Use a chart to represent numeric data visually. Charts are created in Microsoft Graph. When you first insert a chart, it appears in the default chart style (3-D bar) with sample data. Chart data is contained in a table called a datasheet. You enter your own data in the datasheet, and then click outside the chart to deactivate Microsoft Graph and return to Word.

➤ Click a chart to select it (for deleting, moving, or resizing). Double-click a chart to activate Microsoft Graph (for editing the datasheet, changing the chart type, or changing other chart options).

➤ To navigate in the datasheet, use (Enter), (Tab), or the arrow keys. To select all data in the datasheet, press (Ctrl)+(A).

➤ To enter new data in the datasheet, delete the sample data. In the first row under "A," "B," "C," and so on, key your column headings (the category axis of the chart). In the first column to the right of "1," "2," "3," and so on, key your row headings (the value axis of the chart). Key the corresponding numeric values in the datasheet.

➤ Edit chart data by clicking a datasheet cell and keying the new data.

➤ Create a chart from a Word table by selecting the data in the table, and then using the Insert, Picture, Chart command. Edit the datasheet as necessary.

➤ Microsoft Graph has its own Standard and Formatting toolbars for modifying a chart. For example, switch data series from rows to columns by clicking the By Column button ▥. Reverse this procedure by clicking the By Row button ▤. See Table 18-1.

➤ There are 14 different types of charts, and each type has 2 or more sub-types. Use the Chart Type button ![icon] on the toolbar to choose from a group of chart types, or open the Chart Type dialog box (open the Chart menu and choose Chart Type) to see all types and sub-types.

➤ Add options such as a chart title or gridlines by selecting the chart, choosing Chart Options from the Chart menu, and clicking the appropriate tab in the Chart Options dialog box.

➤ Change the chart position in the Word document by clicking to select the chart, and then using the alignment buttons on the Formatting toolbar. Resize a selected chart by dragging a sizing handle to increase or decrease the chart size.

LESSON 18 Command Summary

FEATURE	BUTTON	MENU	KEYBOARD
Insert clip art		Insert, Picture, Clip Art	
Format picture		Format, Picture	
Crop picture		Format, Picture, Picture	
Resize picture		Format, Picture, Size	
Reset picture		Format, Picture, Size	
Change wrapping style		Format, Picture, Layout	
Insert WordArt		Insert, Picture, WordArt	
Select drawing objects			
AutoShapes	AutoShapes ▾	Insert, Picture, AutoShapes	
Line		Insert, Picture, AutoShapes, Lines	
Arrow		Insert, Picture, AutoShapes, Lines	

continues

LESSON 18 Command Summary *continued*

FEATURE	BUTTON	MENU	KEYBOARD
Rectangle		Insert, Picture, AutoShapes, Basic Shapes	
Oval		Insert, Picture, AutoShapes, Basic Shapes	
Text Box		Insert, Text Box	
Change Text Direction		Format, Text Direction	
Fill Color		Format, Autoshape Format, Text Box Format, WordArt Format, Picture	
Line Color		Format, Autoshape Format, Text Box Format, WordArt Format, Picture	
Shadow Style			
3-D Style			
Text Wrapping		Draw, Text Wrapping	
Insert Diagram		Insert, Diagram	
Insert Chart		Insert, Picture, Chart	
View Datasheet		View, Datasheet	
Chart Type		Chart, Chart Type	
Plot data series by rows		Data, Series in Rows	
Plot data series by columns		Data, Series in Columns	
Add/delete chart options		Chart, Chart Options	

Concepts Review

TRUE/FALSE QUESTIONS

Each of the following statements is either true or false. Indicate your choice by circling
T or F.

T F   **1.** You can search for clip art by keyword in the Clip Art task pane.

T F   **2.** After clicking the Oval button ⬭, hold down Alt to draw
a circle.

T F   **3.** The terms "size" and "crop" are used interchangeably in Word.

T F   **4.** After you size or crop a picture, you can easily reset it to its
original size.

T F   **5.** The organization chart option is included in the Diagram
Gallery dialog box.

T F   **6.** To select a WordArt object, point to the object and click.

T F   **7.** You right-click a chart to open the graph editing screen.

T F   **8.** You can edit chart data without changing a chart's datasheet.

SHORT ANSWER QUESTIONS

Write the correct answer in the space provided.

1. Which toolbar appears on the screen when you insert clip art?

_____

2. Which feature do you use to combine two colors in a gradient fashion and
use them as the background in a shape?

_____

3. When you insert a picture, what is its default wrapping style?

_____

4. Name one way to insert WordArt.

_____

5. What is the table called in Microsoft Graph where a chart's values are entered?

_____

6. What is the purpose of 🖼?

_____

7. After a chart is inserted in Word, how do you activate Microsoft Graph to make changes to the chart?

_____

8. How do you use the Microsoft Graph toolbar to display or hide a datasheet?

_____

CRITICAL THINKING

Answer these questions on a separate page. There are no right or wrong answers. Support your answers with examples from your own experience, if possible.

1. Find three examples of clip art in newsletters, advertisements, letters, or other publications. Explain how the clip art helps to communicate the message of the text.

2. When would you use charts rather than tables in a document? Create a few sample charts and explain their potential use.

Skills Review

EXERCISE 18-28

Insert, size, move, restore, and change the wrapping style of clip art.

1. Open the file **Euro**. Change the top margin to 2 inches. Format the title as 14-point bold, centered, all caps. Add an extra paragraph mark below the title.

2. Insert clip art by following these steps:

   a. Position the insertion point at the blank paragraph above "Countries now using."

   b. Open the Insert menu and choose Picture, Clip Art.

   c. In the Clip Art task pane, key **currency** in the Search for text box and press (Enter).

   d. Click the clip with the euro symbol (€) to insert it in the document. (See Figure 18-40 on the next page.)

3. Resize the clip by following these steps:

   a. Click the clip to select it.

   b. Drag a corner handle slightly toward the middle of the clip to make the clip proportionately smaller.

   c. Drag the right middle handle slightly toward the middle of the clip to make the clip narrower.

FIGURE 18-40
Currency clip art

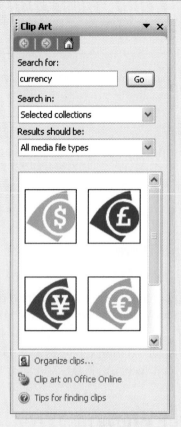

d. Click the Reset Picture button on the Picture toolbar to restore the original size and proportions of the clip.

e. With the clip still selected, click the Format Picture button on the Picture toolbar. Click the Size tab.

f. Under Scale, change the height setting to 50%. Make sure the Lock aspect ratio box is checked. Press Tab. The width should be automatically set to 50%. Click OK.

4. Move and align the clip by following these steps:

   a. With the clip selected, hold down the mouse button and drag the clip to the second paragraph mark below the title. Release the mouse button.

   b. Click the Center button on the Formatting toolbar to center-align the clip.

   c. With the clip still selected, press Enter to insert a blank line below it.

5. Format the line that begins "Countries now using" as bold italic. Format the list below it as a bulleted list. Format the line that begins "Information about using" as bold italic and the paragraph below it as a bulleted list.

6. Position the insertion point to the left of "Information about using" and insert the clip with the dollar sign (the clip that's the same style as the euro clip you inserted).

7. Close the Clip Art task pane.

8. Resize the dollar sign clip to 30% of its original size.

9. Change the wrapping style of the clip to Square by clicking the Text Wrapping button on the Picture toolbar and choosing Square.

10. With the clip now changed from an inline graphic to a floating graphic, drag the clip so it is under the text "Information about using" and to the left of the bulleted text.

11. Save the document as *[your initials]*18-28 in your Lesson 18 folder.

12. Print and close the document.

EXERCISE 18-29

Insert, size, and position clip art, insert and format a shape, and insert and modify WordArt.

1. Open the file **Business**. Change the top margin to 2 inches.

2. Insert WordArt by following these steps:

   a. Select the title.

   b. Open the Insert menu and choose Picture, WordArt.

   c. Click the first sample in the second row and click OK.

   d. In the Edit WordArt Text dialog box, decrease the point size to 32 and change the text to uppercase and lowercase ("Corporate Travel"). Click OK.

3. Modify the WordArt object by following these steps:

   a. Right-click the WordArt and choose Format WordArt from the shortcut menu.

   b. Display the Colors and Lines tab. Choose either a fill color or a different line color. Click OK.

   c. Resize the clip to the width of the document text by pointing to the middle right sizing handle and dragging to the right. (Display the ruler and use it as a guide.)

4. Add two blank lines after the title.

5. Insert a shape by following these steps:

   a. Position the insertion point on the first paragraph mark after the title.

   b. Click the Line button on the Drawing toolbar. Press (Esc) if necessary to close the drawing canvas. Draw a line below the WordArt title and equal to the length of the title. Remember to hold (Shift) while dragging to create a straight line.

   c. Click the line to select it if necessary, and click the Line Style button and change the width to 3 pt.

   d. Click the Line Color button and change the color to coordinate with the WordArt title.

6. Use the Microsoft Clip Organizer to find clip art by following these steps:

   a. Display the Clip Art task pane and click Organize clips at the bottom of the task pane.

   b. Under Collection List, click the plus sign next to Office Collections to expand the list of categories.

   c. Double-click the Business category. Review the pictures in this category and in the Concepts subcategory.

   d. When you find an appropriate clip, right-click it and copy it to the Clipboard. (You might want to move the dialog box out of the way to review the document text before choosing your clip.)

7. Click the document and close the Clip Art task pane.

8. Paste the clip at the end of the document.

9. Change the wrapping style of the clip to Square. Make the clip a reasonable size (no larger than about 1.5 inches wide or high) and position it attractively in the document. You might want to change the color you applied to your WordArt to pick up a color from your clip art.

10. Save the document as *[your initials]*18-29 in your Lesson 18 folder.

11. Print and close the document.

12. Close the Microsoft Clip Organizer. Click No when asked if you want the clip to remain on the Clipboard.

EXERCISE 18-30

Create a diagram and insert a text box.

1. Open the file **Music**.

2. Insert a Radial diagram by following these steps:

   a. Insert a page break at the bottom of page 1.

   b. Open the Insert menu and choose Diagram. Double-click the Radial diagram, which is the last sample in row one.

   c. On the Radial Diagram toolbar, click Insert Shape twice to add two more circles to the diagram.

3. Add and format text in the diagram by following these steps:

   a. Click in the innermost circle and key **Classical Music in Santa Fe**.

   b. In the outer circles, key the following types of music, one type per circle:

      Opera
      Chamber
      Choral
      Symphony
      Musical Theater and Dance

   c. Click the AutoFormat button 🔄 on the Radial Diagram toolbar and choose Square Shadows. Click OK.

   d. Select the text in the center square and format it as bold with 30-point spacing before the paragraph.

   e. Copy this formatting to the text in the other circles.

   REVIEW: To copy formatting repeatedly, click within the formatted text you want to copy, and then double-click the Format Painter button 🖌. Using the Format Painter pointer, click or select the text you want to format. Press Esc to restore the normal pointer.

4. Format the first line on page 1 as 16-point bold italic. Change the line spacing for all the text on page 1 to 1.5 lines.

5. In the paragraph that begins "July 1," format the title "Santa Fe Times," "Madame Butterfly," and "The Marriage of Figaro" as italic.

6. Add and format a text box by following these steps:

   a. Select the last paragraph that begins "For reservations," and click the Text Box button 🔲 on the Drawing toolbar.

   b. Select the text box and click the Line Color button 🔳 on the Drawing toolbar and apply the dark blue color.

   c. Change the fill color of the text box by clicking the Fill Color button 🔳 on the Drawing toolbar, and choosing Pale Blue.

   d. Click the Line Style button ≡ and choose 2 1/4 pt.

   e. Change the left and right indents to .5 and change the line spacing to single. Apply bold formatting to the paragraph.

7. Change the page layout for the whole document so it is centered vertically.

8. Add a page border to the whole document.

9. Save the document as *[your initials]*18-30 in your Lesson 18 folder.

10. Print and close the document.

EXERCISE 18-31

Create a chart, change chart types, add options, and edit chart data.

1. Start a new document. Create a memo to Frank Youngblood from you. The subject is **Climate Charts**. For the body of the memo, key the following text: **We are designing a group of climate charts for our Southwest Web page. What do you think of this type of chart for showing high and low temperatures?**

2. Two lines below the last paragraph, insert a chart by following these steps:

   a. Open the Insert menu and choose Picture, Chart.

   b. Press Ctrl +A to select all the sample chart data. Press Delete.

   c. In the datasheet, key the data from Figure 18-41.

3. Change the chart type and add chart options by following these steps:

   a. Click the Chart Type button 🔳 and choose the 3-D Line chart (second option in the fourth row).

   b. Open the Chart menu and choose Chart Type.

   c. Under Chart sub-type, click the fourth sub-type (line with markers). Use the Press and Hold to View Sample button to see a preview of the chart. Click OK. (See Figure 18-41 on the next page.)

FIGURE 18-41
Temperatures
in Santa Fe

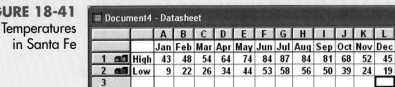

			A	B	C	D	E	F	G	H	I	J	K	L
			Jan	Feb	Mar	Apr	May	Jun	Jul	Aug	Sep	Oct	Nov	Dec
1		High	43	48	54	64	74	84	87	84	81	68	52	45
2		Low	9	22	26	34	44	53	58	56	50	39	24	19
3														
4														

  d. Click the View Datasheet button on the toolbar to hide the datasheet.

  e. Drag the chart's bottom right sizing handle to increase the chart size until you can see all 12 months displayed on the x-axis. Click outside the chart to view the chart in the document. (The chart's right edge should not go beyond the document's right margin.)

  f. Double-click the chart to reactivate Microsoft Graph. Open the Chart menu and choose Chart Options.

  g. Key **Santa Fe, New Mexico: Average Monthly Temperatures** as the Chart title. Key **Temperature** as the Value (Y) axis.

  h. Click the Gridlines tab. Under Category (X) axis, click Major gridlines. Click OK.

  i. Click the chart title. Select the colon after "Mexico" and the space following it, and press [Enter] to create a two-line title.

4. Edit the chart data by following these steps:

  a. Display the datasheet.

  b. Change the January low figure to **19**.

  c. Click outside the chart to return to the Word document.

5. Save the document as *[your initials]***18-31** in your Lesson 18 folder.

6. Print and close the document.

Lesson Applications

EXERCISE 18-32

Insert, size, position, and format clip art and WordArt.

1. Open the file **Special2**. Set a 1.75-inch top margin and 1-25-inch left and right margins.

2. Add two blank lines at the top of the document. At the first blank line, insert a WordArt object. Choose the third style in the third row of the WordArt Gallery dialog box. Use the text **Gateway Travel Tours**.

3. Format the text below the WordArt object (from "Some of our" through the end of the document) as two columns with 0.25 inch between the columns. Balance the length of the columns.

4. In the paragraph that begins "Our local area," change the last words "student tours" to **golf tours**. At the end of the paragraph, insert clip art of a golfer. Search for the clip art by using the keyword "golf." Scale the picture proportionately to 60% and apply square text wrapping. Adjust the picture position to place it attractively to the right of the "Our local area" paragraph.

5. Locate the paragraph that begins "Our naturalist." At the beginning of the paragraph, insert clip art of a backpacker. Size the picture to 1.2 inches high (make sure Lock aspect ratio is checked) and apply square text wrapping. Place the picture attractively to the left of the paragraph.

6. Add a page border to the document. Readjust picture positions after the border is added, if needed.

7. Save the document as *[your initials]***18-32** in your Lesson 18 folder.

8. Print and close the document.

EXERCISE 18-33

Insert a diagram and insert a chart with added options.

1. Start a new document. Create a memo to Staff from Alexis Johnson. The subject is **Sales Analysis**. For the body of the memo, key the following text:

   An analysis of our customer base over a two-year period shows that repeat business and referrals are the foundation of our business. Additional sources of new business are the Internet, advertising (print ads and direct mail), and walk-in traffic to the office. See the attached chart.

2. Two lines below this paragraph, insert a pyramid diagram. Use the Insert Shape button twice so you have a total of five layers. Add the following

text, one per layer, from top to bottom. You'll have to reduce the font size for the top text item to fit.

Walk-ins
Advertising
Internet
Referrals
Repeat Business

3. To make the diagram fit on page 1 of the memo, right-click the diagram border, open the Format Diagram dialog box, and scale the diagram to 85 percent of its original size.

4. Change the text wrapping to square and center align the diagram.

5. Use the AutoFormat button on the Diagram toolbar to choose another style. Make the diagram text bold and change the font to Arial Narrow.

6. Change the paragraph spacing before for all levels except the first layer to center align the text vertically.

7. Deselect the diagram and insert a page break at the end of the page.

8. On page 2, insert a standard column chart (clustered column sub-type) with the data shown in Figure 18-42.

FIGURE 18-42

	2003	2004
Repeat Business	.93	1.41
Referrals	.38	.4
Internet	.24	.26
Advertising	.2	.22
Walk-ins	.11	.07

9. Add the chart title **Sales Analysis** and the value axis title **Millions of Dollars**.

10. Resize the chart to the width of the document text area (6 inches)— use the ruler as a guide. Make the chart taller, as well.

11. Add a continuation page header to page 2.

12. Save the document as *[your initials]***18-33** in your Lesson 18 folder.

13. Print and close the document.

EXERCISE 18-34

Insert a picture and work with text boxes and advanced layout settings.

1. Start a new document. Change the left and right margins to 1.5 inches.

2. Create a text box measuring 3 inches wide by 2 inches high. If the drawing canvas appears before you draw the text box, press Esc to delete the drawing canvas, draw the text box, and turn off the automatic drawing canvas feature.

3. Key **Big Apple Gateway Travel** in the text box and format the text as 48-point bold, small caps.

4. Click the Change Text Direction button on the Text Box toolbar twice, so the text reads from bottom to top. Center the text.

5. Right-click the text box border and choose Format Text Box from the shortcut menu. On the Size tab, set the height to 8.5 inches and the width to 1 inch.

6. On the Colors and Lines tab of the Format Text Box, open the Fill Color box and choose Fill Effects. On the Gradient tab, choose One color. Using the default color, choose Diagonal down under Shading styles. Drag the slider scale slightly toward Light to lighten the color. Click OK twice.

7. Drag the text box to the left margin and centered between the top and bottom margins.

8. Deselect the text box and open the Clip Art task pane. Search for a picture of New York City and insert it. (Choose one that is taller than it is wide.)

9. Apply Square text wrapping to the picture. Size the picture to 3 inches wide. Drag the picture until its top border aligns with the text box and the right margin of the text box aligns to the right margin of the document.

10. About ¼ inch below the picture, draw a text box that is as wide as the picture and that extends to the bottom of the left text box. Remove the line from the text box.

11. In the new text box, key the text in Figure 18-43. Use single spacing and align and format as shown.

FIGURE 18-43

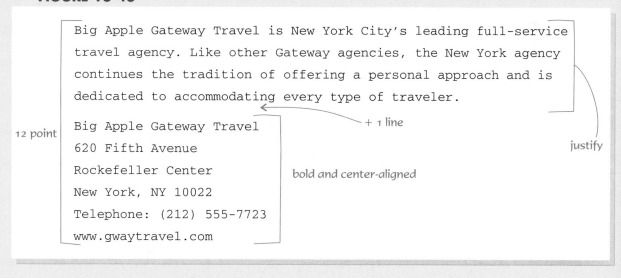

12. Format both instances of "Big Apple Gateway Travel" in the bottom text box as bold small caps.

13. Position the insertion point in the first line of the address and change the paragraph spacing before to 48 points.

14. Save the document as *[your initials]*18-34 in your Lesson 18 folder.

15. Print and close the document.

EXERCISE 18-35 *Challenge Yourself*

Insert, position, and modify AutoShapes, a graphic, and a text box; work with the drawing canvas.

1. Start a new document and make sure the automatic drawing canvas feature is turned on. Change all margins to 1 inch and switch to landscape orientation. Zoom to Page Width.

2. Add a 1-point box page border to the document.

3. From the Stars and Banners AutoShape category, draw the Explosion 2 shape across the top of the drawing canvas.

4. To make the shape look like the one in Figure 18-44, do the following:

   ● Right-click the shape and choose Add Text. Add the text shown in the shape. Format the text as 36-point Arial bold, italic, centered. Size the shape to fit the text, as shown.

   ● Rotate the shape to the right until it's not tilted up.

   ● Add light yellow fill color and apply Shadow Style 6. Exaggerate the shadow by nudging it right and down. (Click Shadow Settings at the bottom of the Shadow menu and use the Nudge buttons on the Shadow Settings toolbar.)

FIGURE 18-44

5. From the Block Arrows AutoShape category, draw a Striped Right Arrow inside the drawing canvas. Apply black fill color, copy the arrow, and then place both as shown in Figure 18-44.

6. Size the drawing canvas to fit the objects.

7. Below the drawing canvas, draw a text box with the text as shown. Format the text as 36-point Arial. Position the text box horizontally so it is left-aligned with the margin and vertically so it is bottom-aligned with the margin.

8. For the logo shown at the bottom right of the figure, insert the picture file **Dukelogo**.

> **REVIEW:** Choose Insert, Picture, From File. Locate the file (in the folder that contains the other files you use in this course) and click Insert.

9. Position the logo so it is horizontally aligned at the right margin and vertically aligned at the bottom margin.

10. Save the document as *[your initials]***18-35** in your Lesson 18 folder.

11. Print and close the document.

On Your Own

In these exercises, you work on your own, as you would in a real-life work environment. Use the skills you've learned to accomplish the task—and be creative.

EXERCISE 18-36
Establish an Internet connection and use Word's Clip Art task pane to search only for photographs with the keyword **sports**. Insert one of the photographs into a blank document, add accompanying text by inserting a text box, and then save the document as *[your initials]***18-36**. Print the document.

EXERCISE 18-37
Insert a target diagram into a blank document. Add text appropriate for this type of diagram. (A target diagram shows steps toward achieving a goal, the goal being the "bull's eye.") Save the document as *[your initials]***18-37** and print it.

EXERCISE 18-38
Create a simple column chart, using your own data. Copy the chart twice in the same document, and apply different chart types or sub-types for the same data (such as 3-D Cone or Pyramid). Save the document as *[your initials]***18-38** and print it.

Unit 6 Applications

Create a new template, create and apply paragraph and character styles, modify styles, and use style options.

1. Create a new document based on the default template. Change the left and right margins to 1 inch.

2. Using WordArt, key the title **North American Travel Update** in 20-point text. Center the WordArt, and press Enter twice.

3. Key **Duke City Gateway Travel Agency** and press Enter three times.

4. Create a paragraph style named Subhead that is 14-point Arial regular, all caps, centered, and apply it to the second line of text.

5. At the last paragraph mark, key **November Events**. Create and apply a paragraph style for this text based on the Normal style and named UpdateHead. Use 12-point Arial bold, all caps.

6. Save the document as a template named *[your initials]***update** in a new Unit 6 Applications folder or in the default Templates folder on the hard disk, whichever your instructor tells you to use. Close the template.

7. Start a new document based on the template *[your initials]***update**.

8. Replace the word "NOVEMBER" with **MAY**.

9. Two blank lines below "MAY EVENTS," key **Boston, Massachusetts**, and apply the Normal style.

10. To the right of "Massachusetts," insert a tab character and key **May 13-15**.

11. To this line of text, create and apply a paragraph style named CityName using 12-point Arial bold with a 3-inch left tab setting.

12. Create a paragraph style named Indentedpara based on the CityName style, using 12-point Arial regular, with 12 points spacing after paragraphs and a 3-inch hanging indent.

13. Assign Indentedpara as the style for the paragraph following CityName.

14. Assign CityName as the style for the paragraph following Indentedpara.

15. Create a character style named EventDate using 11-point Arial bold italic and apply it to the text "May 13-15."

16. Press (Enter) after "15," press (Tab), and key **Annual Chowder Festival: Boston's best clam, corn, and seafood chowders**.

17. Press (Enter) and key the text shown in Figure U6-1. Use single spacing. Press (Tab) before each date and before each description. Apply the EventDate style to all the date text.

FIGURE U6-1

```
New York, New York          May 29
                            The New York City Marathon

Cape May, New Jersey        May 7-9
                            Victorian house tours, lawn parties,
                            and arts & crafts

New Orleans, Louisiana      May 10-15
                            Jazz Festival: A week of jamming with
                            the country's greatest jazz musicians

Tampa, Florida              May 20-22
                            Strawberry Festival: The nation's best
                            strawberries during spring harvest

St. Louis, Missouri         May 5
                            The Annual May Day Parade down Main Street

Flint, Michigan             May 19-22
                            The Greene's Nursery Annual Tulip Festival

Houston, Texas              May 18-19
                            The Annual Amateur Rodeo and Western Show

San Diego, California       May 24-29
                            Pacific Regatta: Hundreds of sailboats
                            decorating the Pacific for a week
```

18. Modify the CityName style to 11 points.

19. Align the entire document vertically on the page.

20. Spell-check the document and save it as *[your initials]***u6-1** in a folder for Unit 6 Applications.

21. Print and close the document.

UNIT APPLICATION 6-2

Insert and resize a chart and a graphic.

1. Open the file **Wonders**.

2. Set a 2-inch top margin and add a memo heading to the document. The memo is to Frank Youngblood from Valerie Grier, regarding the Wonders of the World travel packages. Add your reference initials.

3. Make the following changes to the table:
   - Change the width of column 1 to fit the text and distribute the remaining column widths evenly.
   - Make the column headings bold.
   - Increase row height for the entire table to 0.25 inches and center the text vertically in the cells.

4. Three lines below the table, insert a 3-D column chart that charts just the Total Travel Packages row in the table (and the year column headings). Be sure the years appear along the x-axis. Add the chart title **Total Travel Packages** and hide the legend. Size the chart to about the same size as the table.

5. Above the table, format the text "Number of Packages Sold" to match the chart title. Add a blank line before it.

6. Add a continuation page heading to page 2. Remember to change the header to 1-inch from the edge.

7. In the page 1 header, create a letterhead for Duke City Gateway Travel. Use the following text and the graphic file **Dukelogo** (located where your other student files are stored). Size the graphic to no taller than the header pane, and format and position the text with the graphic attractively.
   15 Montgomery Boulevard
   Albuquerque, NM 87111
   (505) 555-1234
   www.gwaytravel.com

8. Because the "MEMO TO:" line should start about three lines below the letterhead, add a few blank lines, as needed.

9. If your chart appears on page 2, make your table taller (increase row height) to fill up some vertical space.

10. Save the document as *[your initials]***u6-2** in a new folder for Unit 6 Applications.

11. Print and close the document.

UNIT APPLICATION 6-3

Create a main document and a data source, edit the data source, and merge the documents.

1. Start a new document based on the Elegant Merge Letter template. (Tip: The template is on the Mail Merge tab in the Templates dialog box.) The new document already contains the Address Block and Greeting Line merge fields.

2. Key **Duke City Gateway Travel Agency** in the top placeholder. Key the following address information in the bottom placeholders, replacing "FAX:" with "WEB:"

15 Montgomery Boulevard
Albuquerque, NM 87111
(505) 555-1234
www.gwaytravel.com

3. Switch to the Styles and Formatting task pane to make the following changes:

- Modify the paragraph spacing of the Date style to 24 points before and 36 points after.

- Modify the paragraph spacing of the Signature style to 30 points before.

4. Switch back to the Mail Merge task pane.

5. For the data source, key the data in Figure U6-2. Customize the list of field names as needed.

FIGURE U6-2

Title	First Name	Last Name	Address	City	State	ZIP
Mr.	Gary	Pines	1115 Sunrise Bluffs Dr.	Belen	NM	87002
Ms.	Donna	Albert	12 Avon Lane	Fort Sumner	NM	87086
Mr.	Paul	Green	4 Milky Way Dr.	Grants	NM	87020
Ms.	Briana	Chin	48 Lucia St.	Lincoln	NM	87121
Ms.	Gina	Scotto	153 35th Circle	Rio Rancho	NM	87124
Mr.	Raymond	Steel	10 Low St.	Taos	NM	87557

 NOTE: Enter addresses without abbreviating "Drive" and "Street."

6. Save the data source as *[your initials]***u6-3data** in your Unit 6 Applications folder.

7. Check the recipient list against Figure U6-2 for errors, and then sort the list by ZIP Code (click the ZIP heading in the dialog box).

8. For the body of the letter, key the corrected text shown in Figure U6-3.

FIGURE U6-3

As you may have heard, Duke City Gateway Travel is offering a once-in-a-lifetime, four-day visit to Gettysburg, Pennsylvania. During this ~~time~~ visit, you will see the ~~present-day~~ 25 square miles of land where the epic battle (was fought,) ~~which~~ that decided the fate of the United States. Some of the sites ~~that we~~ you will visit are:

Devil's Den

Little Round Top

Culp's Hill

Rock Creek

Wheat Field

Peach Orchard

~~Seminary Ridge~~ (Stet)

McPherson's Ridge

~~We will~~ Walk the ground where, over 140 years ago, men from both the North and the South fought for three long days in the biggest and bloodiest battle ever on American soil.

Please let us know if you are interested in joining us on this trip into America's past. Let us know soon, because space is limited.

9. In the closing text placeholders, key the name **Alice Fung** and the title **Travel Counselor**. On the second line below the title, add your reference initials.

10. Edit the paragraph that begins "Please" to start with merge fields for the title and last name. Include correct punctuation.

11. Preview the letter with the merged data, making sure the letter fits on one page.

12. Notice that the greeting line has a comma instead of a colon. To fix this, right-click the greeting line, choose Edit Greeting Line from the shortcut menu, and change the punctuation to a colon.

13. Turn off the view of merged data (click the View Merged Data button on the Mail Merge toolbar).

14. Sort alphabetically the list of sites (which starts with "Devil's Den") and apply the style List Bullet.

15. Save the main document as *[your initials]***u6-3main** in your Unit 6 Applications folder and print it.

16. Edit the data source to change Mr. Raymond Steel to Mrs. Raymond Steel and add a new record for Mr. Allen Jones, 34 Sky Road, Union, NM 87117. Sort the list by ZIP Code.

17. Complete the merge, merging all data to a new document.

18. Save the new document as *[your initials]***u6-3merged** in your Unit 6 Applications folder.

19. Print the document, four pages per sheet.

20. Use the data source created in this application to create mailing labels. Use the 5160 – Address product number. Save the merged labels to a new document.

21. Format the label text as 11-point Arial bold and then save as *[your initials]***u6-3labels**. Print the labels.

22. Close all documents, saving changes when prompted.

UNIT APPLICATION 6-4   *Using the Internet*

Use the Internet to download a template and clip art; modify graphics; insert text boxes, and create a WordArt object.

1. Connect to the Internet. Display the New Document task pane and click the link Templates on Office Online.

2. In the Templates Gallery, click the heading Holidays and Occasions.

3. Click the For holidays category and then click the Autumn Holidays category. Choose the template Thanksgiving Dinner Menu (casual).

4. Click the Download Now button.

5. Where you would begin typing the menu, key **MENU:**.

6. Delete the graphic(s) at the top of the page, and replace it with WordArt using the same or similar text. You can change the holiday to any other fall celebration, such as a Halloween Feast, Harvest Dinner, or Football Brunch.

7. Below "MENU" insert a text box and key the menu items you would serve at such an event. Below the list of items, include a short paragraph to your guests.

8. Change the wrapping style of the transparent clip at the bottom of the page to Behind Text. Size the clip to as large as you like, making it a background to your page.

9. Make any other improvements to the document; then save it as *[your initials]*u6-4 in your Unit 6 Applications folder.

10. Print and close the document.

Appendixes

APPENDIX A

Proofreaders' Marks

PROOFREADERS' MARK		DRAFT	FINAL COPY
⌗	Start a new paragraph	ridiculous! If that is so	ridiculous! If that is so
⌒	Delete space	to gether	together
# / ∧	Insert space	Itmay be	It may not be
↑ ⌀	Move as shown	it is (not) true	it is true
∿	Transpose	beleivable	believable
		is it so	it is so
◯	Spell out	②years ago	two years ago
		16 Elm St.	16 Elm Street
∧	Insert a word	How much it?	How much is it?
℘ OR —	Delete a word	it may not be true	it may be true
∧ OR ⅄	Insert a letter	temperture	temperature
ℐ OR ⋛	Delete a letter and close up	commitment to buny	commitment to buy
℘ OR —	Change a word	and if you won't	but if you can't
(Stet)	Stet (don't delete)	I was very glad	I was very glad
/	Make letter lowercase	Federal Government	federal government
≡	Capitalize	Janet L. greyston	Janet L. Greyston
∨	Raise above the line	in her new book*	in her new book*
∧	Drop below the line	H2SO4	H_2SO_4

PROOFREADERS' MARK		DRAFT	FINAL COPY
⊙	Insert a period	Mr. Henry Grenada	Mr. Henry Grenada
∧	Insert a comma	a large old house	a large, old house
∨	Insert an apostrophe	my childrens car	my children's car
∨∨	Insert quotation marks	he wants a loan	he wants a "loan"
= OR ̭	Insert a hyphen	a first rate job	a first-rate job
		ask the coowner	ask the co-owner
¦ M	Insert an em-dash	Here it is cash!	Here it is—cash!
¦ N	Insert an en-dash	Pages 1 5	Pages 1–5
———	Insert underscore	an issue of <u>Time</u>	an issue of <u>Time</u>
ital	Set in italic	ital The New York Times	*The New York Times*
bf	Set in boldface	bf the Enter key	the **Enter** key
rom	Set in roman	rom the *most* likely	the most likely
⟨ ⟩	Insert parentheses	left today May 3	left today (May 3)
⊐	Move to the right	$38,367,000	$38,367,000
⊏	Move to the left	Anyone can win!	Anyone can win!
ss	Single-space	ss ⎡I have heard ⎣he is leaving	I have heard he is leaving
ds	Double-space	ds ⎡When will you ⎣have a decision?	When will you have a decision?
(+ 1 line)	Insert one line space	<u>Percent of Change</u> (+ 1 line) 16.25	<u>Percent of Change</u> 16.25
(– 1 line)	Delete (remove) one line space	Northeastern (– 1 line) regional sales	Northeastern regional sales

APPENDIX B

Standard Forms for Business Documents

Reference manuals, such as *The Gregg Reference Manual,* provide a variety of letter and memorandum styles, as well as styles for reports and other documents. Many businesses also have their own styles for documents. This Appendix includes two basic styles—for a business letter and a memorandum. It also shows the most common format for a continuation page (used for either letters or memos).

TABLE B-1   Parts of a Letter

PART OF LETTER	LOCATION/DESCRIPTION
Heading	
Letterhead or return address	Often appears on pre-printed stationery; can also be created in Word. Includes the company name, address, and other contact information.
Date line	Two inches from the top of the page on letterhead stationery or on the third line below a Word letterhead. Use date format shown in Figure B-1.
Opening	
Inside address	Starts on the fourth line below the date; consists of name and address (and possibly company name and job title) of person to whom you are writing.
Salutation	On the second line below the inside address; typically includes a courtesy title (Mr., Mrs., Ms., Miss) and ends with a colon.
Body	
Message	Content of the letter, single-spaced with one blank line between paragraphs.
Closing	
Complimentary closing	On the second line below the last line of the body of the letter. Common closings are "Sincerely" or "Sincerely yours" followed by a comma.
Writer's identification	On the fourth line below the closing, to leave space for a signature; includes the writer's name and job title (and sometimes the department).
Reference initials	On the second line below the writer's name and title; consists of the typist's initials in small letters.
Enclosure notation	On new line below the reference initials if letter has an enclosure. Specify the number of enclosures. Can also use "Attachment" if enclosure is attached.
Optional features	Filename notation—indicates document name for reference purposes; delivery notation—method of delivery (other than regular mail); copy notation—people who will receive copies of the letter (usually begins with "c:" or "cc:").

FIGURE B-1
Business letter style

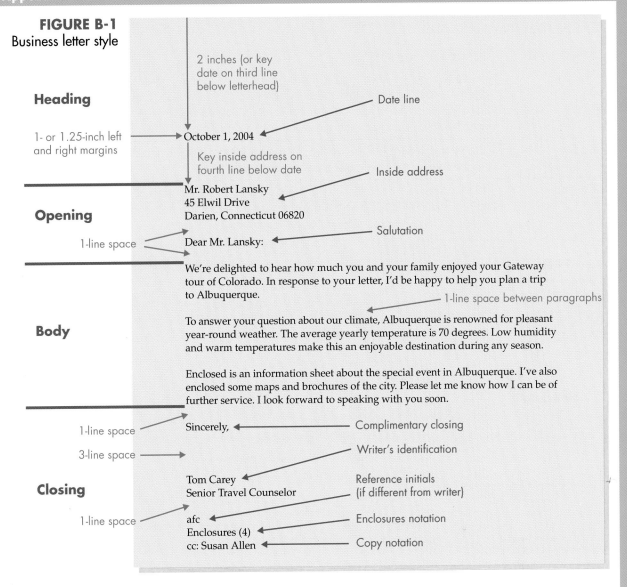

Heading

1- or 1.25-inch left
and right margins

2 inches (or key
date on third line
below letterhead)

Date line

October 1, 2004

Key inside address on
fourth line below date

Inside address

Mr. Robert Lansky
45 Elwil Drive
Darien, Connecticut 06820

Opening

1-line space

Dear Mr. Lansky:

Salutation

We're delighted to hear how much you and your family enjoyed your Gateway
tour of Colorado. In response to your letter, I'd be happy to help you plan a trip
to Albuquerque.

1-line space between paragraphs

Body

To answer your question about our climate, Albuquerque is renowned for pleasant
year-round weather. The average yearly temperature is 70 degrees. Low humidity
and warm temperatures make this an enjoyable destination during any season.

Enclosed is an information sheet about the special event in Albuquerque. I've also
enclosed some maps and brochures of the city. Please let me know how I can be of
further service. I look forward to speaking with you soon.

1-line space

Sincerely,

Complimentary closing

3-line space

Writer's identification

Closing

Tom Carey
Senior Travel Counselor

Reference initials
(if different from writer)

1-line space

afc
Enclosures (4)
cc: Susan Allen

Enclosures notation

Copy notation

FIGURE B-2
Continuation
page header for
2-page (or longer)
letter or memo

1 or 1.25 inches

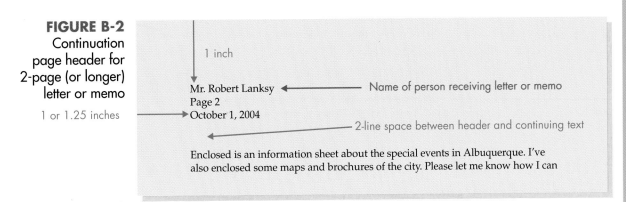

1 inch

Mr. Robert Lanksy
Page 2
October 1, 2004

Name of person receiving letter or memo

2-line space between header and continuing text

Enclosed is an information sheet about the special events in Albuquerque. I've
also enclosed some maps and brochures of the city. Please let me know how I can

TABLE B-2   Parts of a Memo

PART OF MEMO	LOCATION/DESCRIPTION
Heading	Starts two inches from top of page using plain paper or letterhead stationery or on third line below Word letterhead. Consists of guide words ("MEMO TO," "FROM," "DATE," and "SUBJECT") in capital letters followed by a colon. Entries after guide words align at a 1-inch left tab setting. Use date format shown in Figure B-3.
Body	Starts on the third line below the memo heading; contains the message, single-spaced with one blank line between paragraphs.
Closing	On the second line below the last paragraph; includes reference initials (the typist's initials in small letters). Might also include an enclosure notation, a file name notation, and a copy notation or distribution list.

FIGURE B-3
Memorandum style

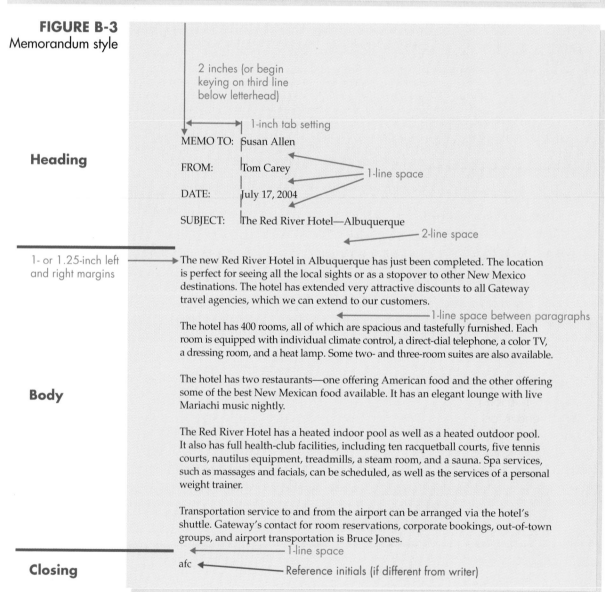

APPENDIX C

Microsoft Office Specialist Certification

TABLE C-1

Microsoft Office Specialist Activities Related to Lessons

CODE	ACTIVITY	LESSON
WW03S-1	**Creating Content**	
WW03S-1-1	Insert and edit text, symbols and special characters	1, 4, 5, 8
WW03S-1-2	Insert frequently used and predefined text	4, 6
WW03S-1-3	Navigate to specific content	9, 10, 11
WW03S-1-4	Insert, position, and size graphics	18
WW03S-1-5	Create and modify diagrams and charts	18
WW03S-1-6	Locate, select, and insert supporting information	4
WW03S-2	**Organizing Content**	
WW03S-2-1	Insert and modify tables	13
WW03S-2-2	Create bulleted lists, numbered lists, and outlines	5
WW03S-2-3	Insert and modify hyperlinks	10
WW03S-3	**Formatting Content**	
WW03S-3-1	Format text	3, 9, 13, 15
WW03S-3-2	Format paragraphs	5, 7
WW03S-3-3	Apply and format columns	14
WW03S-3-4	Insert and modify content in headers and footers	12
WW03S-3-5	Modify document layout and page setup	6, 11
WW03S-4	**Collaborating**	
WW03S-4-1	Circulate documents for review	10
WW03S-4-2	Compare and merge documents	10
WW03S-4-3	Insert, view, and edit comments	10
WW03S-4-4	Track, accept, and reject proposed changes	10
WW03S-5	**Formatting and Managing Documents**	
WW03S-5-1	Create new documents using templates	16
WW03S-5-2	Review and modify document properties	2

continues

TABLE C-1　*continued*

CODE	ACTIVITY	LESSON
WW03S-5-3	Organize documents using file folders	2
WW03S-5-4	Save documents in appropriate formats for different uses	2
WW03S-5-5	Print documents, envelopes, and labels	6
WW03S-5-6	Preview documents and Web pages	6
WW03S-5-7	Change and organize document views and windows	3, 6, 8, 10

TABLE C-2

Lessons Related to Microsoft Office Specialist Activities

LESSON		CODE
1	Creating a Document	WW03S-1-1, WW03S-5-3, WW03S-5-4
2	Selecting and Editing Text	WW03S-5-2, WW03S-5-3, WW03S-5-4, WW03E-4-6
3	Character Formatting	WW03S-3-1, WW03S-5-7
4	Writing Tools	WW03S-1-1, WW03S-1-2, WW03S-1-6
5	Formatting Paragraphs	WW03S-1-1, WW03S-2-2, WW03S-3-2
6	Margins and Printing Options	WW03S-1-2, WW03S-3-5, WW03S-5-5, WW03S-5-6, WW03S-5-7
7	Tabs and Tabbed Columns	WW03S-3-2
8	Moving and Copying Text	WW03S-1-1, WW03S-5-7
9	Find and Replace	WW03S-1-3, WW03S-3-1
10	Sharing Your Work and Hyperlinks	WW03S-1-3, WW03S-2-3, WW03S-4-1, WW03S-4-2, WW03S-4-3, WW03S-4-4, WW03S-5-7, WW03E-4-1, WW03E-4-4
11	Page and Section Breaks	WW03S-1-3, WW03S-3-5, WW03E-1-2
12	Page Numbers, Headers, and Footers	WW03S-3-4
13	Tables	WW03S-2-1, WW03S-3-1
14	Columns	WW03S-3-3
15	Styles	WW03S-3-1, WW03E-1-1
16	Templates and Wizards	WW03S-5-1, WW03E-5-3
17	Mail Merge	WW03E-2-6, WW03E-2-7
18	Graphics and Charts	WW03S-1-4, WW03S-1-5, WW03E-1-3

Glossary

Active window Window in which you are currently working, which shows the title bar highlighted and its taskbar button depressed. (8)

Antonym Word that is opposite in meaning to another word. (4)

Ascending sort order Arrangement that places items in first-to-last or lowest-to-highest order, such as from "A" to "Z" or 0 to 9. (13) (17)

Attribute Setting such as boldface or italics, that affects the appearance of text. (15)

AutoComplete Automatic Word feature that suggests the completed word when you key the first four or more letters of a day, month, or date. (4)

AutoCorrect Automatic Word feature that corrects commonly misspelled words as you key text. (4)

AutoFormat Word feature that automatically changes formatting as you key text or numbers. (3)

AutoRecover Word feature that automatically saves open documents in the background. This backup version of the document can be recovered in case the original is lost or damaged in a power failure or because of a system problem. (2)

AutoShapes Group of ready-made shapes that are grouped by category: lines, connectors, basic shapes, block arrows, flowchart symbols, stars and banners, and callouts. (18)

AutoText Word feature you can use to insert text automatically. (4)

Background repagination Automatic process of updating page breaks and page numbers that occurs while you are creating or editing a document. (12)

Border Line, box, or pattern that you place around text, a graphic, or a page. (5)

Bulleted list List of items, each preceded by a bullet (•). Each item is a paragraph with a hanging indent. (5)

Cell Portion of a table that is formed by the intersection of rows and columns. (13)

Character style Formatting applied to selected text within a paragraph; includes font, font size, and font style. (15)

Click and type Insert text or graphics in any blank area of a document. Position the insertion point anywhere in a document, click, and then type. Word automatically inserts paragraph marks before that point and also inserts tabs, depending on the location of the insertion point. (5)

Clip art Ready-to-use drawings that can be inserted into a document. (18)

Clipboard Temporary storage area in the computer's memory used to hold text or other information that is cut or copied. (8)

Clips Multimedia files that can be clip art, photographs, movies, or sound files. (18)

Comment Electronic note that you add to a document. (10)

Comment reference mark Vertical line that appears where a reviewer has inserted a comment; the mark is color-coded to reviewer. (10)

Contiguous text Any group of characters, words, sentences, or paragraphs that follow one another. (2)

Crop Trim a picture so that only a portion of the original shows. (18)

Cut and paste Method for moving text or other information by removing it from a document, storing it on the Clipboard, and then placing it in a new location. (8)

Cycle diagram Diagram that illustrates a process that has a continuous cycle. (18)

Data series Collections of related data points plotted in a chart. (18)

Data source Variable information in a mail merge, such as names and addresses, to use in personalizing the main document. (17)

Datasheet Table in Microsoft Graph from which data is charted. (18)

Descending sort order Arrangement that places items in last-to-first or highest-to-lowest order, such as from "Z" to "A" or 9 to 0. (13) (17)

Drag-and-drop Method for moving or copying text or other objects short distances by dragging them. (8)

Drawing canvas Rectangle graphic object in which you can draw multiple shapes, and then size and format them as a group. (18)

Drop cap Large letter that appears below the text baseline, usually applied to the first letter in the word of a paragraph. (3)

Em dash Dash twice as wide as an en dash and used in sentences where you would normally insert two hyphens. (5)

En dash Dash slightly wider than a hyphen. (5)

End mark Short horizontal bar on the Word screen that moves as you key text and indicates the end of the document. (1)

End-of-cell markers Character that indicates the end of the content of each cell in a table. (13)

End-of-row markers Character to the right of the gridline of each row in a table; indicates the end of the row. (13)

Facing pages Document with a two-page spread. Right-hand pages are odd-numbered pages, and left-hand pages are even numbered pages. (6)

Field Hidden codes that tell Word to insert specific information such as a date or page number; in a data source table, each item of information contained in a record. (6) (17)

Field code Merge field that appears in the main document as a placeholder for data from a data source. (17)

Field name In a data source table, the column heading of each category of information. (17)

Filename Unique name given to a document saved in Word. (1)

Find Command used to specify text and formatting in a document. (9)

First-line indent Indents the first line of a paragraph. (5)

First-line indent marker Top triangle on the left side of the ruler. Drag the indent marker to indent or extend the first line of a paragraph. (5)

Floating graphic Graphic inserted in the drawing layer of a document so it can be positioned freely on the page, in front of or behind text and other objects. (18)

Font The design applied to an entire set of characters, including all letters of the alphabet, numerals, punctuation marks, and other keyboard symbols. (3)

Footer Text that appears in the bottom margin of a page throughout a section or document. (12)

Gridlines Lines that mark the boundaries of cells in a table. (13)

Gutter margins Extra space added to the inside margins to allow for binding. (6)

Hanging indent Indents the second and subsequent lines of a paragraph. (5)

Hanging indent marker Bottom triangle on the left side of the ruler. Drag the marker to indent the second and subsequent lines in a paragraph. (5)

Hard page break Page break that you insert manually. Does not move, regardless of changes in the document. (11)

Header Text that appears in the top margin of a page throughout a section or document. (12)

Header row First or second row (if the table has a title row) of a table, in which each cell contains a heading for the column of text beneath it. (13)

HTML Represents Hypertext Markup Language. File format used to make a document readable in a browser, on an intranet, or on the Internet. (2)

Hyperlink Text or graphic you click to move to another location. (10)

Hyphenation Division of words that cannot fit at the end of a line. (6)

I-beam Shape of the mouse pointer when it is positioned in the text area. (1)

Indent Increases the distance between the sides of a paragraph and the two side margins (left and right). (5)

Indent marker On Word's horizontal ruler, a small box or triangle that you drag to control a paragraph's indents. (5)

Inline graphic Graphic inserted in a line of text, on the same layer as the text. (18)

Insert mode Mode of text entry that inserts text without overwriting existing text. The Insert key on a keyboard is a toggle key used to switch between Insert mode and Overtype mode. (1)

Insertion point Vertical blinking bar on the Word screen that indicates where an action will begin. (1)

Landscape Page orientation setting in which the page is wider than it is tall. (6)

Leader characters Patterns of dots or dashes that lead the reader's eye from one tabbed column to the next. (7)

Left and right indent Indents left and right sides of a paragraph (often used for quotes beyond three lines). (5)

Left indent Indents a paragraph from the left margin. (5)

Left indent marker Small rectangle on the left side of the ruler. Drag the marker to indent all lines in a paragraph simultaneously. (5)

Legal blackline Word feature that lets you compare and merge two documents. The differences between the two documents appear in a new document as revision marks. (10)

Line space Amount of vertical space between lines of text in a paragraph. (5)

Line-break characters Used to start a new line within the same paragraph. Insert by pressing [Shift]+[Enter]. (2)

List style Formatting instructions applied to a list, such as numbering or bullet characters, alignment, and fonts. (15)

Mail merge Process of using information from two documents (a main document and a data source) to produce a set of personalized documents, such as form letters or mailing labels and envelopes. (17)

Main document Document in a mail merge to be merged with a data source and sent to many people or printed on envelopes or labels. The main document is information that is constant (it does not change). (17)

Margins Spaces at the top, bottom, left, and right of the document between the edges of text and the edges of the paper. (6)

Markup Comments and tracked changes such as insertions, deletions, and formatting changes. (10)

Markup balloon Contains comments or tracked changes in the margin of a document in Print Layout view. (10)

Merge fields In a mail merge, placeholders you insert in a main document that indicate where to insert information from the data source. (17)

Mirror margins Inside and outside margins on facing pages that mirror one another. (6)

Negative indent Extends a paragraph into the left or right margin areas. (5)

Nonbreaking hyphen Hyphen used in a hyphenated word that should not be divided at a line break. (6)

Nonbreaking space Space between two words, defined by a special character that prevents Word from separating two words. Insert by pressing [Ctrl]+[Shift]+[Spacebar]. (2)

Noncontiguous text Text items (characters, words, sentences, or paragraphs) that do not follow one another, but each appear in different parts of a document. (2)

Nonprinting characters Symbol for a tab, paragraph, space, or another special character that appears on the screen but not in the printed document. (2)

Normal style Default paragraph style with the formatting specifications 12-point Times New Roman, English language, left-aligned, single-spaced, and widow/orphan control. (15)

Numbered list List of items preceded by sequential numbers or letters. Each item is a paragraph with a hanging indent. (5)

Optional hyphen Indicates where a word should be divided if the word falls at the end of a line. (6)

Ordinal number Number indicating an order or position, for example 1st, 2nd, or 3rd. (3)

Organizational chart Diagram that illustrates the top-down relationship of members of an organization. (18)

Organizer Feature that lets you copy styles from one document or template to another, or copy AutoText entries from one template to another. (16)

Orientation Setting to format a document with a tall, vertical format or a wide, horizontal format. (6)

Orphan First line of a paragraph that remains at the bottom of a page. (11)

Outline numbered list Numbering sequence used primarily for legal and technical documents. (5)

Overtype mode Mode of text entry that lets you key over existing text. The Insert key on the keyboard is a toggle key, used to switch between Insert mode and Overtype mode. (1)

Pagination Process of determining how and when text flows from the bottom of one page to the top of the next page in a document. (11)

Pane Section of a window that is formed when the window is split. A split window contains two panes. (8)

Paragraph Unique block of text or data that is always followed by a paragraph mark. (5)

Paragraph alignment Determines how the edges of a paragraph appear horizontally. (5)

Paragraph mark On-screen symbol (¶) that marks the end of a paragraph and stores all formatting for the paragraph. (1) (5)

Paragraph space Amount of space (measured in points) before and after a paragraph; replaces pressing Enter to add space between paragraphs. (5)

Paragraph style Formatting instructions applied to a paragraph; includes alignment, line and paragraph spacing, indents, tab settings, borders and shading, and character formatting. (15)

Placeholder text In a template, (or a new document based on a template), text containing the correct formatting, which you replace with your own information. (16)

Point Measure of type size; 72 points equals one inch. (3)

Portrait Page orientation setting in which the page is taller than it is wide. (6)

Positive indent Indents between the left and right margins. (5)

Proofreaders' marks Handwritten corrections to text, often using specialized symbols. (Appendix A)

Property Any information, such as the filename, date created, or file size that describes a document. (2)

Proportional sizing Resizing an image while maintaining its relative height and width. (18)

Pyramid diagram Diagram that illustrates foundation-based relationships. (18)

Radial diagram Diagram that illustrates the relationships of objects to a main object. (18)

Record In a data source table, a row of related information (such as name, address, city, state, and ZIP Code) for one person or business. (17)

Replace Command used to replace text and formatting automatically with specified alternatives. (9)

Reviewer Person who adds a comment or tracked change to a document. (10)

Reviewing pane Narrow horizontal pane that opens at the bottom of the screen to display revisions and comments. (10)

Revision bar Black vertical line that appears to the left of each line containing a revision. (10)

Revision mark Mark that Word applies to text that has been changed while the Track Changes feature is turned on. (10)

Right indent Indents paragraph from right margin. (5)

Right indent marker Triangle on the right side of the ruler; drag the marker to indent the right side of a paragraph. (5)

Ruler Part of the Word screen that shows placement of indents, margins, and tabs. (1)

Sans serif Font characteristics in which the font has no decorative lines, or serifs, projecting from its characters, such as Arial. (3)

Scale Change the size of an image as a percentage of its original size. (18)

ScreenTip Brief explanation or identification of an on-screen item, such as a menu command, dialog box option, or toolbar button. (1)

Scroll bar Used with the mouse to move right or left and up or down within a document to view text not currently visible on-screen. (1)

Section Portion of a document that has its own formatting. (6)

Section breaks Double-dotted lines that appear on-screen to indicate the beginning and end of a section. (6) (11)

Selection Area of a document that appears as a highlighted block of text. Selections can be formatted, moved, copied, deleted, or printed. (2)

Serif Font characteristics in which the font has decorative lines projecting from its characters, such as Times New Roman. (3)

Shading Shades of gray, a pattern, or a color applied to the background of a paragraph. (5)

Shortcut menu Menu that opens and shows a list of commands relevant to a particular item that you right-click. (3)

Size Change the height and/or width of an object. (18)

Sizing handles Squares or circles that appear around the border of a selected object, used for resizing the object. (18)

Smart tags Feature that recognizes dates, addresses, and user-defined data types, all of which you can use to perform actions in Word that you'd normally open other programs to do, such as Microsoft Outlook. (4)

SmartQuotes Quotation marks that curl in one direction (") to open a quote and curl in the opposite direction (") to close a quote. (5)

Soft page break Page break automatically inserted by Word and continually adjusted to reflect changes in the document. (11)

Sort To arrange items in a particular order, such as alphabetical or numerical order. Sorting is often done on tables and lists, but can also be performed on text paragraphs within a document. (13) (17)

Special characters Characters such as the trademark ™ symbol or those used in foreign languages. (5)

Spike Used to store multiple items that you delete, and then paste as a group in a new location. (8)

Split bar Horizontal line that divides a document into panes. (8)

Split box Small gray rectangle located just above the vertical scroll bar. You can drag it down to split a document into two panes. (8)

Status bar Located at the bottom of the Word screen, it displays information about the task you are performing, shows the position of the insertion point, and shows the current mode of operation. (1)

Style Set of formatting instructions that you apply to text. (15)

Style sheet List of style names and their formatting specifications. (15)

Symbol Font that includes special characters, such as the copyright symbol ©. (5)

Synonym Word that is similar in meaning to another word. (4)

Tab Paragraph-formatting feature used to align text. (7)

Tab characters Nonprinting characters used to indent text. (7)

Tab marker Symbol on the horizontal ruler that indicates a custom tab setting. (7)

Tab stop Position of a tab setting. (7)

Table Grid of rows and columns that intersect to form cells. (13)

Table style Formatting instructions applied to a table, such as borders, shading, alignment, and fonts. (13)

Target diagram Diagram that illustrates steps toward a goal. (18)

Task pane Pane to the right of the text area that gives easy access to a variety of functions. (1)

Template File that contains formatting information, styles, and text for a particular type of document or for a Web page. (16)

Text box Free-floating rectangle that contains text. (18)

Text wrapping Graphic option that lets text flow around an object or that positions the object behind or in front of text. (18)

Thesaurus Tool you can use to look up synonyms for a selected word. (4)

Thumbnail Miniature representation of a page in Reading Layout view. (10)

Title bar Displays the name of the current document at the top of the Word screen. (1)

Toolbar Contains buttons you click to initiate a wide range of commands. Each button is represented by an icon. (1)

Tracked changes Revision marks that show where an insertion, deletion, or other editing change occurred. (10)

Venn diagram Diagram that illustrates where areas overlap between objects. (18)

Widow Last line of a paragraph that remains at the top of a page. (11)

Wildcard Symbol that stands for missing or unknown text. (9)

Wingding Font that includes special characters, such as arrows. (5)

Wizard Automated and interactive template that helps you choose formatting options and prompts for text entries to be included in the resulting document. (16)

Index

INTERNATIONAL CONTACT INFORMATION

AUSTRALIA
McGraw-Hill Book Company
Australia Pty. Ltd.
TEL +61-2-9900-1800
FAX +61-2-9878-8881
http://www.mcgraw-hill.com.au
books-it_sydney@mcgraw-hill.com

CANADA
McGraw-Hill Ryerson Ltd.
TEL +905-430-5000
FAX +905-430-5020
http://www.mcgraw-hill.ca

GREECE, MIDDLE EAST, & AFRICA
(Excluding South Africa)
McGraw-Hill Hellas
TEL +30-210-6560-990
TEL +30-210-6560-993
TEL +30-210-6560-994
FAX +30-210-6545-525

MEXICO (Also serving Latin America)
McGraw-Hill Interamericana Editores
S.A. de C.V.
TEL +525-1500-5108
FAX +525-117-1589
http://www.mcgraw-hill.com.mx
carlos_ruiz@mcgraw-hill.com

SINGAPORE (Serving Asia)
McGraw-Hill Book Company
TEL +65-6863-1580
FAX +65-6862-3354
http://www.mcgraw-hill.com.sg
mghasia@mcgraw-hill.com

SOUTH AFRICA
McGraw-Hill South Africa
TEL +27-11-622-7512
FAX +27-11-622-9045
robyn_swanepoel@mcgraw-hill.com

SPAIN
McGraw-Hill/
Interamericana de España, S.A.U.
TEL +34-91-180-3000
FAX +34-91-372-8513
http://www.mcgraw-hill.es
professional@mcgraw-hill.es

UNITED KINGDOM, NORTHERN,
EASTERN, & CENTRAL EUROPE
McGraw-Hill Education Europe
TEL +44-1-628-502500
FAX +44-1-628-770224
http://www.mcgraw-hill.co.uk
emea_queries@mcgraw-hill.com

ALL OTHER INQUIRIES Contact:
McGraw-Hill Technology Education
TEL +1-630-789-4000
FAX +1-630-789-5226
http://www.mhteched.com
omg_international@mcgraw-hill.com